ARCH BRIDGES AND THEIR BUILDERS
1735-1835

TED RUDDOCK

ARCH BRIDGES AND THEIR BUILDERS 1735–1835

CAMBRIDGE UNIVERSITY PRESS

CAMBRIDGE
LONDON   NEW YORK   MELBOURNE

CAMBRIDGE UNIVERSITY PRESS
Cambridge, New York, Melbourne, Madrid, Cape Town, Singapore, São Paulo, Delhi

Cambridge University Press
The Edinburgh Building, Cambridge CB2 8RU, UK

Published in the United States of America by Cambridge University Press, New York

www.cambridge.org
Information on this title: www.cambridge.org/9780521218160

First published 1979
This digitally printed version 2008

A catalogue record for this publication is available from the British Library

Library of Congress Cataloguing in Publication data
Ruddock, Ted, 1930–
Arch bridges and their builders, 1735–1835.
Bibliography: p.
Includes indexes.
1. Bridges, Arched – Great Britain. 2. Engineers –
Great Britain. I. Title.
TG327.R8   624.6'0941   77-82514

ISBN 978-0-521-21816-0 hardback
ISBN 978-0-521-09021-6 paperback

CONTENTS

ILLUSTRATIONS

ACKNOWLEDGEMENTS

Documents in the Public Record Office and Session Records at County Record Offices are Crown Copyright and are quoted by permission of the Controller of HM Stationery Office. Permission to quote from all other manuscript and unpublished sources has been granted by the owners named in the notes at the end of the book. Acknowledgement for permission to reproduce illustrations is due to the following:

Captain C. K. Adam (fig. 9)
The British Architectural Library (figs 36, 66, 67, 68, 71)
The Duke of Buccleuch (fig. 113)
The Marquess of Bute (fig. 115)
The Churchwardens of St Chad's, Shrewsbury (figs 150, 161)
Church Commission Durham Bishopric Estates (fig. 88)
The Corporation of London Records Office (fig. 178)
The Crown Estate Commissioners (fig. 15)
Mr E. W. Currer (fig. 100)
Dublin Port and Docks Board (fig. 98)
Edinburgh University Library (figs 36, 37, 53, 56, 60, 116, 141, 152, 153, 164, 166, 190, 195, 196, 203)
Messrs W. A. Fairhurst and Partners (figs 158, 159, 160)
The Guildhall Library, Corporation of London (figs 28, 61, 156, 180, 182)
The Institution of Civil Engineers (figs 1, 11, 14, 35, 42, 43, 46, 63, 72, 124, 140, 155, 172, 184)

The Trustees of the Lady Lever Art Gallery (fig. 145)
The Director of Highways, Lothian Regional Council (fig. 193)
Messrs Mott, Hay and Anderson (figs 142, 181)
Miss J. M. H. Mylne (figs 70, 92)
The Trustees of the National Library of Scotland (figs 44, 45, 54, 57, 58, 106, 143, 177)
The National Portrait Gallery, London (figs 109, 133, 135, 165)
North Yorkshire County Council (figs 110, 111)
Mr R. A. Paxton (figs 179, 194)
The Earl of Pembroke (figs 8, 9)
The Keeper of the Records of Scotland (figs 16, 171)
The Royal Commission on Ancient Monuments, Scotland (figs 80, 130)
The Royal Dublin Society (fig. 202)
The Royal Society of London (figs 6, 10, 12, 13, 29, 31, 32, 33, 55, 64, 74, 75, 76, 78, 83, 84, 85, 86, 87, 89, 90, 91, 93, 94)
The Scottish National Portrait Gallery (figs 112, 114)
Shrewsbury and Atcham Borough Council (fig. 24)
The late Edwin Smith (fig. 81)
The Trustees of Sir John Soane's Museum (figs 69, 134)
The Society of Antiquaries of London (fig. 139)
The Controller of HM Stationery Office (fig. 170)
The Director of Engineering Services, West Yorkshire Metropolitan County Council (figs 23, 24)

PREFACE

On a family walk on a Sunday afternoon in September 1969 this book was born. Passing under Dean Bridge for perhaps the fiftieth time I wondered afresh why Telford made it such an unusual shape. I thought I might spend a little time in the library the next day to see if I could find any clue to his motive. It was the first historical research I had ever done and I expected it to take a few hours. I have actually written the answer to the question about Dean Bridge, for the first time, eight years later, and at the end of this book, not the beginning. For I found that it was built at the end of a century of great developments in arch bridges. I have now specified the century as 1735–1835: 1735 marks the starting of Westminster Bridge, the first modern bridge over the Thames and the first bridge to be built in London for over five hundred years; 1835 marks the completion of the Broomielaw Bridge in Glasgow and thus ends the career of Thomas Telford, the last great bridge designer of the pre-railway age.

The story of bridge-building is primarily a story of men, from lords to labourers and architects to astronomers, so I have tried to write a book for all men – and women – to read. I have had to use the technical terms of bridge-building, but many of them are explained when they first occur in the narrative and I have also included a glossary at the end of the book. In addition I would recommend the inexpensive *Illustrated glossary of architecture 850–1830* by J. Harris and J. Lever (Faber and Faber, London, 1966 and 1969) in which every term listed is illustrated by a photograph.

For specialist readers I have endeavoured to make the book a thorough reference work by providing extensive notes and bibliography, four appendices, and a tabulated index of bridges as well as the general index. All dates have been converted to new style, but the few dimensions of foreign bridges are quoted directly from the sources noted without converting them to English measure.

This book could not have been written without the assistance of many people, including official bodies, the staffs of libraries and archives, experts who shared their knowledge with me and friends who gave me hospitality on my travels. I hope that many will accept this tribute as their own and forgive my naming only a few. I have benefited repeatedly from discussions with three men who never refused me their time or access to their own researches. They are Professor Alec Skempton of Imperial College, London, Roland Paxton of the Lothian Region Department of Highways, and my colleague at Edinburgh University, Malcolm Higgs. Eddie Mac-Parland of Trinity College, Dublin, made my work in Ireland very much quicker by directing me to interesting bridges and records as soon as I arrived there. Miss Jean Mylne of Great Amwell has welcomed me to her home many times to work on the precious books and manuscripts of her family. Amongst the staffs of libraries I owe most to those of the Institution of Civil Engineers, the Royal Society of London, the National Library of Scotland and the University of Edinburgh. I have also learned, from calls at many of them, to expect excellent service at the County Record Offices of England and Wales.

My first financial help was in two grants from the Edinburgh University Travel and Research Fund in 1970 and 1971. The Nuffield Foundation made me a larger grant in 1973 and the University then allowed me two terms' leave of absence for study and travel.

I must also express my gratitude to Mrs Marjory Waterston who typed the manuscript, and to several members of the Cambridge University Press for their encouragement, particularly Mr Anthony Parker who advised me at the outset to write the book I wanted to write, and then persuaded the Press to publish it.

I have left until last my wife and three sons. They have tolerated my obsession cheerfully, shared it occasionally and now express relief at its fulfilment in this book. I share their relief.

TED RUDDOCK

Edinburgh
September 1977

xiii

PART 1

Labelye and his contemporaries
1735–1759

1

WESTMINSTER BRIDGE

... since I have lived in *Westminster*, I am grown very much out of love with this word *sinking*. It was in *Westminster*, how many years ago I forget, that the Church of St. John's *sunk* ... it was in *Westminster*, about thirty years ago, that his late Honour invented the *Sinking Fund* ... What can we infer ... except that Westminster is a very bad place to lay foundations in?

A writer in the *Westminster journal* (12 September 1747)

The movement which led to the construction of the first Westminster Bridge began in 1734.[1] The fact that the construction did not start until 1738 shows that there were serious obstacles and the chief obstacle, for several reasons, was the old London Bridge. In the first place, Londoners were convinced that the decrepit old structure drew all the trade of the south of England to their town and they therefore resisted all the attempts to build other bridges across the Thames. Until 1729 they succeeded and there was no other bridge below Kingston; after 1729 the nearest was at Fulham. Both of these were timber pile bridges, rather narrow and very prone to damage and decay, but London Bridge, the only stone bridge over the tidal reaches of the river, was worse than either. It was narrow and dirty, its roadway contracted by houses and fouled by horses and cattle, and it was a major obstacle to the flow of the river and the movement of the numerous river craft (fig. 1). To the craft it was often a lethal hazard because of the narrowness of the passages between its piers and their further obstruction by 'starlings', which surrounded the piers to a little above low-water level but were covered and hidden, and therefore doubly dangerous, when the tide rose (see fig. 54). Such a bad bridge was an example ready to be cited by everybody who opposed the Westminster project, and even its owners, the magistrates and Common Council of London, tacitly admitted the great difficulties London Bridge made for boatmen when they claimed that the proposed new bridge would be 'a great prejudice to the navigation'.[2] The watermen of London and Westminster made similar objections, though their first concern was clearly the risk of losing cross-river trade. Uncommitted Members of Parliament were bound to find the argument plausible, and if London Bridge represented the best skill of English bridge-builders Parliament could not be expected to allow another stone bridge to be built at Westminster.

Many architects and tradesmen drew designs with the intention of showing that the new bridge could be made convenient for traffic and yet harmless to navigation, and five pamphlets published in 1735-7 all gave the same assurance. One of these pamphlets[3] was clearly written at the request of the men who led the movement for the bridge. They had worked in the customary way, first calling meetings of leading citizens with sympathetic noblemen and Members of Parliament, then raising subscriptions to pay for surveys of the river, for the drafting of a petition to Parliament and later a Bill, and for legal and technical witnesses to support the Bill when committees were formed to examine it in each of the Houses of Parliament. Their pamphlet was written by Nicholas Hawksmoor, one of the best-known architects of the day and a resident of Westminster. He came near to admitting that British architects were unfit to design or build the bridge, for when he sought instruction, both in books and in standing bridges, about the design and construction of such a large bridge he had found little in Britain and nothing near London to help him. Yet, paradoxically, his pamphlet contained a new and valid answer to the most serious question of debate.

He described the history and the wretched condition of London Bridge and referred to the medieval bridges of Rochester, Bristol and Burton-on-Trent, as well as several on the Severn, but for examples of arches of long span, such as he knew to be necessary in the projected new bridge, he had to look to the north of England at the old bridges of York, Durham (Framwellgate) and Bishop Auckland (Newton Cap), and the new waggon-way bridge south of Newcastle-upon-Tyne which has been called, at different times, Tanfield Arch and Causey Arch. He also noted the Brig' o' Balgownie at Aberdeen, an old arch of 80 ft span, and was aware of the ten years' work of road- and bridge-building which was just being completed in Scotland by General Wade,

3

1   London Bridge 1749. Engraving by S. and N. Buck.

a man who would soon be an active member of the Westminster Bridge Commission.[4] The only arch of long span in the southern half of England which he was able to mention was the one of 101½ ft span in the park at Blenheim Palace, in the building of which he himself must have played some part as assistant to Vanbrugh, the architect.[5]

All the books about bridge design and construction which Hawksmoor found were from the Continent. The Italian writers, Palladio, Scamozzi and Serlio, gave some dimensions and ornamental details of the bridges of ancient Rome and of Trajan's great bridge over the Danube. Hawksmoor admired as 'exceedingly beautiful' the bridge which Palladio, a sixteenth-century architectural theorist, praised and copied, namely the Roman bridge at Rimini (fig. 2) – a bridge which, through Palladio's advocacy, was to influence many designs in Britain and Ireland. He quoted in summary Palladio's rules for the form, proportions and materials of arches, piers, cutwaters and abutments (see appendix 1A) and both Palladio's and Scamozzi's advice on the construction of foundations. Study of Palladio's book was greatly aided by the fact that an English translation had been published in 1715–16.

Hawksmoor's French authority, Gautier (see appendix 1B), was much more recent and gave him full specifications of two large modern bridges,[6] the Pont Royal at Paris, built in 1685, and the bridge of Blois over the Loire, an eleven-arch bridge finished in 1724 and designed by Jacques Jules Gabriel, the Premier Ingénieur of the Corps des Ponts et Chaussées since its inception in 1716. The bridge of Blois was of a similar type and length to what must be built at Westminster and it kept within Palladio's rule for the proportions of arch span to pier thickness, with ratios varying from 3.5 to 5.0, but it had elliptical arches, which Palladio had never considered. The foundations were laid on oak

piles and the piers built up to water level in cofferdams, but Hawksmoor gave no details of these operations.

His own design for Westminster Bridge was of nine semicircular arches, the shape considered strongest since Roman times and by both Palladio and Gautier. His largest arch was to be 100 ft in span and the largest ratio of span to pier thickness 4.4. He followed the common practice of adding up the thicknesses of the piers as a measure of the obstruction offered to the river and also adding up the spans of the arches and comparing the total, 660 ft, with the existing width of the river, 830 ft. But as well as this he offered an entirely new kind of assessment of the effect of the bridge on the flow of the river, which had been made by Charles Labelye. Labelye was a Frenchman resident in Westminster who had been employed by the men who set the project in motion to supervise their soundings of the river and boring of the bed at the possible sites for the bridge.[7] He had been born in Switzerland, son of a Huguenot refugee, and had come to England about 1720 at an age between fifteen and twenty. He appears to have had some experience of construction work before 1736 but he was no ordinary tradesman, and probably not a tradesman at all. A survey map of the coast near Sandwich printed in 1736 is inscribed 'By Charles Labelye, Engineer, late Teacher of Mathematics in the Royal Navy';[8] and he had also assisted Dr John Theophilus Desaguliers either with scientific experiments or with his teaching. Desaguliers was himself the son of a forcibly exiled French protestant pastor but was renowned in London for his lectures on scientific subjects and especially 'experimental philosophy',[9] the subject which laid the foundations of modern engineering theory. Labelye was therefore fully competent in contemporary mathematics and mechanics and he had deduced or invented a method of calculating the 'fall', or difference of level between the surfaces of the water upstream and downstream of a bridge, caused by the piers obstructing the flow. His method, which was at least similar to that given in appendix 2, was, he said,

founded upon the doctrine of falling bodies, the principles of hydrostatics, a great number of observations made in several streams by the Reverend Dr. J. T. Desaguliers, several observations made last year on the greatest velocity of the stream near the place intended for the bridge, the greatest rise of the tide in the same place, and lastly, by a true and exact section of the river, obtained by three actual mensurations of the breadth of the river, and several boreings [i.e. soundings?].[10]

To introduce what is now called a 'factor of safety' or, as Labelye himself put it, 'to obviate or remove all difficulties and objections', he entered in his calculation a tidal rise one and a half times the greatest rise he had measured at the site and a velocity of water almost twice the greatest value he had measured. The result of the calculation for the bridge designed by Hawksmoor was that there would be a fall through the bridge of $4\frac{7}{10}$ in.; and he checked the method by applying it to the existing London Bridge when the answer agreed very closely with the fall observed there. At the worst the fall through London Bridge could be 4 ft 9 in.,[11] the sum of the widths of the openings being only 450 ft when measured above the starlings and 194 ft between the starlings, the latter being less than a quarter of the

width of the river.[12] Labelye asserted, probably quite fairly, that a fall of less than 5 in. through the new bridge would 'never hinder any vessel or boat from rowing through, even against the swiftest tides'.[13] Hawksmoor printed Labelye's short report in his pamphlet and noted that the full calculations were available for inspection and were 'approved of by the Rev. Dr. Desaguliers, and others'. The 'others' included the mathematics master of Christ's Hospital, James Hodgson, who appears to have been acting as an adviser to the Common Council of London, and William Jones, Esq., who later proposed a simpler way of making the calculation.[14] It is Jones's method which has survived and is given in appendix 2.

This calculation by Labelye was the first important contribution by scientific theory to the art of bridge-building in Britain, and it must have made some impression on the committee in Parliament when the result was stated by Labelye and supported by Desaguliers;[15] but it did not go unchallenged. It was challenged by other calculations so simple that they could be understood by every member of the committee, but based on erroneous views of the flow of a river through a bridge. About a year before Hawksmoor's pamphlet appeared with Labelye's report, John Price the elder had addressed a pamphlet to the members of the House of Commons.[16] He offered a design of nine arches with the piers obstructing approximately one-sixth of the

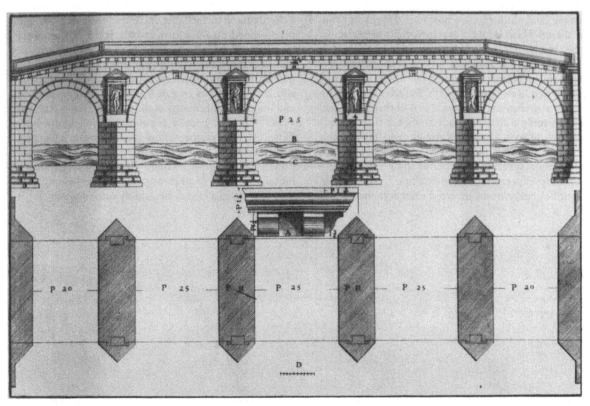

2   Bridge of Rimini. Drawing by A. Palladio, 1570 (reprinted 1965).

width of the river (fig. 3); he said that since one-sixth was obstructed the depth of water on the upstream side during the ebb of tide must increase by one-sixth, and as the depth was 8 ft at low-water there would be a fall through the bridge during the ebb of only one-sixth of 8 ft, that is, 1 ft 4 in. Even this fall he thought to be 'so inconsiderable, that boats may pass and repass at all times, without hazard or hind'rance', though one cannot imagine the boatmen agreeing. Were it not for the terrors of London Bridge, such a statement must have been thought imbecilic. The absurdity of the calculation was easy to prove, since the same rule could be applied at the turn of high tide when the depth could be as much as 24 ft and the fall therefore 4 ft; but there would be no flow at all at the turn of the tide and, logically, no fall. Such an argument does not appear in the parliamentary records and was not printed until July 1736 when Labelye's method had already gained a seal of acceptance by the passing of the Act.[17] It was used in a pamphlet by John James, architect, of Greenwich,[18] in which he reviewed the pamphlets of Price and Hawksmoor and another published in the previous month by Batty Langley.[19] Langley had given a much longer treatment of the question of the fall, but on the same premise as Price and with similar results, and he ridiculed Labelye's method. James's argument exposed the absurdity of Price's and Langley's calculations and he seemed disposed to believe Labelye's result, though he was clearly unable to follow the calculation.

James also offered a design, though without a drawing, and so there were now four designs in print. He shared Hawksmoor's preference for semicircular arches but he was satisfied with a maximum span of 80 ft and so required fifteen arches. His design was therefore the nearest in form to the bridge eventually built. Price was content (with Palladio) to let his segmental arches rise only one-third of their span, and so had a middle span of 100 ft no higher than James's semicircle of 80 ft. Langley, a prolific and often irascible writer on architecture and the building trades and apparently a successful teacher of those subjects, but with little experience of construction,[20] offered the 'scientific' curve of an inverted catenary, that is, the

shape which a hanging chain would adopt naturally, turned over to form an arch which he believed would be stressed in pure compression. He proposed also to equalise the loads over the two haunches of each arch by forming cylindrical voids of unequal size through the spandrels; and to reduce the load on the foundations by another larger void directly over the middle of each pier[21] (fig. 4). The question of whether each arch could stand independently, that is, without the balancing thrusts of its two neighbour arches, was important, and Price intended that all the arches of his bridge should be built before any of the supporting centres were removed. But Langley argued that in his design, and only his, the arches could each stand alone, saying that 'their abutments are within the bases of their piers'. His diagrams (fig. 4) show that he derived this concept from the writings of French authors but without understanding them fully. For decoration Price used Palladian motifs (fig. 3), Hawksmoor a restrained classical form with heavily rusticated stonework, and Langley drew no ornament at all.

On the vital question of how to lay firm foundations in the bottom of a river which rose and fell by 15 ft at spring tides and was up to 24 ft deep at the highest spring tides, the four men also differed. John Price offered practical methods for founding the piers, but they were rather outdated and certainly difficult to carry out in such deep water; Hawksmoor and James made no firm proposals. Langley was original, proposing two alternative methods. If the bottom were found to be 'a strong clay' he would make a double-walled cofferdam round each pier and found the pier on driven piles capped by a platform of oak. If the bottom were 'firm gravel', which was actually much more likely, he believed the water would seep into a cofferdam too quickly for pumps to clear it and so he proposed to dig and level the gravel to 5 or 6 ft below the existing bed and build the first courses of the piers in a flat-bottomed vessel which he named a 'concave parallelopipedon'. (The use of a mathematical description instead of a nautical one was typical of Langley, and James saw fit to translate: 'in English, a hollow chest or coffer'. Langley himself wrote in another work that a parallelopipedon was 'also called a long cube, and by some a prism'.)[22] It

3   Design for Westminster Bridge by John Price, 1735.

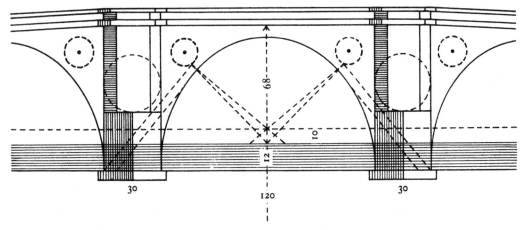

4   Batty Langley's arch.

was to be an open-topped box which would sink by the weight of the masonry and rest on the prepared gravel surface. Whether the sides were to be removed when the top of the stonework reached water level Langley did not say.

These pamphlets did nothing to establish their authors as contenders for the position of architect of the bridge. Both Price and Hawksmoor were dead before the Act was passed and James was well over sixty.[23] Langley may have seemed too clever. But the distrust of native talent, which the pamphlets may well have inspired, caused some of the promoters to write to the British ambassador in Paris.[24] He was asked to obtain designs for the foundation works from Chevalier d'Hermand, an engineer at the French Court, and to persuade the builder of the bridge of Blois, M. Lambotte, to come to London to contract for the construction. This seems to have come to the ears of some master tradesmen in London, for six of them signed a petition presented to the first meeting of the Westminster Bridge Commissioners on 22 June 1736 which said: 'with great concern they have heard it is apprehended there are not people in England capable of building the said bridge'; and begged to be allowed to build a stone pier at their own expense in whatever part

of the river was considered most difficult, to prove their competence. Among the six were Andrews Jelfe and Samuel Tuffnell, masons, Thomas Phillips, carpenter, and George Devall, plumber, all of whom had been contractors in the building of the timber bridge at Fulham in 1729, Phillips holding the main contract.[25]

This offer was allowed to lie unanswered for a year because the lottery intended to finance the building of Westminster Bridge proved a failure and no action could be taken until another Act of Parliament had authorised a second lottery with better terms.[26] Then the tradesmen renewed their offer but the Commissioners were advised by Thomas Ripley, a former carpenter who was now Comptroller of the King's Works and was also a leading figure among the proprietors of Fulham Bridge, that they could have a timber bridge for £35,000 while a bridge of stone would cost £70,000. They decided in July 1737 to build in timber, being still unsure how much money they would have to spend. The surviving pamphleteers, James and Langley, submitted designs of timber along with at least six others, but Charles Labelye offered a design all of stone and declined to make a design of timber. Ripley introduced designs of his own for stone, timber and part stone and part timber, but after repeated discussions

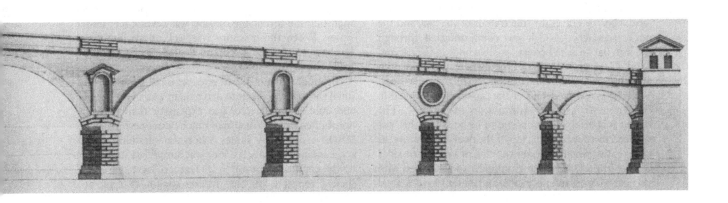

and partly on Ripley's advice the Commissioners decided that they must adopt a step-by-step approach to the work. They asked Labelye twice, in September 1737 and May 1738, to explain to them a method he proposed for founding piers, and he in turn offered to found a trial pier by this method at his own expense. He admitted later that he had taken advice from Andrews Jelfe and a master carpenter named James King on this and other occasions[27] and Jelfe, with Samuel Tuffnell in partnership, and King appeared with him at his second explanation with offers to construct some piers by his method at stated unit prices. The Commissioners accepted.

The method was the same in principle as that which Batty Langley had proposed in his pamphlet for use on a gravel river-bed. Labelye called the open-topped floating box a 'caisson' and it was sometimes referred to later as a 'case' or simply as 'the vessel'. Another tradesman collaborator was Robert Smith, a 'ballast-man' whose normal business was the supply of gravel dredged from the bed of the Thames; he had assured the Commissioners that he had experience of dredging to make a perfectly level bottom such as the caisson would need to rest on. Labelye did not claim complete novelty for the method, and boxes of masonry had been sunk before as the bases of structures, but the use of a box designed to be sunk and floated again as necessary, and of such a large size, was quite new.[28]

Labelye was appointed 'engineer' for the laying of the foundations in June 1738, and the title is significant. Nobody who called himself architect took any further part in the design or construction of the bridge and it seems likely that Labelye himself had suggested a title which was already common in France. He acknowledged early in the work that piles might be used to strengthen the ground under some of the piers, but as work went on he bored the bed of the river at the exact site of each pier and decided in every instance that the gravel was firm enough without piling. James King, however, held that they should have been piled and offered to supply and drive piles for all the piers for only £5,000 or £6,000.[29] Labelye judged both the nature and the strength of the strata by their resistance to the driving of a sharp solid bar, 'the point of which was not unlike that of a watchmaker's drill', after a few initial borings in which a different tool was used to bring up samples of the materials. He could distinguish 'dirt, sand, clay, or gravel' by 'the tremor of the bar, and the noise it made . . . which was communicated through the iron bar so as to be very sensible both to the ear and hand'.[30] Most, if not all, of the piers were founded on gravel several feet below the surface of the river-bed.

Fig. 5 gives the details of the caisson as engraved in Bélidor's *Architecture hydraulique* in 1752.[31] The height was originally proposed to be 30 ft but this was reduced to 16 ft from the top of the floor, or 'grating' as it was called when it became the base of a pier. The sides were of fir timbers laid longitudinally and sheathed with vertical 3-in. planking inside and out to give a total

thickness of 1 ft 6 in. at the bottom and 1 ft 3 in. at the top. Vertical timbers, forty-six in number and 8 in. × 3 in. in section, were fixed to the inside and outside faces and had dovetailed ends at the bottom held in mortises in the floor timbers by iron wedges; the wedges were knocked out when a pier had been founded and the sides were then free of the bottom and could be floated off on a high tide. There was a sluice in the side for letting in the water to sink the caisson when required. The bottom was of 2 ft total thickness, made up of longitudinal timbers 12 in. square with 3-in. planking across their undersides and 9-in. square timbers across their upper sides. Both the 12-in. and the 9-in. timbers were spaced apart about twice their own widths so as to form an open 'grating'; what was used to fill the spaces between them is not recorded. To resist the inward pressure of the water on the sides there were braces across the top and these also carried a floor for stacking stones and other materials. A timber kerb 14 in. × 7 in. in section fitted tight between the bottom of the sides of the caisson and the face of the bottom course of masonry.

The caisson was built on twelve trestles which stood vertically until it was ready for launching but could then be made free, by the removal of braces between them, to rotate on their bottom members and drop the caisson into the water (fig. 5). This was done at high tide. It was then towed out to the site of the pier and anchored within 'guard works'. The caisson for the first pier to be founded was kept afloat while three courses of masonry were laid and cramped, but it was sunk several times by opening the sluice to check the regularity of the bottom of the prepared foundation pit. The pit had been dug by the dredging scoops of the ballastman Smith but needed some trimming where material had since fallen in. Planking attached to piles was used to prevent such falls and later on troughs filled with gravel were placed along the top of the slopes which surrounded the pits, as shown at Y in fig. 5. The arrangements at the first pier are shown correctly in the figure but they were simplified for later piers. When three courses of stone had been laid the caisson was grounded permanently and afterwards pumped dry at each low tide both day and night for the masons to work, until at least six courses had been laid. Every time the tide rose the sluice was opened to avoid danger of the caisson floating, and the masons stopped work; at high tide the top of the caisson was several feet under water. When the masonry was well above low-water level the sides were freed and floated away to be used again for the next pier.

Although the number of piles required was relatively small the machine built to drive them was a very special one, more complicated and expensive than conventional pile-drivers but also of much better performance. It was designed by James Valoué, a watchmaker and acquaintance of Labelye, but was simplified somewhat by the engineer himself.[32] Its features were the use of sprung tongs to lift the driving weight of 1,700 lb,

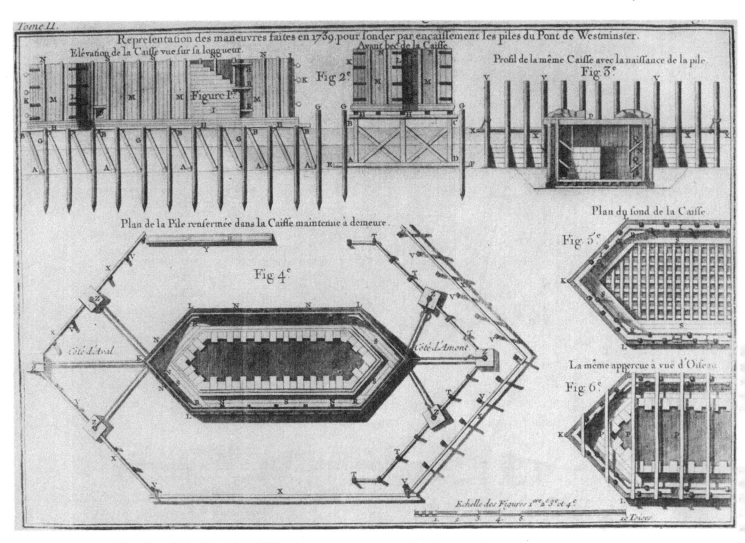

Tome II.

Representation des maneuvres faites en 1739.pour fonder par encaissement les piles du Pont de Westminster.

Elévation de la Caisse vue sur sa longueur.

Fig 2.º

Avant bec de la Caisse.

Profil de la même Caisse avec la naissance de la pile.

Fig 3.º

Figure I.ʳᵉ

Plan de la Pile renfermée dans la Caisse maintenue à demeure.

Fig 4.º

Côté d'Aval.

Côté d'Amont.

Plan du fond de la Caisse

Fig 5.º

La même apperçue à vue d'Oiseau

Fig 6.º

Echelle des Figures 1.ʳᵉ 2.ᵉ 3.ᵉ et 4.ᵉ

10 Toises

5 Caisson of Westminster Bridge. Engraving in Bélidor, 1752.

automatic opening of the tongs to release the weight at the top of its lift, which was up to 20 ft above the pile-head, and the use of a clutch between the drum on which the lifting rope was wound and the shaft of the capstan which turned the drum. Valoué's print of the machine (fig. 6) shows the clutch operating automatically and it may be that Labelye's alteration was to make a simpler clutch for hand operation by the workman in charge, as explained by Bélidor in 1750 in a full description of the machine with good illustrations of the details (fig. 7).[33] The release of the clutch allowed the drum to reverse and the rope to run out, so that the tongs dropped after the weight to re-engage it and lift it again, without need for the capstan to reverse. As compared with other piling machines, this one had three advantages: first, the weight, or 'monkey', was large because it was lifted by a capstan instead of by men hauling on a rope over a pulley – the usual monkeys raised by that method being about 800 lb; second, the height of fall was much greater than the men could obtain, being limited by the length of their

arms' reach, about 4 ft; third, the return of the tongs to the monkey after a blow was much quicker than in other capstan-driven machines where the capstan and its horses or men had to reverse their movement to pay out the rope after the monkey fell. The machine was erected on two barges fitted with a timber floor for the horses to walk on and both well ballasted to keep it as steady as possible in the water. Once it was well run in it could raise the monkey 150 times per hour to a height of 8 to 10 ft, using three horses 'going at a common pace'. Dr Desaguliers claimed in his *Course of experimental philosophy* that it did five times as much work (in equivalent time) as an ordinary piling machine.[34]

Labelye was ordered at the start of the work to make the foundations capable of bearing the weight of 'a stone bridge agreeable to his own plan'; but at the same time the Board of Commissioners made a contract with James King and John Barnard, carpenters, to build a superstructure of timber which was of the same overall dimensions as Labelye's design. It was designed by King and was to consist of thirteen trussed arches and

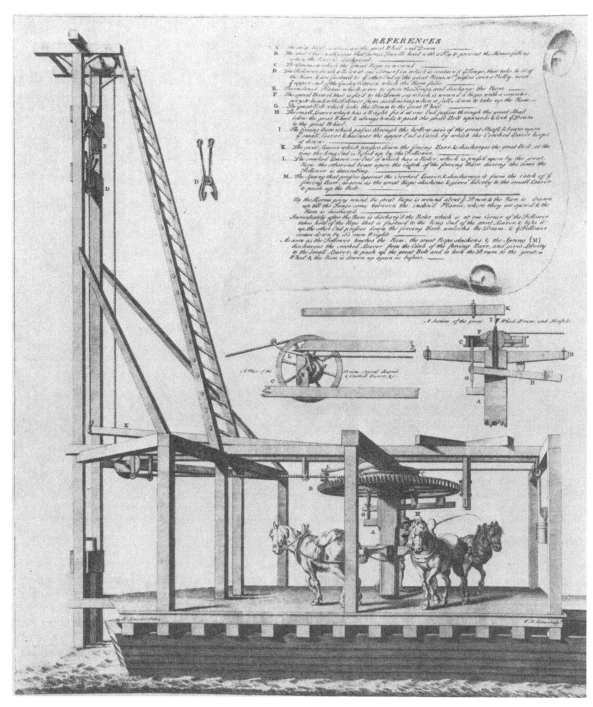

6   Pile-driving machine. Design by J. Valoué.

cost £28,000, but in the winter of 1739–40 when the Thames froze over for several weeks the damage caused by ice to the guard works prompted a reconsideration of this decision; and in February 1740 the Commissioners made a bargain with King and Barnard for cancellation of their contract and started to build the stone bridge designed by Labelye. They paid to King and Barnard £1,797 in compensation and £5,674 for timber already purchased. A little of the timber was used in the

construction of the bridge and valued at £271; the remainder had to be sold for only £1,872.[35]

Labelye became 'engineer' for the whole construction, with power to suspend incompetent workmen and to direct all the work and certify its satisfactory completion for the purpose of payment. For this he was paid only £100 per year with a subsistence allowance of ten shillings a day when in attendance, but he received a gratuity of £2,000 when the bridge was finished.[36] The

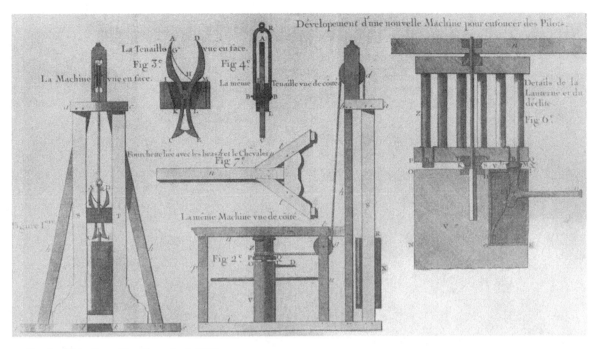

7 Details of Valoué's machine. Engraving in Bélidor, 1750.

Commissioners had other salaried officers: a clerk, with one assistant at least, to keep records and deal with legal matters, a treasurer to handle money, a 'surveyor of streets' concerned with survey and purchase of property, and a 'surveyor and comptroller of the works', Richard Graham, whose duties came closest to the engineer's. Graham was responsible for drawing up contracts, measuring work done, checking accounts, and all financial aspects of the construction of the bridge. He was paid £300 per annum, but without the daily subsistence allowance and with only a small gratuity at the end; the other officers received similar but generally smaller pay. It was a remarkably complex organisation and it allowed the Commissioners to contract for the work in a piecemeal way, as they seem to have felt compelled to do, whether from fear of making mistakes or from uncertainty as to the continuance of the supply of funds. Their funds came from lotteries until 1741 and thereafter by direct grants from Parliament.

The most active Commissioner at most times between 1736 and 1750 was the wealthy and boisterous Ninth Earl of Pembroke (fig. 8). He was a very accomplished gentleman architect who had designed and was, in 1735–7, erecting the most famous of all estate bridges, the covered bridge in Palladian style in his own grounds at Wilton House.[37] An undated memorandum among his private papers lists the officers considered necessary for the construction of Westminster Bridge at the time when the foundations were started, foreseeing the division of responsibility between Labelye and Graham and using the term 'engineer'; there is also a drawing of the elevation of the middle arch which may represent his thoughts on the architecture of the bridge (fig. 9). He has actually been described as the architect[38] and it is not unlikely that he contributed some of the motifs on the facade of Labelye's design. He was clearly Labelye's chief supporter in all controversy and was given credit by both Labelye and Graham for the survival of the project and by Graham for his own employment there. He also gave security for James King's contract with the Commissioners for construction of the first centres. The letters of various officers to him suggest that he was allowed by the Commissioners to act almost as the bridge's patron.[39]

8 The Ninth Earl of Pembroke. Bust by L. F. Roubiliac.

9   The Earl of Pembroke's drawing.

When Labelye was ordered to direct the building of a stone superstructure in February 1740 he altered the elevations of his former design but none of its structural forms or dimensions.[40] His new design (fig. 10) had a pilaster or 'turret', as he called it, on the top of each cutwater, of trapezoidal shape in plan and extending up to the parapet where it enclosed a refuge or 'recess' from the footway of the bridge. Similar but taller turrets were shown in the drawing among Lord Pembroke's papers, which was probably earlier in date (fig. 9). The outsides of the turrets and the arch rings had deeply recessed V-joints which were extended right across the soffits of the arches. The outside faces of the spandrels were designed to be rubble masonry, but of well-squared stones and laid in courses which radiated from the centre of the circle of the arch – another feature which agrees with Pembroke's drawing. Inside the spandrels, each arch was to be backed with a 'secondary arch'

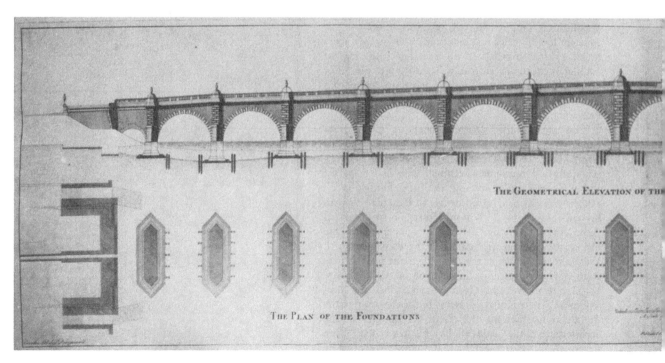

10   Westminster Bridge as built.

tapering from about 2½ ft thick at the crown to a very large thickness at the springings; this was to be built of Purbeck stone which was much cheaper than the best Portland but sufficiently square in cleavage and equally hard and strong. Over the secondary arches the space up to road level was to be filled with gravel but it was to be first divided into nine compartments by four walls of dry (i.e. unmortared) stone (fig. 11). This was a type of construction so unusual in London that when Richard Graham inquired among tradesmen about the proper rate to be paid for it he found nobody with any experience of it. Over four of the recesses on each side of the bridge, namely those next to the ends and to the middle arch, there were semi-domes of stone; domed recesses had been introduced in James King's timber design as stations for the watchmen who would be on the bridge at night when it was finished[41] and domes may have been intended in Lord Pembroke's drawing also. There was a classical cornice and stone balustrade for the whole length. The width was 30 ft for carriageway and 6 ft each for two footways, and the width across the arches 44 ft.

The principal reasons which Labelye gave for all the dimensions and forms of the design were either practical or scientific. He considered the semicircular shape of arch to be the strongest but noted that it was also generally thought 'the most graceful'. The use of V-joints would prevent the breaking or 'flushing' of the edges of arch stones, but they also gave a good rustic appearance. The arches were to spring from a low level, just 1 ft above low-water, which improved the stability of the piers. The proportion of arch span to pier thickness was about 4½ which was smaller, and therefore more stable, than 'the most approved authors' advised.

The whole waterway through the bridge was 880 ft which exceeded the width of the river at the Horseferry a short distance upstream. Labelye thought the radial courses gave the spandrels extra strength and that the weight of the 'turrets' over the cutwaters would act to neutralise the lateral thrust of the arches. He made calculations by the arch theory which French academicians[42] had published and proved to himself that the combined masonry of the semicircular arch and the tapering 'secondary arch' was 'in equilibrio in all its parts' and each arch would therefore stand independently. Lastly, he applied to his own design the method of calculating the 'fall' which he had originally applied to Hawksmoor's, with the same factor of safety in the assumed velocity of flow, and found that the fall would not exceed 3¼ in. He now tried Jones's method too and got the same answer.

The foundations and piers were already under construction and the next step was to erect centring for the arches. Designs and estimates for centres were first made separately by Labelye, King and Barnard, but when the Commissioners came to decide between them Labelye expressed a preference for King's design and King contracted for making the centres, first for three arches[43] and later for all the others. The first three, for the middle three arches, were erected in the summer and autumn of 1741 and the three arches were built before June 1742. The centres were of a trussed kind, supported only near the piers so as to leave a passage for boats to pass under them (see fig. 12).

The contracts made with King for the centres are the most extreme example of the Commissioners' step-by-step proceedings, for they covered the supply and erection of the centres, but not their removal, and the

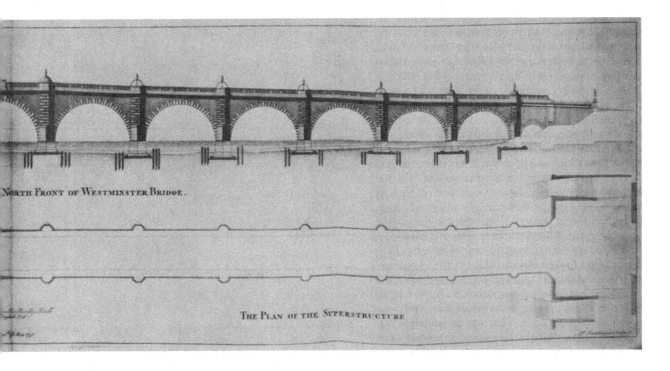

NORTH FRONT OF WESTMINSTER BRIDGE.

THE PLAN OF THE SUPERSTRUCTURE

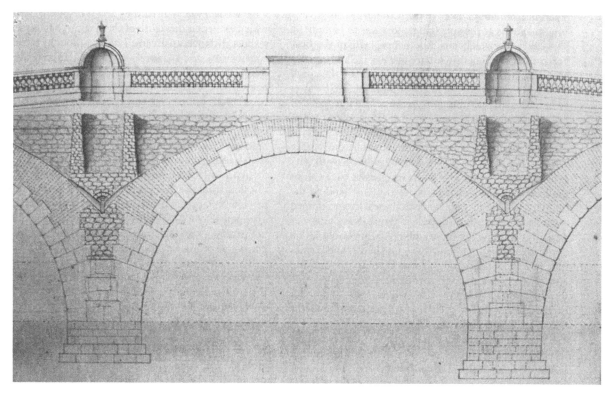

11   Sections of piers, arch and spandrels, drawn by T. Gayfere.

method of removing them was still under discussion
when the three arches were finished. As fig. 12 shows,
each of the five ribs of a centre was supported at each
end on an oak plate or bearer; and that plate was
separated by four timber blocks (called 'blockings')
from a similar and parallel plate which was laid on the
tops of three piles and a post or 'puncheon' which stood
on one of the steps of the pier base. The centre had to be
lowered by removing the blockings, but the removal
should be done slowly and evenly and preferably
without anybody standing underneath. The first
method tried was devised by King. It required one or
more 'circular wedges' of cast iron under the end of
each rib and, although no full description survives, it
can be guessed that each wedge was placed between the
two oak plates with an axle through its centre; that the
wedges were all rotated to a tight position for the
blockings to be taken out, and then turned back in
unison to lower the centre uniformly and slowly from
the arch. However, there was difficulty in the casting
and delay in the delivery of the wedges and they were
then found very difficult to turn, some of the levers used
for this being broken. All through 1743 the Commissioners kept asking for quicker progress and when the
first centre had finally been struck Labelye had to
report that it took thirteen weeks' preparation and three
or four weeks to strike it. That was in January 1744,
eighteen months after the first three arches had been
finished. Centres were standing in the river for six or

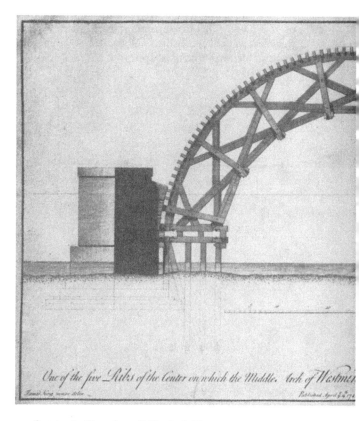

One of the five Ribs of the Center on which the Middle Arch of Westmr

12   Centring for Westminster Bridge, by J. King.

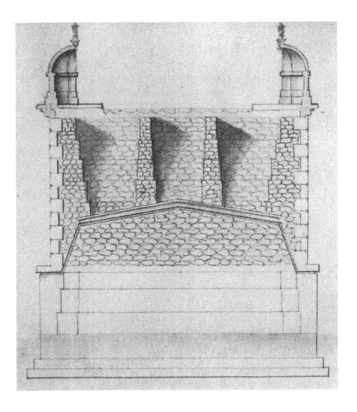

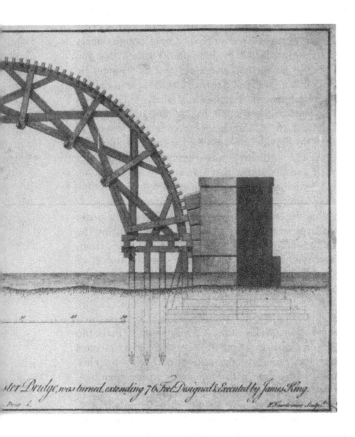

ster Bridge, was turned, extending 76 Feet Designed & Executed by James King.

seven arches all of which would soon be finished, and a better method of striking was needed urgently. King being confined to his room and soon to die, Labelye considered several proposals and opted for one suggested by William Etheridge, King's foreman. It was to insert long multiple wedges stretching the whole width of the bridge between the parallel oak plates, drive them tight enough to let the blockings be removed and then drive them back so as to lower the ribs of the centre all at once (fig. 13). The driving was done with what Labelye later called a 'battering-ram', and was presumably a heavy balk swinging from a shear-legs or scaffolding. It was safe in the sense that nobody need be underneath when the centre was struck and it eased all the ribs at once, but it can hardly have been gentle.

On several different occasions cracks and visible movements of the masonry were reported to the Commissioners by members of the public, and each time Labelye allayed their fears, usually with the support of Richard Graham and the contractors. In September 1744 some of the radial joints on the faces of two arches were opening but Labelye, Jelfe and Tuffnell all insisted that opening of joints was normal and was not to be called cracking. At the same time it was observed that some of the puncheons which stood on the steps of the piers and should have supported the oak plates under the centres were loose, because the piers had settled more than the piles which also supported the plates (fig. 12). Labelye explained that the piers carried a high proportion of the weight once the keystones of an arch had been driven since the tightening of the keystones tended to lift the rest of the arch off the centre; the load on the piers was simultaneously and suddenly increased and this could cause the settlement observed. To estimate how much more settlement might take place when the centres were finally lowered he made long calculations 'founded not only upon mathematical and mechanical principles (which are in no wise sufficient for the purpose) but also upon observations and experiments of his own, relating to the force necessary to overcome the friction or the roughness of the stones and the tenacity of the mortar or tarris used in those joints'. His conclusion was that the centres, while they were in place after keying of the arches, would support only 600 tons of the middle arch and 550 tons of the next arch; therefore the extra load which would come on the intervening pier when both centres were struck would be half of (600 + 550), or 575 tons, and this was only an addition of one-twelfth to the load it was already carrying. As similar loads of stones had previously been stacked over the piers Labelye expected no further settlement when the centres were struck.

There is evidence in the minutes of the Commissioners and their committees that some concern was felt about the foundations at several other times. For instance, it was resolved after some discussion to cut off the piles which had been driven near the piers to support the centres and for other purposes, rather than

13   Drawing owned by J. Smeaton, inscribed 'This draft describing the manner of striking the centers of Westminster Bridge was copied from one given to Richd. Hazard Esqr. by a person that had worked there under Mr. Etheridge.' There were actually only five ribs, not seven.

risk disturbing the ground by withdrawing them.[44] William Etheridge designed an underwater sawing machine which worked well and the Commissioners later paid him a gratuity of £200 for the 'invention' of this and his method of striking the centres.[45] It would seem that the settlement of all the piers was enough to be visible but the fourth pier from the western (Westminster) end settled more. Reporting on 8 September 1747 Labelye referred to 'the former settlings' of 7 in. but said that since the beginning of August it had begun to settle again and had gone down a further 13 in. He had been absent from London, and while he was away Richard Graham had ordered the removal of the parapets to prevent them being cracked by the movement. Some of the spandrel filling had also been taken out. All the master tradesmen advised that the two adjacent arches were in danger and should be propped with centres but opinions differed about what should be done to remedy the trouble. Labelye himself designed these centres and Etheridge was ordered to make and erect them; but it was over six months before the first one was in place. The settlement had slowed down and eventually stopped but by May 1748 when both the centres were in place it was 2 ft 5 in. at one side of the pier and 2 ft 2 in. at the other. Labelye's proposal, which was supported by Jelfe and Tuffnell, was to apply to the pier a load greater than the 700 tons which had been removed from it, let it settle to a stop and then build it up to its proper height again and repair the arches. 'Iron ordnance', or cannon barrels, was borrowed from the Army's store at Woolwich and brought up river to be laid on the pier. By July the load of ordnance was 700 tons, the pier had settled further and the arches were bearing so heavily on the centres that the latter were badly distorted. The Commissioners were strongly advised by Jelfe and Tuffnell that the load should be further increased, and as strongly by Etheridge and Graham that the centres were in danger and the load should be removed. Labelye inclined to the masons' view, though much less strongly than before, and the Commissioners gave him an order contrary to his advice – he said for the first and only time.[46] The ordnance was removed, the spandrel emptied and the arches taken down.

In December 1748 Labelye reported on possible schemes of repair. He rejected as impractical suggestions that the two arches might be replaced by a single one of large span, and also any proposals requiring the complete removal of the pier and its foundation, because after its settlement its base was 4 ft lower than those of the adjacent piers, which he therefore considered to be placed at risk by the digging. Instead, he offered two alternatives: to replace the two arches by three, founding two new piers in caissons sunk on piled bottoms one each side of the settled pier; or to make the settled pier as firm as possible by three rows of piles driven close round its edge and rebuild the two arches thinner than before with large voids in the spandrels over them, the voids roofed by segmental arches. He favoured the latter and so did the Commissioners. William Etheridge then raised the question whether, when the two damaged arches were taken down, there would be danger from the thrust of the next arches on the two unbalanced piers. He claimed to have 'experimentally found a considerable lateral pressure in all the arches' but Labelye opposed to this his claim to have designed the secondary arches with their tapering thicknesses conforming to 'a curve which (as it has been demonstrated ever since the year 1695) brings all the parts of those arches to an equilibrium, and consequently they act no other way on the piers but by their whole natural weight'.[47] Therefore no arch would be in danger from the lack of a balancing thrust from its neighbour. This disagreement of experts was a very difficult thing for a lay Board of Commissioners to resolve but after a month's delay and long discussion they voted to accept Labelye's opinion by seven to one.

The repair took another two years and was expensive. The scheme was simplified by omission of two of the three rows of piles when the first row was found to compact the ground much more than Labelye had expected.[48] Fig. 14 is Gayfere's drawing of the repair. Etheridge left the site and Edward Rubie, introduced as a 'water carpenter', took over the carpenter work, including piling. The bridge was finally opened to traffic at about 2 a.m. on Sunday 18 November 1750, more than fourteen years after the passing of the enabling Act, twelve after the start of work on the foundations and three after the first completion of the structure. It was just in time for its 'patron', who died on 9 January 1751. Horace Walpole reported to a friend:

Lord Pembroke died last night: he had been at the Bridge Committee in the morning, where, according to custom, he fell into an outrageous passion; as my Lord Chesterfield told him, that ever since the pier sunk he has been constantly *damning* and *sinking* . . . He was one of the lucky English madmen who get people to say, that whatever extravagance they commit, 'Oh, it is his way'.[49]

Labelye had been subject to some public criticism and had been castigated in a pamphlet by Batty Langley both for the sinking of the pier and for 'pirating' Langley's invention of the caisson;[50] but he had retained the confidence of the Board of Commissioners and was now appointed at a reduced salary of £150 per annum to undertake any surveys or repairs the bridge might need.[51] But he informed them that he had 'contracted an asthma (daily growing upon me) by my constant attendance on the works, especially on the water in winterly and rainy weather' and in April 1752 he left for France, intending to spend twelve months at Béziers in the south and hoping thereby to recover his health. In fact he never returned.

What sort of man he was we can only guess from his formal reports in the Commissioners' minutes, two or three letters to the Earl of Pembroke and a few published reports and pamphlets. There is just enough to convey an impression that he disliked publicity and was over-sensitive to criticism. We may guess that he lacked humour. But his protestant honesty is beyond question and his professional judgement was always confident and generally very sound. And he brought

14 Repair of the 'sunken pier', drawn by T. Gayfere.

something absolutely new into British bridge-building: the direct use of science to decide questions about construction. In this he moved architecture a step further than had Sir Christopher Wren. His most important scientific prediction proved to be safe but not very accurate, for the greatest 'fall' of water level he ever measured through his finished bridge was only $\frac{1}{2}$ in., compared with his prediction of $3\frac{1}{4}$ in. Had he not used his 'factor of safety' in the calculation the prediction would have been much smaller, but still probably twice the fall which he measured.[52]

The architecture of the bridge is as enigmatic as the architect. It seems to have only one possible source, a design for the improvement of London Bridge by Wren (fig. 53), which has Gothic turrets up the spandrels over the piers; but the similarity is not enough to establish Wren's design as more than an influence on Labelye's. Labelye's 'style' was taken as a model for two or three important bridges and then forgotten. It was never copied again after about 1765. Of its appeal to artists, however, there is no doubt, for drawings and engravings of Westminster Bridge are legion. It appears in no less than eleven drawings and paintings by Canaletto.[53]

So much will be said about the faults of the bridge as seen by engineers of the following century that the reactions of Labelye's contemporaries must be set down first. An outstanding tribute was paid by Bélidor, the greatest engineering writer of the day, when he described Westminster Bridge in 1752 as 'the most magnificent monument of our times'.[54] He considered the method of founding piers to be a real innovation and the piling machine designed by Valoué 'the most advantageous which has yet been used'. A copy of the machine had been used at the reconstruction of the bridge of Sèvres across the Seine within a year or two of the publication of Valoué's print in 1738.[55] The caisson method of founding piers was first adopted by French engineers on the deep gravel bottom of the Loire at Saumur in 1757.[56] The influence of the bridge on British design and construction was more diffused, but greater. Many tradesmen like Tuffnell, Etheridge, Rubie and their foremen had learned to build in deep water; young professionals in London, John Smeaton among them,[57] had observed all the operations; and the successful completion of the bridge inspired thoughts of other new bridges in various parts of the country, not least on the Thames itself. Westminster Bridge is a point of reference to which we shall often return in studying the bridge-building of the ensuing century.

2

BRIDGES IN THE COUNTRY

In Britain . . . these useful works have been constructed at the expense, and under the direction of particular, and frequently very limited districts, communities, or individuals, whose chief object has been, in general, economy.

Thomas Telford, 'Bridge' (1812)

Neither the special techniques nor the complicated organisation devised for the building of Westminster Bridge were required for the great majority of bridge projects. Most rivers were much smaller than the Thames and non-tidal, and the volume of road and river traffic was seldom so great. This chapter is a record of *ordinary* bridges and bridge-building in several parts of the country remote from London and where only one or two extraordinary bridges were built during the first half of the eighteenth century. It is a random sample chosen for interest and makes no claim to represent the whole of British bridge-building at the time.

Bridges in the country were the responsibility of landowners, parish vestries and county Justices of the Peace; some town bridges were built and maintained by the borough councils. Most of the members of the Westminster Bridge Commission must have had previous experience of bridge administration as members of some of these bodies, but one very active Commissioner could recall an intimate experience of actual construction. This was General George Wade (fig. 15).[1] Wade was an Irishman and a distinguished soldier of fifty-one when King George I appointed him in 1724 commander of the forces in 'North Britain', the name by which it was hoped to make Scotland forget its separate identity after the Union of Parliaments in 1707. His task was to establish law and order in the Highlands, an area rife with armed robbery, cattle-rustling and what would today be called protection rackets. In his first report to the King in 1724 Wade instanced 'the want of roads and bridges' as one of the great difficulties for a peace-keeping garrison, and during the next ten years he was given grants by Parliament which he used to make 250 miles of roads. The roads were intended to speed the movement of the Army about the Highlands but they were also used freely by the public, and they were the first long roads designed for vehicular traffic in the whole Highland region.

In the accounts of expenditure which Wade submitted to the government he mentioned the construction of thirty-five bridges. Of these two were of wood and thirty-one were of the simplest and cheapest type of masonry. Only about five of the thirty-one were of more than a single arch and many of the single-arch bridges were only 10 to 20 ft in span. Twenty-seven bridges were stated to have been built for £2,001.[2] From the separation of their costs in the accounts it may be assumed that they were all built by contract, whereas

15 General George Wade. Portrait by J. B. van Loo.

19

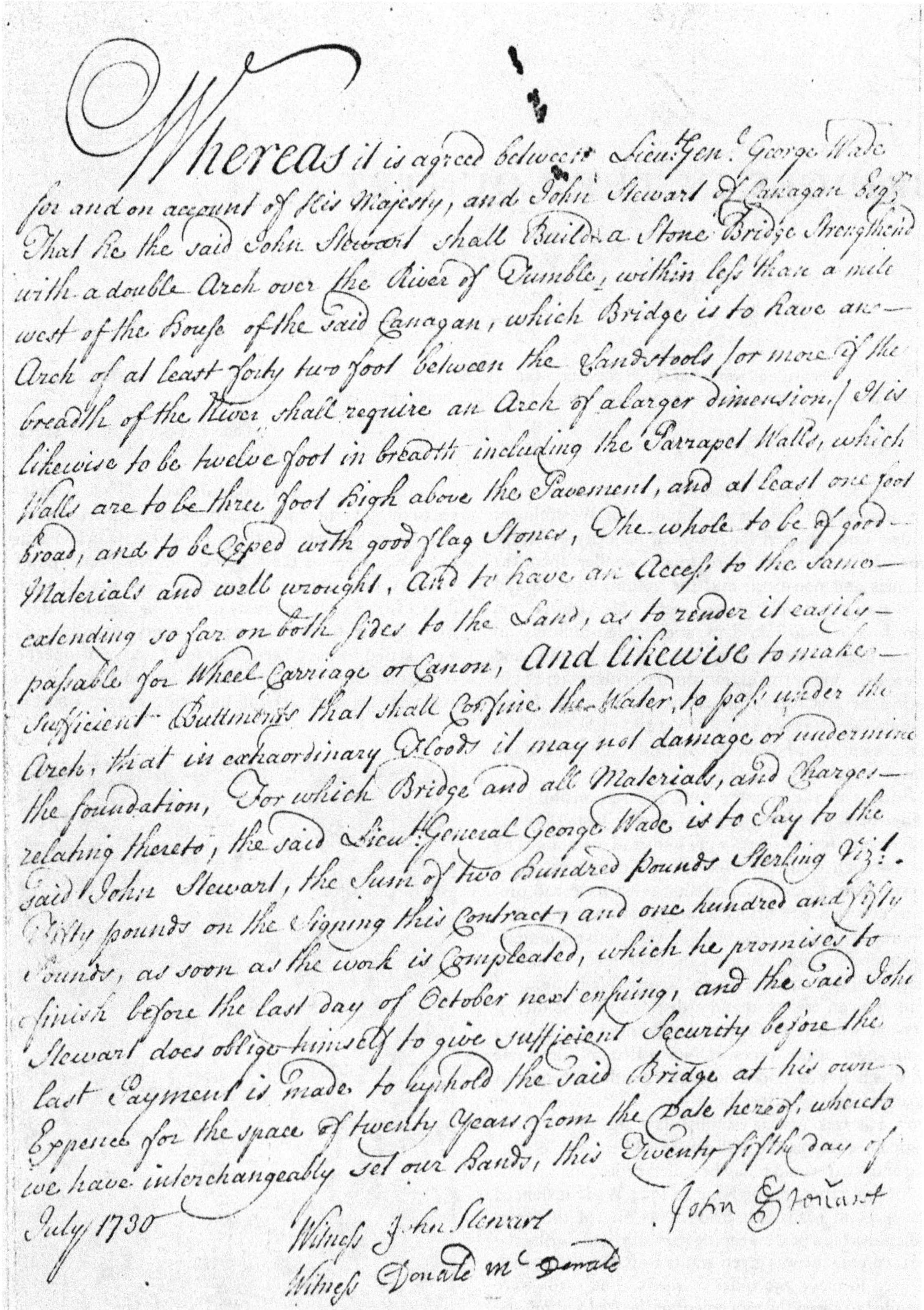

Whereas it is agreed between Lieut Gen. George Wade for and on account of His Majesty, and John Stewart of Canagan Esq. That he the said John Stewart shall Build a Stone Bridge Strengthened with a double Arch over the River of Tumble, within less than a mile west of the house of the said Canagan, which Bridge is to have an Arch of at least forty two foot between the Land stools, for more if the breadth of the River shall require an Arch of a larger dimension. It is likewise to be twelve foot in breadth including the Parrapet Walls, which Walls are to be three foot high above the Pavement, and at least one foot broad, and to be Coped with good flag Stones, The whole to be of good Materials and well wrought, And to have an Access to the same extending so far on both sides to the Land, as to render it easily passable for Wheel Carriage or Canon, *and likewise* to make Sufficient Buttments that shall Confine the Water, to pass under the Arch, that in extraordinary Floods it may not damage or undermine the foundation, For which Bridge and all Materials, and Charges relating thereto, the said Lieu. General George Wade is to Pay to the said John Stewart, the Sum of two hundred pounds Sterling Viz. Fifty pounds on the Signing this Contract, and one hundred and fifty Pounds, as soon as the work is Compleated, which he promises to finish before the last day of October next ensuing, and the said John Stewart does oblige himself to give sufficient Security before the last Payment is made to uphold the said Bridge at his own Expence for the space of twenty Years from the Date hereof, whereto we have interchangeably set our hands, this Twenty fifth day of July 1730

Witness John Stewart

Witness Donald mc Donald

John Stewart

military labour was used to make the roads; but whether Wade was ever able to contract with a local master mason is a matter of some doubt. The only one of his contracts which is now to be found was made with a local gentleman near Tummel Bridge and one has to assume that the gentleman employed journeyman masons to do the building, possibly importing them from lowland towns.[3] In the contract for Tummel Bridge (fig. 16) Wade does not enter into the details of either design or construction; even the precise site of the bridge is left to the contractor, within a space of a mile. A minimum span for the arch is specified, together with easy approaches (but 'easy' is not defined), a 3-ft high parapet and 12-ft wide roadway, the latter being apparently a standard width adopted for all Wade's bridges. In return for so much freedom with regard to construction the contractor is bound 'to uphold the said bridge at his own expense for the space of twenty years', a fearful commitment which could never have been wrung from any tradesman in a more civilised part of the country. The price to be paid is only £200, although this is one of the larger bridges on Wade's roads, a single arch of 'at least forty-two foot between the landstools' – which is very likely the arch of 55 ft still standing, though a second arch has been added.

The use of contracts which specified nothing but the main dimensions may explain why General Wade's bridges were built of traditional forms and materials which were well established before he came to Scotland and continued to be used under his successors Major Caulfeild and Colonel Skene. These two officers extended the military road system until by 1785 it comprised 682 miles of roads and 938 bridges.[4] The continuity of style has resulted in bridges all over Scotland, many in places where Wade never set foot, being known today as 'General Wade bridges'. A well-known example which predates Wade is the old Carr Bridge, built in 1717 with a width between the parapets of only 7 ft,[5] and a very late example is Sluggan Bridge about 3 miles further up the Dulnain, where a bridge was built by Wade but the present one was only erected after the great floods of 1829.[6] By then the traditional building had given way to the written specifications of Thomas Telford, but Sluggan Bridge is very much in the old style.

The style, if it can be called a style, was derived from the methods of building and these can be deduced from a study of the standing bridges. If at all possible a site was chosen which allowed the bridge to be founded on rock above the summer water level and this has resulted in quite a number of bridges, such as Dulsie Bridge in

17   Crubenbeg Bridge.

Nairnshire (a post-Wade military bridge), the old Carr Bridge (a landowner's bridge built in answer to a petition of parishioners),[7] Feshie Bridge and the old bridge at Invermoriston (both of unknown origin), springing dramatically from rock to rock over cascades of tumbling water. An example which was probably built by Wade is the single arch crossing the Truim at Crubenbeg 5 miles north of Dalwhinnie (fig. 17). The two-arch bridge nearby at Crubenmore is much later; it required a pier to be founded in flowing water and such piers were very few up to Wade's time and for some decades after him.

The arches were almost invariably built of local schistose stone, commonly called 'whin', which quarries in thin flat pieces quite long enough to form the thickness of an arch but with very irregular edges and surfaces. The most regular of these stones were chosen to make the faces of the arch on the elevations and others, often thinner, were used to make up the rest of the arch. For some, and possibly a majority, of the arches, it is likely that the arch stones were first placed on the centring standing on their ends, with little or no mortar and presumably wedged tight with thin stones or slates from one abutment to the other; mortar was then poured or packed into the irregular voids between the stones. This is suggested by the fact that the thin voussoirs on the facades of many of the bridges stand much nearer to vertical than to the direction radial to the curve of the soffit. However, there are also some

bridges (see fig. 18) in which the stones for some distance up from the springings lie at angles nearer to horizontal than radial, suggesting that these portions of the arches were built by successive corbelling without centres and presumably using mortar as the stones were laid; higher up the same arches some stones stand higher than radial, as can also be seen in fig. 18, suggesting that the other technique was used after the initial corbelling. The Scottish masons had clearly absorbed no inhibiting ideas from geometry or aesthetics about the proper direction of joints between voussoirs; neither had they any notion that semicircular arches were best, for even the shortest arches, which could easily have been made semicircular, are sometimes segments of no more than 90° or 100°.[8] This would be an advantage in the tight-wedged method of forming an arch envisaged here.

The spandrels of all these bridges were filled with gravel or earth and the external walls built of the local rubble masonry, sometimes of whin but equally often of granite. The granite is not squared stone and is probably roughly dressed boulders rather than quarried material. The plane and vertical spandrel walls continued without any break into the wingwalls, a noticeable difference from the Telford bridges which came after them. Many have been buttressed or tied in later years, outward bulging is normal and some of the spandrels and wingwalls have collapsed, leaving bare arches, such as Carr Bridge and the old Invermoriston Bridge, which remain perfectly stable. An alternative process of decay can be observed in bridges where holes have formed in the arches, for instance Crubenmore Bridge (fig. 19, seen before it was repaired in 1975) and the two bridges near Meall Garbha on the road over the Corrieyairack Pass from Lagganbridge to Fort Augustus. Water percolating through the filling material over the arch slowly leaches the lime out of the mortar which then begins to fall out of the soffit, assisted no doubt by the bursting pressure of ice formed within it in winter. The volume of mortar is so great that its loss allows some of the irregularly shaped stones to shake loose and eventually one or more fall out to leave a hole in the arch. Although the bridge is always closed to traffic at this time if not before, extension of the hole is very slow because the original packing and wedging of the arch stones was such that each one is held up by friction at its contacts with two or more of its neighbours. The thrust in the arch is being transmitted, as it always has been, through these contacts and hardly at all through the mortar, which served only to steady the stones. The large forces at the contacts ensure large enough friction forces to support each stone until decay affects the stone itself, and the result is a progressive but slow loss of arch stones, one or a few at a time.

These simple building methods served for all but two of Wade's bridges. The Tay was much too wide and large a river and he chose to cross it at Aberfeldy by a bridge of five arches; while the line of his road through the Great Glen from Fort William to Fort George

19 Hole in arch of Crubenmore Bridge, looking downwards.

crossed the Spean where it ran in a steep-sided gorge about 100 ft deep, and he decided to raise his bridge to near the top of the gorge so that it required three arches, two of 40 ft and one of 50 ft, with two tall piers. The road was 80 ft above the common water level. He christened it High Bridge and had it built with more finesse than was his wont. The piers, which still stand, have granite quoins and cutwaters extending up half of their large height, dressed back to the spandrels in three stages; and an old photograph shows a well-defined arch ring on the face of the middle arch and a projecting

cornice. It cost £1,087, was begun in 1736 and was the last piece of work in Wade's campaign of road-building. Its decay, before many of the smaller and rougher bridges, may be due to neglect after the route was abandoned as a main public road about 1815. There is now only one arch standing.

The Aberfeldy bridge (fig. 20) was the second last in Wade's campaign and was clearly designed to be his memorial. It cost £4,095, more than half his total spending on bridges. The House of Commons was told in 1734 that it would be 'a freestone bridge over the

20 Aberfeldy Bridge.

21   Framing and setting at base of a pier of Aberfeldy Bridge.

Tay, of five arches, nearly 400 feet in length, the middle arch 60 feet wide, the starlings of oak, and the piers and landbreasts founded on piles shod with iron' and Wade reported later that

all proper precautions were taken to render ye work secure and lasting. The best architect in Scotland [William Adam] was employed and master masons and carpenters sent from ye northern countys of England, who were accustomed to works of that nature. These with some of ye best masons of ye country and about 200 artificers and labourers from ye army were employed for a whole year [1733], the winter season for preparing materials and ye summer in laying ye foundation and in building ye bridge . . . There are 1,200 piles shod with iron to support the foundation of the piers and landbreasts, and the starlings are made of the best oak.

Here, it would seem, was the General's real education in practical bridge-building, and he was learning from the tradesmen of the North of England where Nicholas Hawksmoor found examples of large arches, just two or three years later, to include in his pamphlet on the proposed Westminster Bridge[9] (see chapter 1). Probably the Scottish masons and wrights present learned as much if not more. The river-bed was almost certainly

gravel and the water shallow in summer; low coffer-dams would probably have been used while starting the masonry on top of the piles. But in fact the foundations were not laid deep enough, for they were endangered immediately by scour at the tails of the piers and a 'dam' was made a few yards downstream to raise and steady the water as it passed through the bridge. It was almost certainly a line of sheet piles driven close together right across the river, probably with some stones or gravel dumped against it on one or both sides.[10] The idea of such a dam was that the river itself would be caused to deposit its sediment on the upstream side of it and so raise the bed round the feet of the piers themselves. This principle and several applications of it were discussed in Gautier's book,[11] and it must have been used in Britain before. Wade seems not to have reported the difficulty to Parliament but it must have made him cautious when discussing the much greater foundation problems of Westminster Bridge both in 1738–9 and in 1747. What is meant by the references to 'starlings' of oak is not clear. They must have been some sort of surround to the bases of the piers, but the only such protection which can be seen today (fig. 21) is of the type described below as used under Yorkshire bridges and lying flat on the river-bed. The line of the dam can still be detected when the water is very low, but it is only an irregular bed of stones now.

By his choice of architect (fig. 22) Wade obtained for posterity a bridge of remarkably fulsome decoration (fig. 20). It is all built of greenish-grey chlorite schist quarried locally, the arch rings rusticated but the piers and spandrels smooth, with an archivolt over the ring of the middle arch and plain projecting cornices and copings; all the arch rings have triple keystones. Triangular cutwaters are dressed back to form buttresses of trapezoidal plan running up the spandrels to enclose refuges from the roadway; and surmounting the refuges at the middle arch are four tall stone obelisks. There are other rusticated quoins, inscribed plaques and moulded pedestals, but the most important and most copied architectural feature is the profile of the parapet, outlined at its top and bottom by the projections of the coping and the cornice. The line is horizontal across the middle arch and its refuges; just beyond each of these refuges it turns vertically down and then curves round to become a straight line falling at a constant gradient to the end of the bridge. With this arrangement the arches on each side of the middle one have to be much lower, and logically also of much smaller span, than the middle arch itself. The plethora of decoration and mixture of architectural sources is typical of Adam[12] and probably pleased the General, as it pleases visitors today. But it has as much of the art of the wedding cake as of bridge architecture and that is not really appropriate to the landscape of Perthshire and the largest river in Scotland. The views of the Poet Laureate, Robert Southey, doubtless concurring with those of his companion Thomas Telford, when they viewed the bridge in 1819, are understandable. 'At a distance it looks well, but makes a wretched appearance upon close inspection. There are four unmeaning obelisks upon the central arch, and the parapet is so high that you cannot see over it. The foundations also are very insecure, – for we went into the bed of the river and examined them.'[13]

22 William Adam, architect of Aberfeldy Bridge. Portrait by William Aikman.

Between traditional Scottish bridges and those of Wales there are some similarities and some differences.[14] They were similar in the sites chosen, the Welsh masons having the same preference for single arches founded on rock above the summer water level, and in the form of the arches, segmental arches being as common as semicircular. The differences were in the materials used. Firstly, the Welsh limestone quarries included those of Aberthaw and Penarth which were known to give hydraulic lime as strong as any in Britain. Secondly, there was hard stone in most parts of Wales which quarried easily into oblong blocks, generally of small size but true surfaces and with good square edges and corners. The thicknesses varied but were sometimes very small, as little as 2 to 3 in. As a result the Welsh masonry is very much neater than Scottish

23 Survey drawing of Wakefield Bridge, 1752, with façade of widening added 1758.

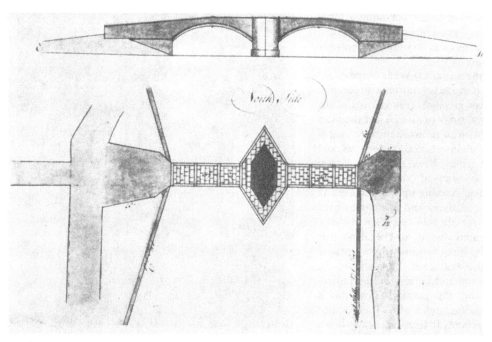

24   Survey drawing of Newing Bridge, 1752.

rubble and looks almost like hewn work, although very little stone-dressing was done (see figs 49 and 50). The arches were built with joints strictly radial to the soffit curve and with constant thicknesses of mortar, at least in the regular arch rings on the faces. The largest stones available were chosen for these rings, the rest of the arch being usually of thinner stones in less regular courses, and they were doubtless 'flushed' with mortar of pouring consistency although they were probably bedded in stiffer mortar first. There is reason to believe, at least in some instances, that the thickness of the arch through these inner parts is less than that of the arch rings on the faces,[15] and if this is the case generally the Welsh arches must often have been much thinner than the textbook rules of proportion demanded. The high strength of the lime in use, together with ignorance of the textbook languages – even ignorance of English was common – could do much to explain the growth of such practice.

The description above would apply to most of the bridges built and maintained by the Justices of Welsh counties during the eighteenth century. The Justices of the Peace throughout Britain were empowered by law to maintain and improve all bridges within their boundaries which were deemed to be of use to the public,[16] that is, the population of the Kingdom as a whole; but their response to the law varied enormously from county to county. As late as 1826 the Justices of Middlesex recognised only four bridges within their borders as 'county bridges', while Devon had 247 county bridges by 1809 and the West Riding of Yorkshire had 112 in 1702 and 120 by 1752. The North

Riding had very few county bridges in 1700 but adopted or built eighty-one more before 1760. By 1806 it had 115.[17] The West Riding Justices ordered a survey in 1752 of all their county bridges, and the drawings which were made for them[18] give an accurate impression of the bridge stock of one county. Full statistics are given in appendix 4 and only some generalisations here.

Twenty-six of the 115 bridges surveyed displayed some antiquity by their arches being either ribbed or pointed. It is worthy of note that only one had arches *both* pointed and ribbed, though both techniques are commonly regarded as medieval. Two of the twenty-six, the bridges of Rotherham and Wakefield, had chapels at midstream and these, together with a small house on Castleford Bridge (possibly a tollhouse), were the only buildings on any of the 115 bridges. The largest arch span was 68 ft (at Keighley) and only seven bridges had arches of more than 50 ft in span; but even for short spans, semicircular arches were not common and the largest arches of most of the bridges were segments between 90° and 120°. Most of the old ribbed arches were also segments within this range. Only fifteen bridges were said to be founded on rock and thirty-nine had some paving of the river-bed to resist scour. The technique used was called 'framing and setting' (fig. 21) and it consisted of frames of timber balks laid flat on the river-bed and usually fixed to short piles driven into the bed, the frames then packed with large squared stones or 'setts' which would not be easily moved by the current and so defended the river-bed against the violence of floods. It was a method used throughout the country to prevent scour near piers whose foundations were at a shallow depth below the surface of the bed. It might be used as a surround to each pier (as at Wakefield Bridge, see fig. 23) or any pier or abutment especially threatened, but on bottoms considered to be very unstable the whole bed from bank to bank (as at Newing Bridge, fig. 24) and sometimes for some feet up and downstream was framed and sett. In the West Riding this had been done at fourteen bridges, which, with twenty-five others partially framed and sett, suggests a high proportion of shallow foundations.

The most unsatisfactory feature revealed by the 1752 survey, however, was the narrowness of the bridges. Only one small bridge, recently adopted, was more than 18½ ft wide and only eleven were wider than 15 ft, both these figures including the thickness of the parapets.

The great majority were only 10 to 14 ft wide, which was insufficient for any two-way traffic by wheeled vehicles. At least thirty-six also had narrow or angled approaches, including Newing Bridge (fig. 24). Between 1758 and 1768 five old bridges at important river crossings, all consisting of strings of short arches, were widened or rebuilt to widths of 23 to 27 ft. These were the busy town bridges of Wakefield (fig. 23), Leeds, Rotherham and Sheffield, and Ferry Bridge on the Great North Road. They were the first of many widenings and reconstructions the total effect of which can be seen by comparing the statistics for the West Riding bridges in 1752 with those from a similar survey of North Riding bridges about 1805 which are also given in appendix 4.

3

TIMBER BRIDGES

Many persons of judgment and taste were extremely disgusted at the thoughts of a wooden bridge.

Short narrative of the proceedings . . . in obtaining the Act for building a bridge at Westminster (1737)

Designers of timber bridges from 1715 onwards leaned just as heavily on the books of Palladio and Gautier as those who designed in stone. Both writers offered designs which would extend the spans of timber bridges beyond the 20 or 25 ft which was possible in a bridge of horizontal floor beams supported on rows of piles; and this was the improvement most needed in British bridges of timber.

Palladio[1] described only one timber bridge which he had actually built and it was a pile bridge of several spans not unlike the multi-span bridges at Kingston, Chertsey, Staines and other towns up the Thames, except that it was roofed like many Alpine bridges. For longer spans he proposed four designs, all drawn for a single span. One (fig. 25, bottom left) was a type of which he had seen an example spanning the river Cismone in northern Italy. It was of two trusses with five cross-beams slung from their bottom chords to carry the road. The span between masonry abutments was over 100 ft. Two other designs (fig. 25, right) were for truss bridges of different form, and the fourth (fig. 25, top left) was an 'arch' of eleven 'voussoirs' each of which consisted of a timber frame with X-bracing. In all four designs the carriageway lay entirely between the trusses or arches which therefore served as guard-rails as well as structural members. Gautier[2] reproduced and described all Palladio's designs and was able also to give a drawing of the recent (in 1714) three-span Pont St Vincent at Lyons with trusses up to 90 ft in span, and to refer to a serious proposal to bridge each of two branches of the Seine at the village of Sèvres near Paris by an arch made up of prefabricated timber 'voussoirs'. In that design, however, made by the architect of the Louvre, Claude Perrault, at some date before 1688,[3] there were five arch ribs, not two, and they had therefore to be placed underneath the road, though the published drawing shows an upward extension of the outside ribs to form guard-rails. The span proposed by Perrault was 150 ft. Although the arches were of low rise, the placing of them under the road occasioned a need for high approaches. Gautier added three other

designs which might be called 'arches'. The first would be best described as a double trussed arch and had only been used for spans up to 36 ft; the second was in part a cantilever-beam design and in part a truss, and it had been used up to 60 ft in span. The third (fig. 26), which was as yet an untried 'invention', made an approach towards the methods of lamination which were used much later for very large arches. Gautier proposed to make an arch up to 150 ft in span with 'voussoirs' composed of straight pieces of timber only 5 or 6 ft long, using as many pieces as necessary placed parallel and tied and braced together both radially and across the width of the arch, and then to raise timber framing from the arch to carry a horizontal road. He did not explain how the vital junctions between adjacent 'voussoirs' would be made. He considered the design particularly suitable for arches of large rise, unlike later exponents of laminated arches. He was helped by the fact that in a high arch the whole of the spanning structure could lie below the line of the road and there could therefore be as many ribs as were needed.

The brief descriptions of timber bridge designs recorded in the minutes of the Fulham Bridge Commissioners in the years 1726–9 read like a preview of the designs submitted for Westminster Bridge in 1737–8; and so do the names of most of the designers.[4] A design by John Price had trusses in the planes of the side-rails, supported by piles, and the road carried by cross-beams as in Palladio's truss designs. It was rejected on the ground that the cross-beams would sag, the Commissioners thus showing awareness of the difficulty which must beset any of Palladio's truss and arch designs if used for a wide bridge. A later design by Thomas Ripley consisted of eleven or more timber 'arches' on stone piers and was described as 'a new invention', but no details of it are given. After many meetings, all novelty was abandoned in favour of a traditional bridge of timber piles, beams and floor, very little different from the old bridges at Kingston and elsewhere. It was

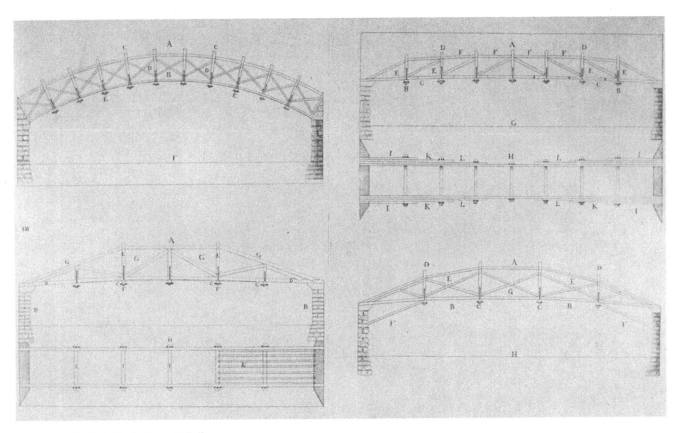

25   Timber bridge designs by Palladio.

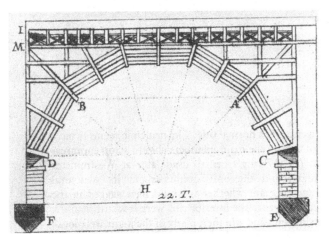

26   Laminated arch design by Gautier, 1714.

designed by an amateur, Sir Jacob Acworth, and built by Thomas Phillips, carpenter, all in about nine months, for less than £14,000. The overall breadth was 24 ft and the spans in general 15 ft with five wider openings for navigation, the largest being 30 ft.

Of the timber designs for Westminster Bridge more detail has survived. One was a direct copy of Gautier's double trussed arch, with spans of 100 ft, by a certain John Westley of Leicester.[5] The boldest design was by

his brother William Westley, a design said to be of 'one arch and two water passages',[6] which apparently meant a single arch of enormous span and two auxiliary openings at the ends. There are no further details and no surviving drawing. Batty Langley offered two designs (fig. 27) which were more realistic and, as was his wont, prophetic of future developments.[7] Both consisted of trusses spanning 100 ft between 'piers' of timber piles and masonry. The whole of the trusses was above high-water, thus avoiding decay, and below the road, allowing them to be at a lateral spacing of only about 10 ft. The height for navigation above high-water was 22 ft in one design and 13 ft in the other. The disadvantage was that the road surface was about 40 ft and 30 ft respectively above high-water and the approaches required would have been very high and expensive. The prophetic aspect was the form of the trusses, each having a series of continuous struts rising at low angles from the piers or near them right to the main king-post at the middle of the span. This arrangement was proved astonishingly effective when Ulrich Grubenmann built his great bridge at Schaff-hausen in Switzerland in 1754 (fig. 36), a bridge which rapidly and rightly became famous all over Europe.

The Westminster Bridge Commissioners rejected all these offers in favour of James King's design of thirteen arches standing on stone piers,[8] the middle arch of 76 ft span (fig. 28). King's trussed arches owed nothing to

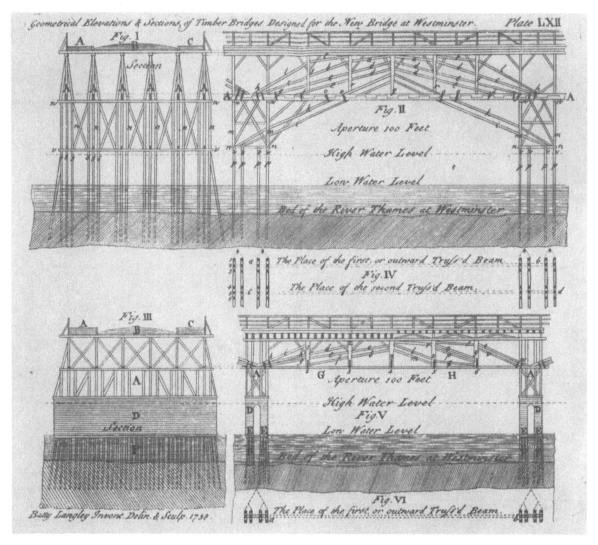

27 Designs for Westminster Bridge by Batty Langley.

Palladio or Gautier; on the contrary, they could be considered an application of the principles used by Batty Langley (and later by Grubenmann) to openings of arched shape. All King's arches were of approximately semicircular 'soffit'. Every arch had five ribs and each rib was made up of straight timbers extending the lines of each of the nine short members which formed the soffit 'curve'; they extended upwards as far as they could go – to the cross-beams under the road in the three interior ribs, to the top of the guard-rail in the two outer ribs – and downwards to the top of the pier or else to meet like struts from the next arch over the middle of the pier, from which point the load was taken down to the pier by vertical posts. The five ribs were linked by ten cross-beams under the soffit, and radial members from these beams went up to the top of the guard-rail. There was also diagonal bracing in the soffits to prevent lateral sway of the ribs. The tops of the piers were at ordinary high-water level, so none of the timber would

be wetted by a normal tide. The principle seems to have been that all the main members should act in compression (as in King's centre used later at Westminster Bridge), and the only doubts about its structural competence are whether it would have sagged in the middle because the interior ribs were less stiff than the outside ones and whether, if any of the joints were loose under tensile force, there might be some longitudinal sway of an arch when it was loaded unequally by traffic. As noted in chapter 1 the contract for construction of this bridge was later annulled, but the principles were adopted by at least three other designers and two of their bridges were built.

The first was William Etheridge, King's foreman at Westminster who after his employer's death took over the carpentry of that bridge on his own account. While still at work there he designed a bridge over the Thames for a wealthy landowner named Samuel Dicker, of Walton in Surrey, and Dicker obtained an Act of

Parliament in 1747 permitting him to build this bridge and charge a toll to reimburse himself.[9] An unusual clause in his Act was the stipulation that no foreigner should be employed in building the bridge, an evident insult to Charles Labelye but an ineffective insurance against his participation because he had become a naturalised Englishman by Act of Parliament in 1746.

In anticipation of opposition from bargemen, which duly arose, it was decided to build a single arch over almost all of the ordinary navigable channel. To keep the road at reasonable gradients this arch had to be of low profile and its rise only 27 ft in a span of 130 ft. These dimensions were written into the Act. There were side arches of timber of 44 ft span and approach ramps faced with brick and pierced by five brick arches to help the discharge of flood water.[10] The timber structure was all above an assumed normal flood level which itself was 7 ft above the normal low-water (see fig. 29). The tide does not run to Walton. The piers up to the assumed flood level were built of stone and founded in caissons only a few feet below the normal

(low) water level. Etheridge took the whole of King's system of trussing and gave it its first practical trial. He had the advantage of needing only three ribs because the road was only 19 ft wide and of these ribs the outside pair had the full stiffness imparted by the height of the guard-rail. Etheridge had so little faith in the strength or stiffness of the middle rib that he strengthened each of the cross-beams between the outside ribs into a trussed beam with two slanting struts notched into its ends and into a king-post on the line of the middle rib (fig. 30).

Walton Bridge was opened in 1750 and extravagant claims were made for it. It was said that any one piece of timber could be replaced without disturbing the remainder, and this may have been true of the arch ribs – it was not true of the floor. Dicker claimed that it could last for two hundred years, but he wished that he had eased the gradients by making the end arches longer and higher. It attracted Canaletto to paint it twice and a number of engravings of it were published. It was the longest arch in Britain for a few years and the

28   Design for Westminster Bridge by James King.

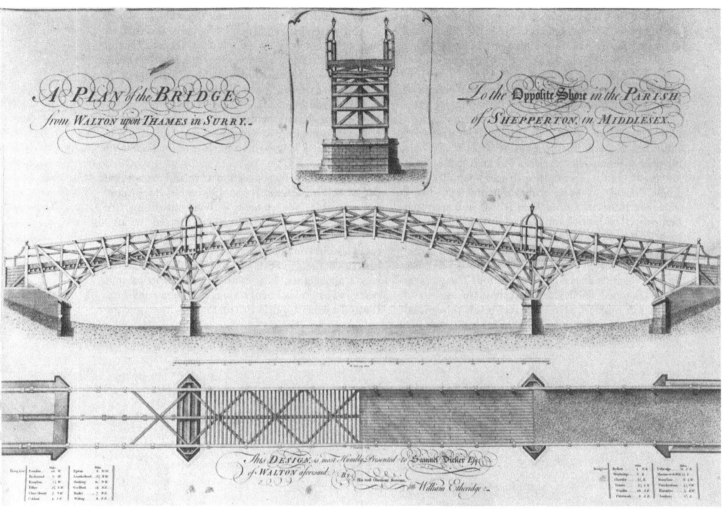

29   Walton Bridge, 1750.

longest timber span until well after 1800. And it proved the strength of the trussing system King had devised. It failed, however, after less than thirty years for what seems a trifling fault. When John Smeaton was called upon to examine it in July 1778 it was already propped by a scaffold erected on piles under the middle of the large arch. Smeaton found the main structure perfect: 'the timber in general was perfectly sound, and much more hard than it was when the bridge was erected', but in a few isolated parts it was 'perfectly rotten and decayed' and these were 'critical points where several pieces were assembled, and thereby loosening the geometrical bond and dependence of one part upon another', so that the whole would have been brought eventually to collapse. The precise location of the initial decay was in the joints where the sloping struts of the trussed cross-beams were notched into the beams themselves (fig. 30), but only in the four nearest the crown of the arch where the top surface of the beams was nearly horizontal and the rainwater which perco-lated through the gravel of the road had lain long enough to seep into the joints. Only one end of one cross-beam had completely given way, pulling away from the iron strap which held it to the outside radial member, but this threw so great a load on the middle arch rib at this cross-beam that one or more of the timbers of the rib had sprung out of line, cracked and splintered, causing the whole of the road near the crown of the arch to sag and twist.

Smeaton's advice, typically economical, was that the props should be left in place and the bridge would probably last for another twenty-five years in that unsightly condition. His report records other interest-ing facts about the structure. The weight and thrust of the large arch had been so great that the lower ends of the main struts had 'perfectly sunk into the bed timbers' laid on top of the piers, and the piers themselves had tilted towards the river-banks; the simultaneous depression of the crown of the arch was more than 2 ft from its original level. Smeaton thought the weight of the gravel had probably caused most of this movement and recommended that it be reduced. In

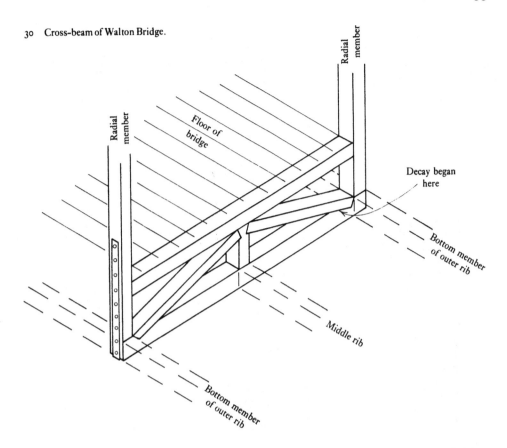

30   Cross-beam of Walton Bridge.

Radial member

Radial member

Floor of bridge

Decay began here

Bottom member of outer rib

Middle rib

Bottom member of outer rib

fact, however, the bridge was taken down a few years later and replaced by a four-arch bridge of stone (see chapter 9).

The second use of King's system of framing was by the man who was his partner in the unexecuted Westminster Bridge contract, John Barnard. He designed a seven-arch bridge to cross the Thames at Kew (fig. 31), the largest arch of 50 ft span and each arch made up of four ribs.[11] The main struts of the outside ribs did not extend up into the guard-rail but the radials did and there was X-bracing in the panels between them. The supports were treble lines of piles

forming 'piers' pointed up and downstream and rising almost 10 ft above the normal water level to the springings of the arches. It was built in 1758–9 and replaced by a stone bridge in 1784–9 (see chapter 9). The details of its decay are not recorded.

The third user of King's trussed arch system was John Smeaton, in an unexecuted design for a bridge at Richmond-on-Thames probably made about 1772.[12] The details of the connections are shown more clearly in his large-scale manuscript drawing (fig. 32) than in the prints of the designs by King, Etheridge and Barnard.

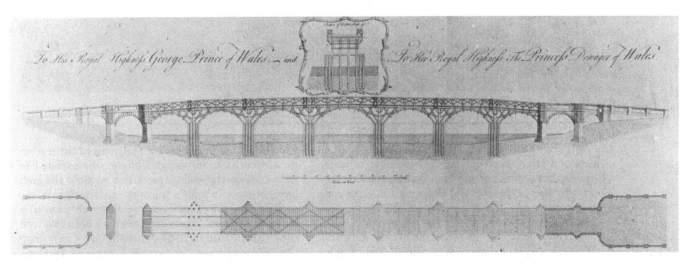

31   Kew Bridge, 1759.

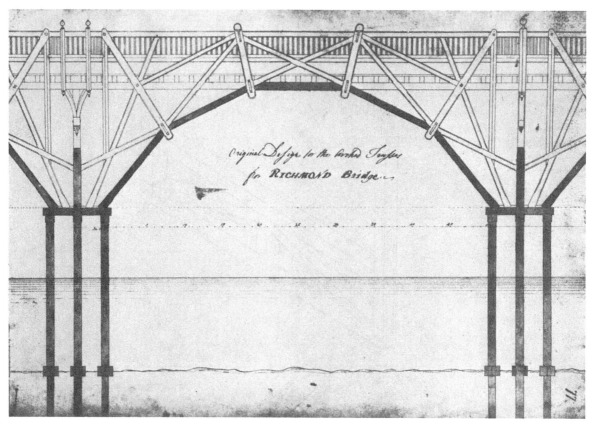

32   Design by Smeaton for a bridge at Richmond-on-Thames.

33   Hampton Court Bridge, 1753.

Between Etheridge's bridge and Barnard's, both in time and place, stood another timber bridge of seven arches. It was designed and built at Hampton Court in 1753 by Samuel Stevens and Benjamin Ludgator and, to a modern eye, the design looks as frivolous as the names of its builders (fig. 33). But for its time it was a large construction, about 20 ft wide like the Kew and Walton bridges, its timber 'piers' pointed, and its style, which was consciously ornamental, derived either from the current taste for things Chinese (the pagoda-like tollhouses on the middle piers and possibly the arch form) or from Palladio's 'voussoir' arch design. It was probably the only application of this arch form to a multi-arch bridge. There can only have been two arch ribs, with the road carried by cross-beams of 20 ft span. One contemporary description[13] explains that the profile of the road did not follow all the ups and downs of the arches, so the thickness of gravel near the piers must have been large and the height of the guard-rail from the road very uneven. There appear to have been pedestrian recesses just where the adjacent arches should have thrust against each other, a mistake which careful reading of Palladio would have avoided. With the large span of cross-beams, the heavy load of gravel and the lack of abutment for the arches it is not surprising that this bridge had to be replaced in 1778. The Palladian pattern-book was not for use by novices. The replacement was another pile bridge of short spans which, in spite of its out-of-date construction, lasted until 1864.

The influence of Palladio was strongest in the parks of gentlemen's houses and many small bridges were built both of the 'voussoir' type and the truss type. Thomas Telford noted one of the 'voussoir' type in the Marquis of Buckingham's estate at Wotton in 1812 which was 87 ft in span, 13 ft rise and 20 ft wide.[14] Being generally quite narrow and lightly loaded, such bridges could be supported adequately by only two ribs. There were also small bridges inspired by King's method of trussing. The best known is a footbridge spanning the Cam at the back of Queens' College, Cambridge, which has been repaired and renewed many times without changing its structural form. It was designed by William Etheridge and built in 1749–50 by a carpenter named James Essex, and Essex is said to have built others of the same type in and about Cambridge and to have called them 'mathematical' bridges.[15] This probably refers to their visible frame-work, which Smeaton would have called 'geometrical construction'. There was a very similar footbridge over the canal at Winchester until 1976 (fig. 34).

Abraham Swan published designs in the 1750s which overcame some of the difficulties in the Palladian models. He was a successful London tradesman and built at least two small timber bridges at Blair Atholl in Perthshire, but his designs published in 1757 and 1759[16] range from small to very large spans. He proposed a single-arch bridge of 220 ft span to cross the River Tay at Dunkeld (fig. 35), with four arch ribs of the 'voussoir' type, but all four entirely under the floor of the road, and he claimed that this type of arch could be made to span even longer openings.

In the same decade Ulrich Grubenmann built his two-span bridge over the Rhine at Schaffhausen (fig. 36) with spans of 171 and 193 ft, and a few years later he and his brother Johannes built a single span at Wettingen near Zurich using laminated arch ribs of 200 ft span and 25 ft rise, the seven laminations made up from pieces 12 to 14 ft long and clamped together by iron straps every 5 ft along the arch (fig. 37).[17] These designs were obviously derived from the live craft of Alpine carpenters who had centuries of experience of bridging rapid rivers with wood. This craft was a more reliable source than any of the books followed in Britain

and a British bishop attempted to tap it for the benefit of his diocese. He was Frederick Hervey, a man of noble birth who opted for a career in the Church but was not willing to devote himself to the duties of a parish.[18] He became Bishop of Derry in 1768 through the influence of his brother who was Lord Lieutenant of Ireland, and he was later to become Earl of Bristol without relin-quishing his wealthy See. He had a passion for architecture and also for travel. Immediately after his appointment as bishop he donated £1,000 to the City of Derry towards the construction of a bridge across the Foyle and by the end of the year (1768) he had been in consultation with Robert Mylne.[19] Early in 1769 Davis Duckart in Dublin made designs both of stone and timber, but the bishop also wrote to a French contact asking for a plan and elevation of Schaffhausen Bridge. There was no bridge in Britain across a tidal water so deep as the Foyle and the use of long spans with very few piers was obviously an attractive idea. In or soon after 1770 the bishop himself went to the Continent, taking with him a young Cork architect named Michael Shanahan who measured and drew for him many of the large timber bridges in the Alps and northern Italy, including Schaffhausen.[20] At the same time he made the drawing for the surviving engraving (fig. 38) of a design which was brought to Derry in 1772; but it

34 Footbridge at Winchester.

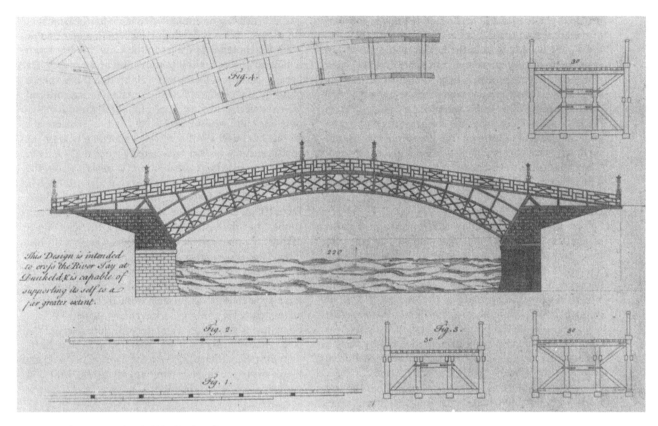

35   Design for a bridge at Dunkeld by Abraham Swan.

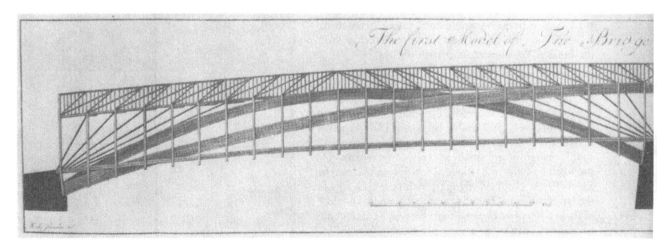

38   Design for a bridge at Derry, 1772.

arrived there not only as a drawing but in the form of a model 19 ft long and built from 11,734 separate pieces of wood. The designer was John Conrad Altheir, a mason of the canton of Appenzell, and he had pushed the model across Europe on a four-wheeled cart with the help of two of his countrymen, the journey taking almost six months at the rate of 6 to 8 miles a day.[21] The design was similar to Schaffhausen Bridge in some respects, having a single pier in the middle and two spanning systems, one a single arch spanning from end to end with its crown over the middle pier, the second a pair of arches from the ends to the middle pier (fig. 38). Each spanning system was presumably of two ribs, one each side of the road as at Schaffhausen. It was unlike Schaffhausen in its arches, which were all curved and laminated, whereas the Schaffhausen spanning systems were really trusses with long struts rising at low angles from the supports to midspan, like Batty Langley's

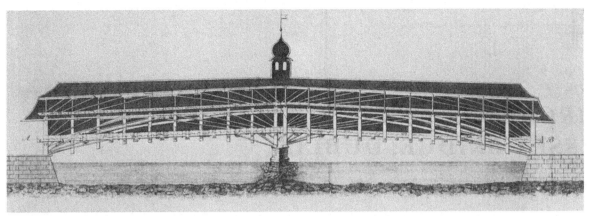

36   Schaffhausen Bridge.

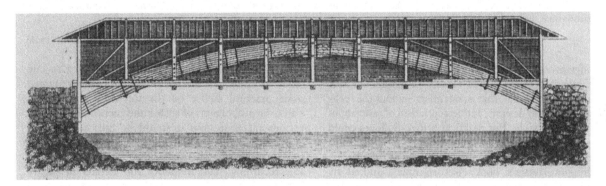

37   Wettingen Bridge.

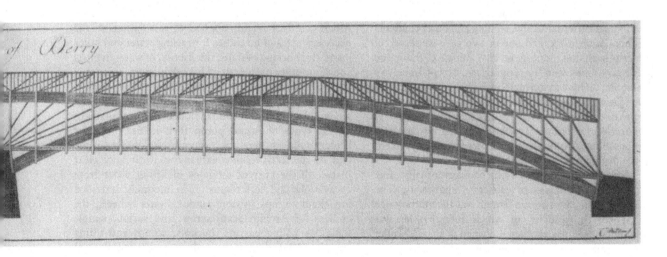

designs for Westminster Bridge (fig. 27). It was also more than twice the size of Schaffhausen, both in total length and the length of the individual spans, each span being over 400 ft. The laminations of the arches were to be held together by some sort of saw-tooth notching, and the whole bridge was to be enclosed by wooden walls and roof in the manner of Swiss bridges, with sixty-two windows in each side.

Why this design was abandoned is not known. The bridge built at Derry in 1789–91 was a variant of the simple piled system with many short spans and was the first of six thrown over wide Irish estuaries and rivers in the short space of seven years; but not being arch bridges they will not be discussed in this book. No further developments of long timber arches for public bridges took place until about 1795 and developments after that date belong to chapter 13.

4

ARCHITECTURE HYDRAULIQUE AND ESSEX BRIDGE, DUBLIN

I had frequent reference to my books . . . They told me, indeed, to make an inclosure; and so might they tell a man, that to measure time, he must make a clock.

George Semple, *A treatise on building in water* (1776)

Bridge promoters and parliamentary committees in the eighteenth century turned readily to mathematicians and 'philosophers' – that is, scientists – about the technical aspects of designs, and to gentlemen of taste about questions of style. But practical bridge-builders could expect little help from such people in their most difficult task, that of laying foundations in tidal rivers. Though Gautier's book[1] was so much quoted by English pamphleteers, it is doubtful if operative tradesmen used it much, as it was written in French and rather inadequately illustrated. Most of the pamphlets concerning Westminster Bridge in 1736–9 and 1747–51 mentioned cofferdams, sheet-piling and other elements of foundation work, and one or two gave some verbal descriptions, but none gave any guidance about the necessary dimensions or proportions, except Labelye's pamphlet on his caisson method.[2] The only drawings of foundation methods in the whole series of pamphlets were those in Charles Marquand's *Remarks on different constructions of bridges* (1749), illustrating a single new proposal for floating a half-built pier from a convenient place of building to its intended site.

More general and more useful books eventually came from the new profession of teachers of engineering. The first school for engineers in Britain was the military one of Woolwich Academy of which John Muller was appointed headmaster when it opened in 1741.[3] Muller was German by birth and must have had a thankless task trying to teach cadets who were generally only twelve to fourteen years old and who had to be restrained by watching corporals 'from leaping upon or running over the desks'.[4] His interests are revealed by his first book, *A mathematical treatise*, which was published in 1736 and deals more with mechanics than with pure mathematics, but has virtually nothing to say about practical construction. The French school of artillery had been established much earlier, in 1720 at La Fère, and one of its first teachers was Bernard Forest de Bélidor.[5] His first book was also on mathematics and

was published in 1725, but he followed it with one called *La science des ingénieurs* in 1729. This contained some practical advice on the thickness of retaining walls, the proportions of arches and piers and the use of various materials, but it would be mostly of use to designers. Bélidor clearly spent years in the field studying construction work of both past and present and recording measurements and details in plenty before he was ready to lay down firm rules of procedure for all the outdoor work comprehended by the title of his great work *Architecture hydraulique*. In the first two volumes of this book he gave all the theory of mechanics which could be readily applied to calculations of the necessary sizes of machines for raising water and other loads, the working of mills, the friction of water in pipes and some other problems, all his calculations being true to the theory and generally simple enough for practical use. The first volume was published in 1737 and the second in 1739. The second included excellent plates of all the types of piston pump then in use,[6] but for bridge-builders there was a more useful chapter at the end of the first volume explaining, with very clear plates, all the current methods of lifting water from excavations and cofferdams.[7] The methods described are chain pumps, piston pumps, water wheels, the endless (or Archimedean) screw, and various simple tools like baling baskets, buckets, scoops and tilting troughs. Bélidor warns the tradesman that mistakes are easily made in the design of piston pumps, water wheels and the endless screw unless hydrostatic and hydraulic principles are fully understood, and gives several instances of machines well made but very inefficient. For a lift of about 8 ft he considers that chain pumps give good service, with output of at least 2,600 cu. ft per hour from the constant work of four men turning crank-handles (see fig. 39). For higher lifts two or three pumps can be used, the first discharging the water into a basin from which the second lifts it again, and so on. The pump may consist of a string of buckets on an

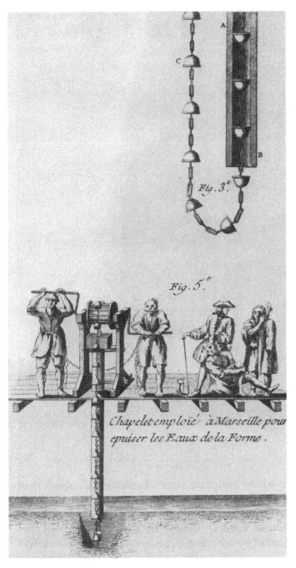

39   Chain pump worked by convicts (from Bélidor, 1737).

moving end dipped in the water to fill it and then lifted by two men to shoot the water over the dam. A two-ended see-saw trough (in the same figure) has a similar height of lift. For really fast baling, however, as required in very leaky ground or to empty an excavation for a short period at low tide, he prefers to do without machines and simply employ a large gang of men with buckets, paddles and other hand tools. (Gautier had preferred men with buckets at all times – see appendix 1B.)

The third and fourth volumes of Bélidor's monumental treatise deal with large works as a whole, including ports, canals and bridges, but the volumes were long delayed and only published in 1750 and 1752 respectively. In volume III there is a chapter on piling and battardeaux with several plates of machines for driving and extracting piles and one showing sections of several battardeaux.[8] In the fourth volume is a long abstract of Labelye's description of his caissons together with a plate (see fig. 5) which must have been made from the drawings intended to be published with his *Description of Westminster Bridge* in 1751 – for they are the only drawings absent from the collection made by Thomas Gayfere for that purpose[9] and Telford copied the plate exactly when he wrote his account of the caissons in 1812 with the help of Gayfere's drawings and conversation with Gayfere himself. Gayfere was then aged ninety but still living in Abingdon Street, Westminster,[10] the street where Labelye had lived at number 24, a house Telford himself was to purchase some years later.

Almost the last thing in Bélidor's book was a chapter on masonry bridges in general.[11] His advice to designers on form and proportions, summarised below in appendix 1C, allows thinner arches and piers than either Gautier's rules or the rules he himself had proposed in 1729.[12] His advice on construction is all

endless chain but more often of paddles or pistons moving up a tube which dips into the water at its bottom end and discharges it at its top; it may be either on an incline or vertical. Several of the chain pumps Bélidor describes can be carried from one site to another as portable equipment.

He has little to say of the endless screw (fig. 40), but its height of lift could not be very great because it worked at a fairly low angle of inclination. Water wheels are too large and complex for temporary work and require a rapid stream to drive them. The 'hollandoise', a large long-handled scoop suspended from a tripod and swung by one man (see fig. 41, top right), can be used for lifts of 3–4 ft with an output which Bélidor estimates at about 450 cu. ft per hour and compares unfavourably with the output of each man on a chain pump. For similar heights of lift he describes a long trough (also seen in fig. 41) hinged on the top of the cofferdam, the

40   Endless or Archimedean screw (from Bélidor, 1737).

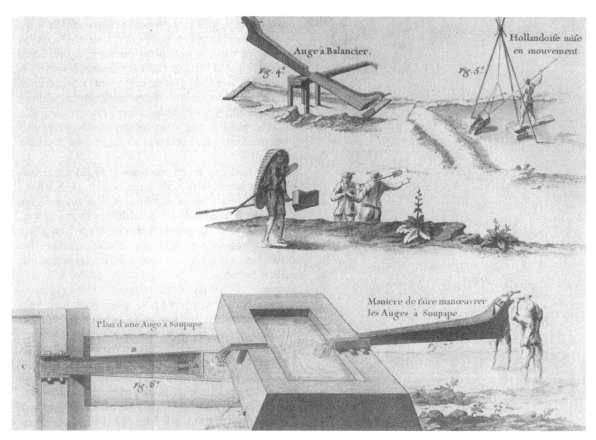

41   Baling out of excavations (from Bélidor, 1737).

derived from practical examples, chiefly those of the Pont Royal at Paris (1685) and the bridge of Compiègne (1735) where he had watched the work closely. The emphasis of the chapter is on the foundations and masonry of the piers and he adds in a further chapter a proposed method of founding a pier a few feet below water level without a cofferdam by sinking a grating, driving piles through it and cutting them off under water.[13] This invention had come from Jean-Rodolphe Perronet, the coming master of French bridge-builders.

Until Perronet wrote a book in 1783[14] Bélidor's four volumes were by far the most useful treatise on construction in water in any language and many men used them as a practical manual. George Semple in Dublin[15] had actually been searching for such a book for almost a year before Bélidor's last volume was printed. He was an architect and builder, designing and/or building five houses and a hospital, when he boasted in conversation with the Archbishop of Dublin and another leading citizen that he could make the damaged Essex Bridge over the Liffey fit for traffic in ten days for a hundred guineas – one-fifth of the estimate the corporation were considering. That was in May 1751 and, having made a repair with timber between 20 and 30 July and for less than the hundred guineas, he was induced to undertake the rebuilding of

the whole bridge in stone. Looking back more than twenty years afterwards he gave himself credit for 'a great desire to acquire knowledge in difficult matters of arts and science', though he also said later that 'the whole of my scholarship, except what little I got whilst I was a meer child, was acquired within the compass of six winter weeks, in the thirteenth year of my age'.[16] He searched his many books and went to London to buy more books and engravings which cost him £40, but though he reached the decision that he must form an 'inclosure' or cofferdam to remove the old foundations and make new ones he found no description of an 'inclosure' which could possibly keep out the water of the Liffey, whose depth would be 27 ft at a high spring tide. He insisted on an 'inclosure', as against the use of caissons, because his twenty-eight borings all discovered rock at depths less than 30 ft below high water, through less than 20 ft, and often less than 10 ft, of sand, mud, gravel or stones, much of the material being loose or soft. He believed he must remove that material and found his bridge 'on or tolerably near the rock'.

After making a design in which he copied every line of the architecture of Westminster Bridge (fig. 10), he went again to England and consulted Charles Labelye, William Etheridge and 'Mr Preston', the latter two working at Ramsgate Harbour. Preston had worked as foreman for Andrews Jelfe at Westminster Bridge, as

Etheridge had for James King. Labelye told Semple that laying the foundations in cofferdams 'would not answer the purpose' and, as neither Preston nor Etheridge had ever seen a cofferdam in use, they could not give any advice. But when Semple returned to Dublin a bookseller offered to ask his son who was in Paris to buy any books or prints which might help him; and in response there came all four volumes of Bélidor's book and a copy of an engraving of the cofferdam in use just then at the new bridge of Orléans, showing men at work inside the dam.[17] 'The language I was a stranger to', wrote Semple, 'but on turning over the plates, I quickly perceived the construction of coffer-dams . . . My drooping spirits then instantly revived, and I . . . entertained the most sanguine hopes of success.' Till the day he wrote his account of the event in 1776, he never knew what was in the French text. Neither, we must assume, did he ever appreciate the irony that amongst the hundreds of diagrams in Bélidor's book which were drawn from actual engineering works, the one which taught Semple most and on which the design of his cofferdam was based was nothing but a tracing of a plate in Gautier's *Traité des ponts* and had therefore been in print since 1714.[18]

Semple boldly planned his cofferdam to enclose half the length of the bridge and connect with the banks at a considerable distance up and downstream. Semple's drawing of the plan (fig. 42) is diagrammatic, the shaded band A representing the cofferdam and DD pipes passing through it. C is a sump dug deeper than the excavation (B) for the bridge foundations so that when work was in progress on the foundations, water would run from B to C and from there be pumped out into the river, leaving the work area dry. The cross-section of the cofferdam is shown in the upper drawing of fig. 43, where the timbers labelled C are braces at wide spacing, the general cross-section being two large steps on each side of a middle compartment, all five compartments being filled with puddled clay and each compartment faced with vertical planking fixed to the piles. The planking is indicated by the parallel lines in the plan (fig. 43, lower drawing). The width of the dam at the river-bed varied from 12 to 17 ft according to the depth below the water. Some of the piles for the upstream wing of the dam, where the water was shallowest, were driven simply with a 'two-handed maul', but for piling in deeper water Semple made a piling machine containing some hints from Bélidor's plates and others from Valoué's design used by Labelye at Westminster Bridge (fig. 6). Fig. 44 is Semple's plate of the front elevation of the frame and details of the monkey and the lifting tongs. In constructing the downstream wing of the cofferdam he met difficulties of a kind which troubled all who set about replacing old bridges in rapid rivers. The increase in kinetic energy caused by the contraction of the flow through narrow arches, and the eddies formed at the downstream ends of the piers, always caused deep scour just a little further downstream. In his initial borings Semple had

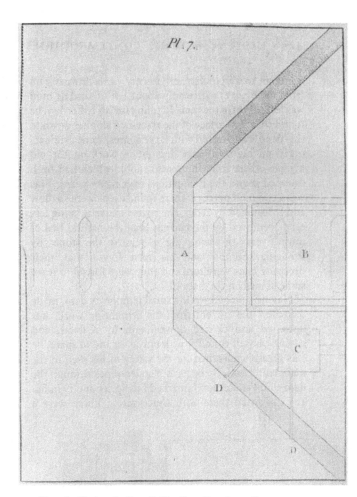

42 Plan of cofferdam for Essex Bridge (from Semple, 1776).

43 Cross-section of cofferdam for Essex Bridge (from Semple, 1776).

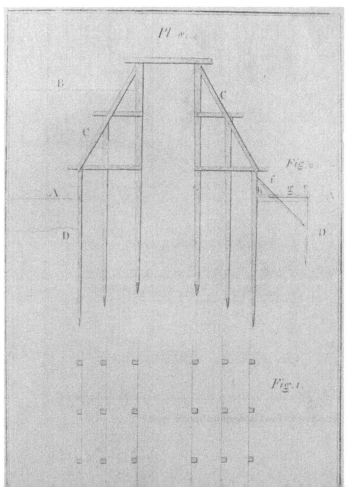

found an area in midstream, twenty yards downstream from the bridge, where only about 1 ft of sand or mud was left on top of the rock. In piling for his cofferdam he found that the surface of the river-bed sloping down to this deep was covered with large stones, some originating from his own early demolition work on the old bridge, others from the coarse rubble which had been dumped round the foundations over many years. This was a common practice where foundations were shallow and threatened by scour, the coarse material being less easily removed by the current than the natural bed of most rivers. In attempting to remove the stones by dredging, Semple and his men found that their dredging bags were torn and they were forced to leave some stones on the bottom.

Having watched while a small temporary dam, made to deflect the current from his demolition work, was uprooted and carried downstream by a flood, and having himself got off it with only a second to spare, he was greatly concerned for the safety of his men in the main cofferdam. So he fixed wooden pipes through the dam at low-water level (marked f in fig. 45 and D in fig. 42) and fitted them with sluice-gates. These were a

device he did not find in Bélidor. By closing the gates he began to test the cofferdam early in 1753 with small differences of water level from outside to inside, and also from inside to out. When the difference of level approached the full height of the tide (10 ft) the water began to leak under the dam 'after a very furious manner' at the place where the coarse stony bottom had been left under it. To prevent a complete outwash the sluices were quickly opened and the tide allowed in.

At low water [he wrote] I set all the drudge [i.e. dredge] men and water-men to that corner, both without and within, and in short, we took out and removed from that corner every stone that was in our power, and filled the vacancies with clay, etc. Aug 3d, the sluices were screwed down at low water, and when there was nine feet flood without, the water had rose only thirteen inches within.

5th, after an exceeding heavy rain there came down a sudden rapid flood . . . but we were prepared for it, and kept open our sluices, so that the water then rose as fast within as it did without; but it did us no other damage than delay us a little, and wash away some of our clay; and after all this, the next day we proved the dam again, and during the whole tide, the water rose only four inches.

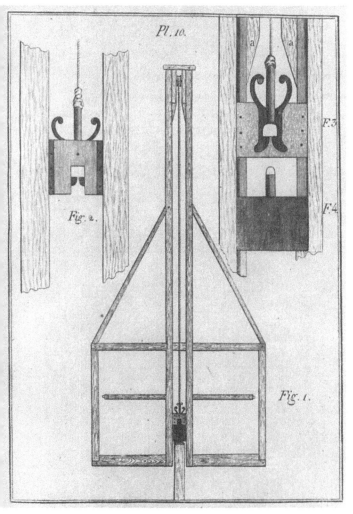

44   Piling machine for Essex Bridge (from Semple, 1776).

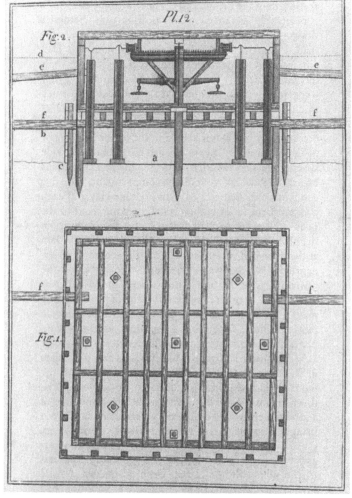

45   Pumping machine at Essex Bridge (from Semple, 1776).

The removal of the old foundations and excavation down towards the rock was already under way and by mid-September the excavation was 10 ft below low-water, but the pumping machine was coping easily with the water which seeped up in the floor. It was a simple capstan-driven machine (fig. 45) located over the sump labelled C in fig. 42, apparently worked by men, not horses, and driving eight piston pumps. Work within the dam was continuous twenty-four hours a day and also on Sundays whenever it was necessary, though there were only two teams of workmen alternating in eight-hour shifts ('September 28th', said Semple's diary, 'at 3 o'clock in the morning we set the first cut-stone of the land abutment, without any parade, which time would not admit of'). At the depth of 10 ft below low-water and still 3½ ft above the rock, however, the floor of the pit being a compact sandy loam, a spring boiled up at one point. Semple controlled it by driving a timber tube 6 in. in diameter, previously used when making borings, down to the rock. Water rose up this tube when the tide rose and when it flowed out over the top he plugged it. 'Gentlemen' were called to examine the phenomenon, and they smelt and tasted the water and discussed it as a possible mineral spring, or the leakage of a sewer. But the following day one, a Dr Rutty, returned and opined, as Semple already feared, that 'it was no species of spa water, but he believed, a large body of subterraneous water, that ran along on the surface of the rock, and communicated with the sea-water'. It seems certain that it was, in modern terms, a body of water which was subject to artesian pressure at high tide. All the men were informed and urged to work faster, and the intention to lay the foundations on or near rock was abandoned. The ground was simply levelled and masonry laid as fast as possible. At one point the pressure below opened a crack in the ground 10 ft long and 3 or 4 in. wide. Semple drove another tube which eased the pressure, but water filled the excavation to 18 in. deep. When the tide fell the largest and flattest stones available were laid over the crack and masonry quickly built over them several feet high. Semple was convinced that he had averted by only the tiniest margin a total disaster to his bridge.

The piers of the old bridge were on timber gratings floored with oak planks, each whole timber platform having apparently been fabricated on shore, floated to its place and simply sunk to the bed of the river without any piles. The bed between these platforms had been scoured easily by the floods and, in Semple's own words, 'the only surprising thing is, that [the bridge] stood so long'. He decided, in sharp contrast, to lay a continuous floor of masonry, which he called his 'thorough foundation' (possibly a mis-spelling of 'through'), across the whole river, piers and arch openings alike. At the upstream and downstream faces of this masonry, which were several feet beyond the points of the cutwaters, there was sheet-piling, 6 in. thick, with interlocking V-shaped edges and driven to

rock. This had formed the walls of the excavation pit and was later cut off at the level of the top of the 'thorough foundation'. The two piers at the northern end had bearing piles driven to rock, as shown in fig. 46, but for those in the southern cofferdam, where the rock rose nearer to the bottom of the 'thorough foundation', no piles were used. Semple's description of the piling and masonry of the foundations, although somewhat confusing, is very full. He is writing chiefly of the northern half of the bridge:

After we had drove the oak piles down to the rock ... we spread a plentiful coat of roach-lime and sharp gravel over the ground, and laid a course of large flat stones, and filled and hearted them in close about the pile for about a foot high; then we covered that course with another plentiful coat of the dry grout (i.e. the roach-lime and sharp gravel) and the next course and all the rest was laid with mortar after the usual manner, only with this difference, that every stone swam in mortar, and each course was grouted as above; and so we went on till we came to the level for the caps that were laid over each row of piles, and then wrought up and levelled them, on which we laid three beams stretching the whole length of the pier from sterling to sterling, and filled up about them with the masonry; but the thorough foundation that lay under the arches, was laid and securely bedded, and many of the stones were rather pitched upon their ends and wrought close together, so that there is a substantial stone floor for the bed of the river between the piers.

The thickness of this body of masonry laid right across the river was as much as 9 ft in places and probably nowhere less than 5 ft. It was therefore a much more complete protection for the feet of the piers and the river-bed than the framing and setting practised by Yorkshire bridge-builders and others. There was a plate and description of a bridge with a similar foundation in Bélidor,[19] but as it was designed for closure of the arches as sluice-gates Semple is unlikely to have copied it as a standard method; we must attribute his 'thorough foundation' largely to his own intense resolve to build a bridge as permanent as a mountain. The first ashlar course of each pier, which was laid over the three beams mentioned above and the masonry packed between them, had a continuous bar of iron 3 in. square laid in a groove about 1 ft from the edge all round and run with lead; and the second and third courses, each stepped in from the course below, had similar bars 2 in. square.

He followed the unusual course of erecting the centring for the two northern arches of his bridge on the dry foundation within the cofferdam, which allowed him to support the centres on many posts standing vertically on the surface of the 'thorough foundation', but he had to remove the cofferdam and take down the centres quickly in the spring of 1754 to open the arches to river traffic, because his second cofferdam, enclosing the southern half of the old bridge, was nearing completion. An abstract of dates in his diary reveals a remarkable speed of construction, both of arches and foundations:

1754

19 March	First piles driven for southern cofferdam.
29 March	Second pier from north end complete.
30 March	Masonry of second arch from north end started.
8 April	First arch at north end opened to river traffic (showing that part of northern cofferdam was already removed).
23 May	Second arch from north end opened to traffic.
25 May–5 June	Constant floods delaying progress.
14–15 June	Similar floods.
21–7 June	More floods.
5 August	First ashlar course of south pier of middle arch.
25 August	Pumps in southern cofferdam stopped 'as we had no further occasion for them' (presumably means that all foundations were complete and piers above low-water).
7 October	South abutment up to springing of arch.
1 November	Erection of centres for first and second southern arches begun.

1755

5 March	Both southern arches complete.
8 April	Middle arch complete and a temporary carriageway 20 ft broad ready for use.
10 April	Temporary carriageway opened to the public.

While the façade of Semple's bridge (fig. 46) was a smaller replica of Labelye's (fig. 10), Essex Bridge surpassed Westminster in several ways. It replaced an old bridge built in 1676 and widened only just before 1750, of seven arches, with piers from 9 to 13 ft thick and a total waterway of 170 ft; the new bridge was of five arches, two piers of 8 ft and two of 9 ft, leaving a waterway of 200 ft. On the old bridge there had been clear evidence of irregular settlement of the northern piers to a maximum of 3 ft 6 in., with a difference of almost 2 ft between the upstream and downstream ends

of the northmost pier, causing a split 8 in. wide between the original road of the bridge and the new part recently added. The performance of Westminster Bridge, on a virgin foundation, compares more readily with these settlements of the old Essex Bridge than with the firmness of the new one, and moreover Semple's task, with the old bridge to remove, was more difficult than Labelye's. His progress was also much faster. In width, too, Essex Bridge surpassed Westminster, for he had chosen to make it as wide as the Pont Royal in Paris (51 ft), though he could not have read Bélidor's advice to do so.[20] Semple's description of the Pont Royal was taken from Nicholas Hawksmoor's pamphlet of 1736[21] and his choice of width is the more surprising because the streets which led to the bridge at both ends were much less than 51 ft. The making of an approach on the south side to match this width was proposed by Semple and was the starting point of much wise planning by a body called the Wide Streets Commission later in the century.[22]

Unlike Labelye who kept to his role of engineer and never handled his employers' money,[23] Semple took the whole responsibility of design and construction, but simply on his public declaration of estimates of the cost (£20,500) and time for completion (two years), without any written contract. Such informality was exceedingly rare and speaks volumes for his reputation both for skill and honesty. Essex Bridge was finished very near his estimates; it cost £20,661, by the best report which can be found,[24] and was opened just two years and eighty days after the old one was closed. In two respects the end of the story is similar to that of Westminster Bridge. Semple was by no means young when he started Essex Bridge – his father, he said, was a mason as early as 1675 – and his health, like Labelye's, broke down within about a year of his finishing the bridge. He 'was obliged to retire to the country in a deplorable condition . . . particularly afflicted with the gravel and rheumatism', but he did recover his health some years later.[25] In the meantime, like Labelye, he had received a gratuity, in his case of £500 granted by the Irish Parliament in response to his petition in 1761.[26]

His bridge did not fulfil his boast that it would last as long as the Sugar-Loaf Mountain, but the pier foundations may yet do so, for they were left in place to bear the weight of the new Grattan Bridge built to replace Essex Bridge in 1873–4 and Grattan Bridge is still in use. Semple's bridge was structurally sound in 1873 but a wider bridge of lower gradient was needed. Most of the hewn stones were re-dressed and built into the new bridge.[27]

46   Middle arch of Essex Bridge (from Semple, 1776).

5

THEORY OF ARCHES AND PONTYPRIDD

The true mathematical and mechanical form of all manner of arches for building, with the true butment necessary to each of them. A problem which no architectonick writer hath ever yet attempted, much less performed. abcccddeeeeefggiiiiiiiillmmmmmnnnnnooprrssstttttttuuuuuuuux.

Robert Hooke, *Description of helioscopes* (1676)

Before the dawn of the eighteenth century men of science had twice offered guidance to architects about the design of arches. The two approaches were initially quite different. One of them appeared first in a list of single-sentence solutions of problems in mechanics, inserted by Robert Hooke, with a nonchalance that was quite false, 'to fill up the vacancy' of a page at the end of his printed lecture on *Helioscopes and some other instruments,* in 1676.[1] It was expressed as the Latin anagram which heads this chapter and which reads, when solved and translated: 'As hangs the flexible line, so but inverted will stand the rigid arch.'

Twenty years later another Englishman, David Gregory, derived the equation for the curve of the hanging flexible line or chain and called it the 'catenaria';[2] and in a corollary he expanded Hooke's assertion and added a sentence of his own which has been quoted by later writers nearly as often as Hooke's: 'When an arch of any other figure is supported, it is because in its thickness some catenaria is included.'[3] Nowadays the curve is called a catenary.

The catenary idea became known and understood by architects well before 1735, perhaps because Hooke's statement was in the nature of an axiom confirmed by intuition or demonstration, not by mathematics. It was also easy to draw the catenary curve for any given span and rise of arch by suspending a cord or chain against a vertical sheet of paper or a wall, tracing out the line and then inverting it. But it was only the pedant for science, Batty Langley, who proposed catenary arches for Westminster Bridge in 1736.[4] The experienced architect, John James, admitted that such arches were, 'geometrically considered, the strongest of all others', but he thought them of 'ill appearance' and had 'never read or heard of any architect, who either advised or made use of this kind of arch, where the height has been about half the extent, or under it'[5] – as were the arches of most bridges. Such scientific objections as James

made were muddled and he failed to note the two important differences between a hanging chain and a real bridge arch: first, that a masonry arch was not at all flexible, and, second, that the chain was loaded only with its own weight while a bridge arch carried also the weight of spandrels, parapets, road and traffic. Indeed the mere fact that arches built of many different curves stood safely should have cautioned men of science against proclaiming a single ideal arch form and architects against listening to them. As far as the catenary was concerned, either this argument or the ones about ill appearance and inconvenience of shape did prevail for most of the eighteenth century.

The second approach by scientists could also be criticised for failure to explain the stability of existing arches but it attracted more attention and, towards the end of the century, a measure of respect. It has become known as the theory of equilibration, and the basic geometrical construction and its interpretation are set out in appendix 3. It was first published by Philippe de la Hire[6] in Paris in 1695 and reconsidered by Parent in 1704 and la Hire himself in 1712. Gautier, in 1717, could not understand la Hire because he used algebra which, said Gautier, was not understood by tradesmen and architects.[7] Later French writers, especially Couplet in 1729–30, introduced new concepts in attempts to make the theory correspond with experience, but the earliest books in English ignored all such advances. The first of the English books was John Muller's *Mathematical treatise,* published in 1736, about two years before Labelye used the theory to work out the thickness of the arches of Westminster Bridge. (But Labelye knew the French writings as well and may well have used theory more advanced than Muller's.) Muller's expression of the theory was mostly geometric as la Hire's had been in 1695; but the second English version, though no different in principle, was expressed largely in terms of 'fluxions', or calculus. It was by William Emerson, a man whose admiration of Isaac Newton and the method of fluxions was almost as irrational as his personal behaviour.[8] He was thirty-one or thirty-two when he married the niece of an eminent surgeon and cleric, who had promised her a dowry of £500 but refused after the marriage to pay the money and treated Emerson 'with contempt, as a person of little consequence'. The angry bridegroom packed up all his wife's clothes and sent them back to her uncle, vowing to prove himself a better man – by becoming a famous mathematician! The strange thing is not that he succeeded, for he had already shown some talent, but that he spurned all the social rewards of his fame and clung to a rural life that was aggressively simple and ways of speech and dress that were coarse and eccentric. He lived at Hurworth near Darlington, went only once to London, never rode a horse or went in a carriage and never stopped to eat when working at his books; but if he went to market at Darlington he often spent two whole days in the public houses on his three-mile journey home, talking volubly but never becoming the worse for drink.

In his *Doctrine of fluxions* (1743) Emerson stated the arch problem thus:

The nature of the curve forming an arch being given [fig. 47, top left]; to find the nature of the curve [labelled TSt] bounding the top of the wall supported by that arch; by the pressure or weight of which wall, all the parts of the arch are kept in equilibrio without falling.

The shape of spandrel required in a bridge would correspond to the shape of wall determined, and the figure shows the shape Emerson found to be necessary to 'equilibrate' a semicircular arch. The result was obviously absurd because there were real arches which supported many other shapes of wall 'without falling'. And he found a similar rise of the wall to infinite height to be necessary over an elliptical arch or any other arch with a horizontal joint at the springings. Fig. 47 actually comes from a later book, *The principles of mechanics*, 2nd edn (1758), in which he attempted not to improve the theory but to find shapes of arch for which the result of the calculations would be a wall bounded at its top by a horizontal line. In the case of a bridge this line coincided with the line of the roadway which had to be nearly horizontal for traffic. Some shapes of arch and wall which resulted from these calculations and which Emerson thought to be practicable are shown in fig. 47. The top right drawing is a circular segment of 90° arc with the crown thickness one-sixteenth of the span, at the middle left is a catenary arch with crown thickness almost one-third of the span, and at the bottom is an irregular arch curve with span 60 ft, rise 30 ft and crown thickness 3 ft 6 in. A table of ordinates was given to enable the reader to draw the irregular curve.

It was clear to Emerson, as also to la Hire even in 1695, that arch and spandrel shapes were not so sensitive to collapse as these findings suggested. But he claimed that arches of the ideal curves he derived were the strongest possible and that any others 'must in time

give way, and fall to ruin',[9] and he did not even try to explain why arches which were not of an ideal curve did not collapse. Explanation is easy in retrospect. The curve of arch which Emerson derived for a given shape of spandrel was the curve (inverted) which a flexible cord would take up if loaded with weights corresponding to the shape of the wall. As it was the line of tension force in the cord, so the inverted curve would be the line of compression force or thrust in the arch. In a real masonry arch, where there is frictional contact between

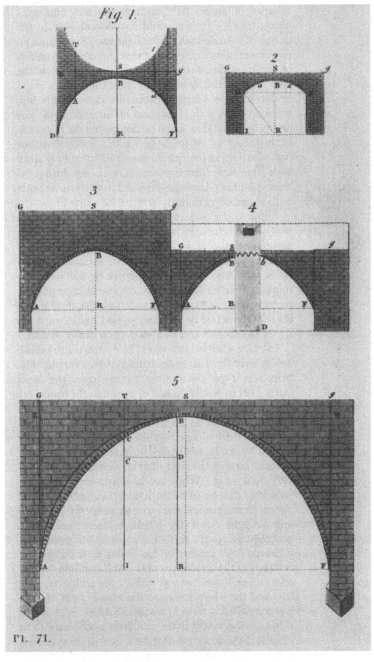

47 Arches of equilibration (from Emerson, 1758).

the voussoirs, there is no need for the line of thrust to be perpendicular to the joints between voussoirs (and therefore parallel to the soffit); it is enough that it lies within the thickness of the arch at every joint. David Gregory came very near to this truth in 1697 but, thinking of a cord loaded only with its own weight, he assumed that the ideal curve was always a catenary.

Other serious omissions from Emerson's theory have yet to be mentioned. He did not attempt to calculate the magnitude of the forces an arch would exert on its piers or abutments. Nor did he consider the effect on the arch of movements of these supports; he was therefore ignoring the most common cause of distortion and collapse of arches. His assumptions and omissions meant that his theory would be relevant only in the design of a single-arch bridge light enough to cause no movement of its supports and thin enough to cause risk of the line of thrust straying outside the thickness of the arch. It happened, however, by sheer coincidence that a bridge built in Wales between the dates of his first writing on the arch (1743) and his second (1758) was just so light and thin; and it demonstrated the relevance of the theory by collapsing first with a wrong distribution of weight in the spandrels and standing ever since with the right distribution. It was the bridge at Pontypridd in Glamorganshire and its story must begin with notice of its builder, William Edwards.[10]

Edwards (fig. 48) was a self-taught mason living on his family's farm in Eglwysilan parish. He first repaired and built dry stone walls, then small buildings and, at the age of about twenty, a large forge in Cardiff, twelve miles from his home. At twenty-six he was ordained minister of an independent church and he continued to be both pastor and farmer, as well as mason, throughout his life. He is said to have known no English until the age of twenty-one and he always preached in Welsh.

In 1746, when twenty-seven, he contracted to build a bridge over the Taff at the hamlet of Pontypridd. The price was £500 and the contract included the usual clause that he would uphold or maintain the bridge for its first seven years. He built the bridge, of two or three arches, and it stood for over two years but was then demolished by a flood. As the failure was obviously caused by scour, he decided to replace it by a bridge without piers in the river, that is, with a single arch of very long span. When he had first convinced his securities that he could do it and then built the arch almost to completion, the centring suddenly collapsed and brought down the whole bridge. There were probably delays for further discussion but eventually Edwards built another bridge of one arch and almost certainly of the dimensions of that still standing; by this time, if not before, he was making the arch extremely thin and the whole structure abnormally light, doubtless compelled to do so by his financial loss. With such a thin arch the weight in the spandrels would have to be close to the ideal weight. But the weight of the spandrels Edwards built was not correct and after standing for six weeks the arch collapsed again, this time by breaking

48   William Edwards.

upwards at the crown. According to one account, the parapets had not been built before this collapse occurred.

There is no full record of Edwards's thoughts or his negotiations with the owners of the bridge after this third failure, but they either encouraged or forced him to try again, providing extra money to cover part of his loss. A story that he consulted John Smeaton, then just becoming well known as an engineer, was passed by William Jessop, Smeaton's first pupil, to Thomas Telford many years later.[11] At any rate, he somehow reached an understanding of the cause of the failure, that the load was too great on the haunches or too little on the crown; and he rebuilt the arch in 1756 with the load corrected by forming three voids through each spandrel of 9 ft, 6 ft and 4 ft diameter (fig. 49). It was said that a fourth semicircular void was made inside the walls and the space over the voids filled with charcoal instead of gravel, but these statements have never been checked, since digging in a spandrel would change the distribution of load and nobody has cared to risk another collapse. It had taken ten years to get a stable bridge but the one which stood safely in 1756 had the longest arch in Britain (140 ft span) and that remained true for forty years. It was, and remains, astonishingly thin, especially considering that it is built of so-called rubble masonry and of very small stones (fig. 50). It is, in fact, in all except its long span and the circular voids, a perfect example of the traditional Welsh bridge-

49   Pontypridd in an old photograph.

50   Arch ring and parapet of Pontypridd.

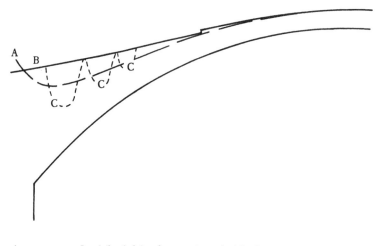

A —— —— Load (i.e. height of masonry) required by theory

B ———————— Actual load if bridge were solid (including parapets)

C − − − − − Reductions of load by circular voids

51   Theory applied to Pontypridd.

building which was described briefly in chapter 2 and the strength of the slaty stone and the mortar of Aberthaw lime could not be better demonstrated. Nevertheless, the story of the third collapse shows that so thin an arch required to be 'equilibrated', that is, subjected to a correct distribution of vertical load, or it would be unstable and collapse. Fig. 51 shows the height of solid masonry which was required above the soffit line according to the theory (line A), and the actual height, including an allowance for the weight of the parapets, provided in Edwards's third bridge which collapsed (line B) and in his fourth which still stands (line C). Such close correspondence to theoretical requirements would suggest a mathematical design, were it not that no mathematician ever claimed credit for it. It would seem that Edwards simply learned from the upward breaking of the third bridge and decided to lighten the spandrels, and Smeaton may have helped him in this; but that the closeness of the design to a mathematical solution happened by chance.

The best-known successor to Emerson in the theory of arches was another North of England mathematician, Charles Hutton, who was stimulated by the collapse of the old bridge at Newcastle-upon-Tyne in 1771 (see chapters 6 and 8) to write his *Principles of bridges* (1772) in which he drew the curves of equilibrium load for most of the common shapes of arch soffit and also solved the inverse problems of finding the ideal soffit curves for given tops of the spandrels. It was more readable but no more practical than Emerson's writing and twenty years more passed before designers began to ask Hutton's advice about real bridge projects (see chapters 11 and 13).

In contrast with their slow acceptance of theoretical methods, designers reacted to Pontypridd as soon as they heard of it. Some of them copied it, making circular voids in the spandrels, especially of single-arch bridges, but sometimes also between spans. Most of these bridges are in South Wales and adjacent counties of England but there are instances in the North and in Scotland and one over the canal in Regent's Park, London. Good examples are Kenmore Bridge at the outlet from Loch Tay in Perthshire, designed by John Baxter in 1774,[12] Middleton-in-Teesdale Bridge, built in 1812, and Hampton Court Bridge over the Lugg in Herefordshire, built by John Gethin in 1826.[13] In South Wales William Edwards himself (probably assisted by his sons) built two more bridges of long span with circular voids, at Wychtree and Dol-au-Hirion, the latter still standing and having a stone in the parapet carved with the name of Thomas Edwards, William's son, and the date 1773. There are more than a dozen other Welsh bridges of shorter span with circular voids and whose dates of construction extend to at least 1829.[14] Most, if not all, of them would not need the voids to equilibrate their arches, and it is obvious that the voids became part of a tradition which in one late instance deserves the name of a style. The three-arch bridge of Pant-y-goytre over the river Usk (fig. 52) is nothing if not stylish.

William Edwards also began a family tradition of bridge-building. Pontypridd was a dismal failure as a river crossing, being only 11 ft wide between the parapets and so steep that wagons and carriages had to use a 'chain and drag' to descend from the crown. The method was described by T. M. Smith at the Institution of Civil Engineers in 1838:

When a carriage reaches the centre of the bridge, one end of the chain is attached to the hinder part of it, the other being secured to the drag, upon which a boy generally places himself, so that as the carriage descends upon one side the drag is pulled up the other, and thus relieves the horse in descending.[15]

The fault was corrected in Edwards's later bridges by adding built-up approaches of reasonable gradient. He had a number of contracts for bridges (table 1) and was consulted as a recognised authority in 1778 to inspect the large bridge at Builth Wells built by James and James Parry, father and son.[16] His son Thomas made many similar inspections for the justices of Breconshire between 1781 and 1791.[17] Thomas and his brother David each designed and built bridges on his own account but each, like their father, followed another vocation as well. David was a farmer at Beaupré near Cowbridge in Glamorgan and Thomas kept the Three Cocks Inn at Aberlunvey near Glasbury, Radnorshire.[18]

The bridges of all the family are listed in table 1, so far as they are known. Most of them might have been deferred for mention in chapter 9, being built in the same years as the bridges described there, but they are so much a growth from older Welsh masonry that they seem to belong to an earlier time. Neat, slender arch rings of constant thickness, small, oblong stones and thin joints, piers with right-angled points and often some stone paving in the river-bed – these are the elements of the Edwards' bridges as they were of earlier and later Welsh masons' work. Circular spandrel voids and a lengthening of arch spans were William Edwards's additions to the local style, and the voids were added to Welsh bridges until well into the twentieth century[19] for various reasons he never thought of.

52 Pant-y-goytre.

TABLE 1. *Bridges by the Edwards family*

Bridge	Date of construction	Description	Designed and built by	Contract price	Notes	References
Pontypridd (R. Taff)	1756	1 arch, 140 ft 3 voids in each spandrel	William	£500	See text	See text
Wychtree (R. Tawe)	?	1 arch, 95 ft 2 voids in each spandrel	William	?	Demolished	(a) B. H. Malkin, *Scenery of South Wales* (1804) pp. 88–9 (b) J. G. Wood, *Rivers of Wales illustrated* (1813)
Dol-au-Hirion near Llandovery (R. Tywi)	1773	1 arch, 84 ft 1 void, 8 ft diameter, in each spandrel	William? and Thomas	?		(a) Malkin, p. 89 (b) Inscription on parapet
Pontardawe (R. Tawe)	?	1 arch, 80 ft	William	?	Widened with similar masonry. Contrary to Malkin's statement, there are no voids	(a) Malkin (b) T. Rees, *Beauties of England and Wales*, vol. XVIII (1815) p. 714
Aberavon (R. Avon)	?	1 arch, 70 ft	William	?	Demolished	Malkin, p. 89
Pont-y-clerc near Ammanford (R. Loughor)	?	1 arch, 45 ft	William	?	Probably the site of the 'Bettws bridge' mentioned by Malkin. Pont-y-clerc rebuilt 1817 and since widened with concrete	(a) Malkin, p. 89 (b) Dynewar MSS 287/2, at Carmarthens. CRO (c) E. Jervoise, *Ancient bridges of Wales and western England* (1936) p. 81
Glasbury (R. Wye)	1776–7	5 'flat segments on high piers'	William and/or Thomas	?	Destroyed by flood 1795	(a) Radnor QS records at NLW (b) Jervoise, p. 114
New Bridge near Tredunmock (R. Usk)	?	4 arches	William?	?	Assumed to be the 'Usk Bridge' named by Malkin as built at the town of Usk	(a) Malkin, p. 88 (b) Wood
Pont Ithel near Hay on Wye (Afon Llynfi)	1783	3 arches, 20 ft, 32 ft and 20 ft	Thomas	£160	Widened with masonry 1952	Brecons. QS records, at NLW
Talybont (R. Usk)	1781–2	?	Thomas	£159	Possibly a timber bridge, rebuilt in 1931	(a) Brecons. QS records (b) Jervoise, p. 103
Forddvawr (Digedi Creek)	1791–2	?	Thomas	£50	Rebuilt 1812	(a) Brecons. QS records (b) Jervoise, p. 114
Brecon (R. Usk)	1793–6	Widening of old bridge on upstream side	Thomas	£1,000	Damaged by flood 1795. David Edwards and the widow of Thomas paid £150 compensation in 1801 for discharge of Thomas's bond for 7 years' maintenance	Brecons. QS records
Llandilo-yr-Ynys (R. Tywi)	1786	3 arches, 45 ft	David	?	Partial collapse in flood 1931. Underpinned and repaired for £4,791, exactly to former appearance	(a) Jervoise, p. 76 (b) Malkin, pp. 93–4 (c) Document RB1057 at Carmarthens. CRO

Table 1 (*cont.*)

Bridge	Date of construction	Description	Designed and built by	Contract price	Notes	References
Pont Loerig (Afon Taf)	?	3 arches, 27 ft	David	?	Widened with concrete on downstream side 1931, reproducing the shapes of the masonry features	(a) Malkin, p. 94 (b) Document RB156 at Carmarthens. CRO
Newport (R. Usk)	1793–1801	5 arches, 62 to 72 ft	David and Thomas (or David and his son Thomas?)	£10,165	Widened 1865 Demolished 1927	(a) Malkin (b) Q/CB 0002 at Gwent CRO (c) Jervoise, p. 108
? (R. Tawe)	?	3 arches	William	?	Mentioned only by Malkin	Malkin, p. 88
Edwinsford (R. Cothy)	?	?	David	?	Replaced by lattice girder before 1936	(a) Malkin, p. 94 (b) Jervoise, p. 77
Bedwas (R. Rhymney)	?	2 arches	David	?	Mentioned only by Malkin	(a) Malkin, p. 94 (b) Jervoise, p. 96

6

FOUR ANCIENT BRIDGES

It is so much a mystery and profession to navigate through this bridge.

Robert Mylne in evidence to the House of Commons committee on London Bridge (1800)

Westminster Bridge caused none of the hindrances to nagivation that its opponents had foreseen, but it may have begun to divert some trade from London. Streets of new houses were built near its western end and by the time it was opened to traffic the citizens of London must have been eyeing it with envy. Before the second modern tidal bridge was opened in Dublin five years later, a movement for improving London Bridge was well under way, and similar movements in other towns with old, narrow bridges were not long delayed. This chapter focuses on four old bridges on tidal rivers, all of which were by 1755 inadequate for their traffic but not in danger of collapse (as Essex Bridge in Dublin had been).

The first to be described is Rochester Bridge.[1] It was built in the years 1383–92 and its dimensions either then or soon afterwards were as follows:

	Dimension (ft)	Ratio of dimension to length between abutments
Total length	560	
Length between abutments	534	1.00
Sum of thicknesses of ten piers	206	0.38
Sum of spans of eleven arches (=waterway measured above top of starlings)	328	0.62
Sum of widths of eleven passages (=waterway measured below top of starlings)	117	0.22
Width	14	

It was of more or less symmetrical design with stone parapets, iron railings on top of them, and a drawbridge over one span, and it changed little in appearance over the centuries although some arches had to be rebuilt

from time to time. A description in 1782[2] said that 'the river has a considerable fall' through the arches and that the bridge was still only 14 ft wide. The fall was due to the very thick piers and starlings, the manner of construction of which was revealed when the bridge was demolished in 1863:

The foundations to support the piers, etc., were constructed by driving piles of elm timber, shod with iron, into the bed of the river, at this part mostly chalk. The piles were about 20 ft. long, driven close together and forming platforms about 45 ft. in length and 20 ft. wide ... the starlings outside the platform, with half timber piles secured by ties, enclosing a space of about 90 ft. long by 40 ft. wide, the space between the piles and those of the platform was filled with chalk, the top and sides of the starling being covered with elm plank. A course of flat bedded stones of Kentish rag, about 8 ins. thick, was laid over the piled platforms and on that the masonry was built – a solid mass, the mortar being nearly as hard as the stone. The number of piles removed from the old bridge was about 10,000; the quantity of timber, about 250,000 cubic feet.

The bridge wardens' accounts during the early years of the life of the bridge (1398–1479) record the annual driving of numbers of piles, sometimes as many as one hundred in a year, and it may be assumed that most of these were used in repairs to the starlings. There may have been some reduction in the width of the starlings before 1736, when Hawksmoor wrote[3] that the waterway was 230 ft at its minimum, but since some starlings at least were still 40 ft wide in 1863 it seems more likely that Hawksmoor was mistaken. The level of the platforms on which the piers proper were founded must have been below low-water, since they would otherwise have decayed. It is worth noting that Hawksmoor could only express an *opinion* that the bridge was founded on piles, since one of the chief difficulties facing architects who tried to improve old bridges was ignorance of whether, within the massive starlings and below low-water, the piers themselves were founded on piles or not.

Rochester Bridge was maintained largely by the rents

received for property, including land, buildings and at least one ferry, which had been gifted for its upkeep at various times. Luckily buildings were never erected on it and so changes could be made in its structure without incurring loss of their rents.

London Bridge[4] (fig. 1) was much more complicated, both in its structure and financial support. It was built in 1176–1209, had nineteen or twenty arches of shapes and spans which varied due to numerous repairs and partial rebuildings, and the piers and starlings showed an equal variety of form and materials. In 1746 the City Lands Committee commissioned a report from Charles Labelye after first submitting ten queries about the existing condition of the bridge to their clerk of works, George Dance (the elder), and their 'tide carpenter', Bartholomew Sparruck.[5] In their answers to these queries Dance and Sparruck emphasised their ignorance of what lay within most of the structure but described what they had seen in a few parts during repairs, several times extending 2 or 3 ft inside the surface and once to a depth of 7 ft. They said that there were piles under the piers themselves, all the piles 'sound at heart' but some at least decayed about 1 in. deep at their surfaces. Some of the piles they had seen were of round rough timber but shod with iron. There was generally no platform of timber between the pile heads and the masonry. The bottom of the masonry was at various depths from 1 ft to 2 ft 4 in. below low-water and from 4 to 6 ft below the tops of the starlings. They were unsure whether the starlings had been increased in width since the bridge was first built but believed them to have been greatly lengthened 'in order to erect houses thereon'. The last question and both men's answer were:

What is the inside of the stone piers made of? whether of the same sort of stone as the outside; cut and laid regular, or only common rubble stones, laid in very bad mortar, as it is in Rochester-bridge?
Answer I have seen, in several breaches, the texture of the piers: and by them it appears to me, that the insides of the said piers are filled with rubble; and the external faces are formed with ashler laid in courses: but the rubble appears to me to be laid with good mortar.

Thus it would seem that the piers were much more solid than the starlings. Dance also made a plan of the piers

and starlings and a list of soundings in all the openings or 'locks', as they were called; and he added as an extra comment: 'In the middle of every arch there are driven down piles, called dripshot piles, in order to prevent the waters from gullying away the ground.'

Labelye showed to the Committee a scheme for improvement of the bridge proposed by Christopher Wren during the period of reconstruction which followed the Great Fire of London in 1666. Wren had wished to remove every second pier and rebuild the whole superstructure with arches of double the existing spans (fig. 53), but Labelye suggested less radical changes. He would increase the free waterway by removing the starlings and 'encasing' the piers with Portland stone 4 ft thick; the waterway at low tide would be increased from 236 ft[6] to 396 ft and the greatest 'fall' reduced to 15 in. He did not propose to remove the houses or widen the roadway, the roadway being generally 20 ft wide and the combined width of the houses on both sides 53 ft. The masonry part of the bridge was much less than 73 ft, however, as the houses overhung the spandrel walls, many being restrained from falling outwards by floors or roofs which spanned over the roadway at second floor level to tie them to those on the other side.

The City authorities failed to act on Labelye's advice and early in 1754 the Common Council appointed a committee to report on London Bridge 'what method is the most proper to make it more safe, commodious and ornamental, and the expense that will attend [it]'.[7] This committee seems to have acted contrary to their instructions[8] in recommending the construction of a new bridge at Blackfriars in September 1754;[9] but they also gave their advice concerning London Bridge. They considered that the houses were 'a public nuisance, long felt, and universally censured and complained of' and should be removed, and the bridge should be widened to give a roadway of 33 ft and two footways of 6 ft each. Dance had examined the pier foundations and found that they were 'likely to stand for ages', allowing for 'the common annual repairs'. He had computed the construction cost of the improvement as £30,000 and that of buying up the leases of the buildings on the bridge about £8,940; also that the City would lose £828 per annum in ground rents paid for the buildings. The committee noted that compensation would have to be

53 London Bridge improvement designed by Christopher Wren (from Maitland, 1760).

54 South end of London Bridge after improvement (from Cooke, 1833).

paid to the two parishes in which the buildings stood, for loss of revenue in the land and other taxes, which they estimated to be £485 per annum.

Such a scale of expense led naturally to a motion in the Council that Parliament should be petitioned for financial aid, but the motion was defeated by four votes.[10] When the Council decided almost a year later to petition for a Bill to build the new bridge at Blackfriars they again rejected a proposal that powers to improve London Bridge be included in the Bill.[11] But almost immediately, in January 1756, another petition for just those powers was presented to Parliament independently by 'divers merchants, tradesmen, citizens and inhabitants of the City of London, the Borough of Southwark and the several Counties of Kent, Surrey and Sussex',[12] and the House of Commons referred both petitions to a single committee which duly recommended that London Bridge should be improved, as well as allowing the Bill for the new bridge.[13] George Dance produced his plans and estimates to this committee and the Common Council undertook the improvement, being clearly unwilling that others should meddle with a bridge which belonged to the City.

The improvement was designed by Dance and Robert (later Sir Robert) Taylor and involved removing all the buildings, rebuilding all the spandrel walls in Portland stone, widening where necessary – since the existing façades were irregular in line[14] – and replacing two of the arches in the middle of the river with a single arch of double the span, the intervening pier to be removed to give a wide passage for river craft. A temporary bridge which had to be built for part of the length was burned down soon after its first erection, causing delay and alarm because the burning was thought to be deliberate; and the improvements were carried out in a piecemeal way, the last houses not being taken down until at least 1761.[15] But the new large arch was built by the end of the summer of 1759. 'It is finished in the Gothic taste', said the *Universal chronicle*,[16] referring to Gothic mouldings on the spandrels over the piers (fig. 54). But every second pier along the bridge supported a new trapezoidal 'turret' complete with semi-dome of stone, like those of Westminster Bridge. This was copying Labelye's architecture, but nothing like so complete a copy as George Semple's Essex Bridge in Dublin. The old pier under the middle of the new arch was used to support the centring (fig.

55), which was designed by Dance and Taylor and contracted for by John Phillips. It was criticised severely by a correspondent in the *London magazine* for the huge quantity of wood which 'had been crammed into it'. From a short distance, he said, 'it seemed as if it had been entirely solid'; and he explained that Phillips had been paid at an agreed rate per cubic foot of timber and the public had therefore paid £1,000 more than a well-designed centre would have cost.[17] The grossness of this centre was also used to discredit Phillips in pamphlets concerning Blackfriars Bridge and Bristol Bridge designs. The centring was struck in 1759 but removal of the pier and starling did not begin until 1762.[18] In the meantime most of the widening and new facings had been completed and the City had successfully petitioned Parliament for grants-in-aid amounting to £71,210.[19] The Act of 1756[20] had granted them increased tolls on the bridge and on boats passing under with cargo but that income was quite insufficient because the Act also required them to compensate all the leaseholders and the parishes. Another requirement was that the flow of water available to turn the wheels of the London Bridge waterworks must not be reduced. The water wheels were placed in the passages under four of the arches at the north end of the bridge and were used to drive pumps which supplied part of the City with water.[21] The waterworks company held leases of no less than 500 years, which would not expire until AD 2082. The force of water which turned their wheels was the force of the famous 'fall', which could not be maintained at the north end but reduced in midstream. The result was that Dance and Taylor's work did only a little to ease navigation through the bridge.

The demolition of the surplus pier had results which we shall describe in chapter 8, but there is one thing to be noted here. It was found that there were no piles within the area of the bottom of the pier (fig. 56). There was simply a wall of piles, four piles thick and about 6 or 7 ft deep, driven round the perimeter, and masonry of some sort dumped to fill the space within up to about low-water, at which level the pier proper was founded on the dumped material. The starling had then been built round the outside.[22] This manner of construction was confirmed in 1826 when William Knight examined another pier during its demolition,[23] but other investigators found different construction in other piers, and at least one pier with no piles at all.[24]

The citizens of Bristol decided very quickly that their bridge must be rebuilt but then engaged in years of controversy about how it should be done and how to raise the money. The rebuilding seemed so probable to an architect named James Bridges that he started making 'observations' at the old bridge in the middle of 1757 at his own expense. In 1758 and 1759 he was employed by the Corporation to attend and serve committees, make designs and surveys and try quarries for stone.[25] In 1760 an Act was obtained and by August of that year designs had been offered to the bridge committee by Bridges (four designs), by John Wood the younger, architect, of Bath (two designs), and by Ferdinando Stratford, engineer, of Gloucester (two designs). Each had written pamphlets in support of his own designs and they continued this practice for over two years until at least twenty pamphlets and letters had been printed, their tone becoming more irritable as time passed.[26]

The bridge to be replaced (fig. 57) was a four-arched structure built in 1247, with the piers at an angle to the flow of the water so that the ebb tides and downstream spates were directed on to the face of a quay below the bridge and tended to dislodge the ships moored there.

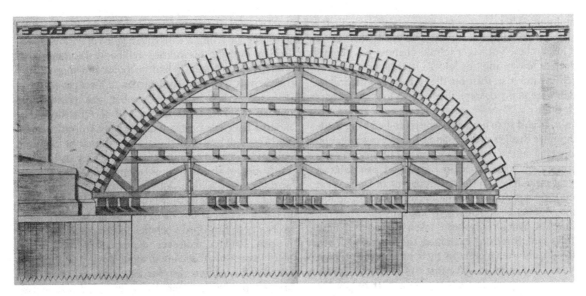

55 Centre for London Bridge arch, 1758.

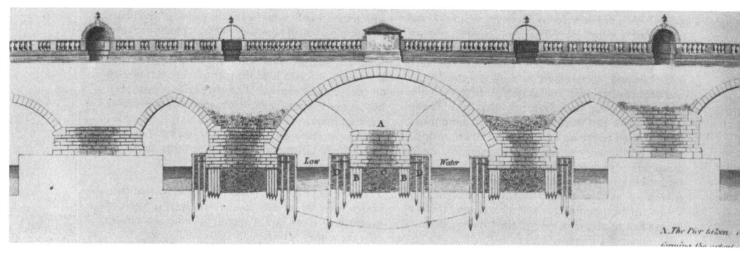

56   Section of old London Bridge, 1762 (from *Commons reports*, 1800).

There were three piers, the middle one 35 ft thick because it had originally supported a chapel, and the others 25 and 26½ ft, while the arches varied in span from 22½ to 26 ft, giving a total waterway of 98½ ft and sum of obstructions of 86½ ft. The bridge itself was less than 20 ft wide but wooden houses up to five storeys high and 24 ft deep from front to back had been built on both sides of it by a curious and seemingly precarious method. On the ends of the piers, which projected 20 ft up and downstream, arches and surmounting walls matching the arches and spandrels of the bridge, but only 4 ft thick, had been erected; and from the top of these walls to the outer faces of the bridge timber beams were laid and the houses built up from the beams (fig. 58). In spite of wind blowing up through the floors and an occasional ship's mast coming up through a kitchen window or a shop floor, these buildings were let at the highest rents in the city until the day of their demolition. There was some navigation through the bridge but probably not a great deal, due to the narrow and low arches. The water in mid-channel was about 5 ft deep at low spring tide and 28 ft at highest water, a much greater rise than on the Thames. The arches, which were pointed, were filled almost to the crowns at high tides.

In 1758 and 1759 borings of the river-bed were made with James Bridges present, and a further investigation by opening an inspection shaft into one of the piers was suggested by John Wood in his second pamphlet in August 1760. After protracted argument and indecision shafts were sunk down the middle of each pier and one of the abutments, apparently in August 1762 when the old arches had already been taken down, and both Bridges and Stratford examined the interiors. They differed quite remarkably in their judgement of what they saw, but they did agree that two of the piers were founded on timber frames only 1 or 2 ft below the river-bed and there were no piles under them. The existence of masonry at such a depth below the level of high tides is explained by the fact, well attested by ancient documents, that the old bridge was founded on dry ground with the river diverted, the construction being more like that of old bridges on shallow inland rivers than that of London and Rochester Bridges. The piers were all of masonry with no surrounding starlings, though they restricted the waterway almost as much as those of London Bridge. James Bridges measured the 'fall' though the bridge at half ebb and found it 1 ft 4 in., but the fall during the first quarter of the ebb was greater. Stratford claimed that the water of 'freshes', or floods, went through the bridge at between 6 and 10 ft per second and up to 14 ft deep, for several hours at the end of the ebb. The facing of the piers was of well-hewn ashlar above low-water but, according to Stratford, it was of small flat stones below low-water and with some parts 'blown' and undermined. There were also some cracks and the body of the piers inside the facing was 'of flat stone, intermix'd with rough stone of various sizes'. Bridges thought this interior well bonded and mortared while Stratford said it had no bond and most of the mortar was 'perished, and turned to a coarse sand'!

A temporary bridge of timber was erected and opened to wheel traffic at the end of 1761, but many months later the bridge trustees were still undecided. Stratford advised that the old piers must be removed and said he could lay new ones on the rock, as much as 16 ft below the old foundations, by enclosing the working space with a double-walled cofferdam. This would allow the new bridge to be better aligned relative to the direction of flow, and the use of cofferdams was necessary to Stratford's scheme for a single arch of 150 ft span. Bridges, however, claimed that the existing piers, cut down to low-water, could be used as foundations for new thin piers to support a three-arch bridge – but he also offered a single-arch design in case the trustees should prefer it. To help them decide they invited an opinion from John Phillips and he came from London in August 1762 and reported in favour of using the old foundations for the sake of cheapness, and against the single-arch designs because he thought them very risky. In reply, John Wood wrote scathing remarks about Phillips's reputation in London and especially his centre for London Bridge, and Wood and Stratford in unison convinced the trustees that the old piers, or what was left of them, were useless. At the end of the year they decided by the slender margin of one vote in thirty-one to adopt Bridges's plan of three arches with new foundations, but eight months more passed while they obtained estimates. These were 'vastly too large' and in November 1763 they again debated at one meeting which design to use and at the next whether to build on new foundations or the old. It was argued at last, and probably correctly, that new foundations could not be sunk to rock without undermining the temporary bridge, and a final vote of forty-five to eighteen settled it in favour of the old foundations. The contract price was £10,300 but the total cost of the project, including purchases of land and buildings, construction of the temporary bridge and payments for the many designs and reports, has been quoted as high as £49,000.[27] All three designers took some part in drafting the contract and specification for the new bridge, but none had any part in the construction, for which the contractor was 'Mr Britton and Co.' and the trustees' supervisor Thomas Patty.[28] The design, as printed in Barrett's *History of Bristol* (1789), borrowed some of its decorative details from Westminster Bridge, and it was the last important bridge in Britain to exhibit such an influence.

The construction of the old Tyne Bridge at Newcastle is chiefly known from records made later than 1760,[29] but it did not differ much at that date from that of London Bridge and Rochester Bridge. There are said to have been grids of piles over the whole area of the bottoms of the piers.[30] The masonry was not in good but not in very bad condition, the starlings were faced with oak piles showing no serious decay, and the widths of all the arches together made up two-thirds of the width of the river, some piers being 23 ft thick; but the widths of the spaces between the starlings (i.e. width of waterway at low tide) was less than half the river's width. The soundness of some of the masonry was proved by the need for some parts to be broken up by blasting in July 1772.[31] The width between parapets was only 15 ft, but in spite of this buildings had been erected on the projections of all the piers or starlings and continuously along the southern part of the bridge which was owned by the Bishop of Durham. Some of the buildings contracted the roadway to only 8 ft wide.

Any hopes of improvement were greatly prejudiced by the bridge's unusual patronage. Three and a half piers and four arches were in the County of Durham and therefore owned and kept in repair by the Bishop, while the remaining five and a half piers and six arches

were the property and the responsibility of the City of Newcastle. Boats were navigated through the bridge in large numbers and John Smeaton's reports of 1769 and 1771 speak of frequent damage to the masonry of the piers and arches by collisions of the coal-carrying 'keels'. In the oldest arches, which were ribbed, many of the ribs had fallen and others had settled and lost contact with the soffits of the arches which they might have been thought to support, one or two appearing to Smeaton to be in danger of falling on the boats and boatmen passing below. Both owners had asked him to survey the bridge, the Bishop in 1769 and the City in 1771, but there was no real attempt at improvement until the river itself took the initiative and demolished the bridge. That story is told in chapter 8.

57 Old Bristol Bridge (from Seyer, 1821).

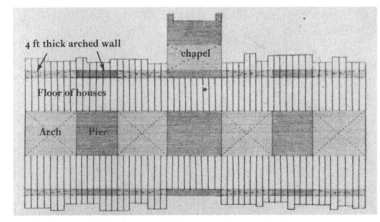

58 Plan of old Bristol Bridge and floors of houses (from Seyer, 1821).

PART 2

The years of Mylne and Smeaton
1759–1796

7

BLACKFRIARS BRIDGE

In respect to the elliptical arches, I must refer to a very easy and familiar experiment, which is, to take an egg, and try whether it will not bear a great deal more weight on the ends than it will on the sides before it burst.

Isaac Gadsdon (1739)

Although the argument between improving London Bridge and building a new bridge had ended in a decision to do both in 1756, the City's Common Council took no steps towards the building of the new bridge for two years more. There was no parliamentary grant for this bridge and with the country at war there was some doubt as to whether the money could be raised as a loan. Materials were also more costly in wartime and labour might be rather scarce.[1] However, the City's credit was so good that when they finally advertised for subscriptions in July 1759, the money to bear interest and be repaid out of the produce of a toll on the bridge, the whole amount of £144,000 was promised within a few weeks. The bridge committee therefore advertised that they would meet on 4 October to receive designs and estimates from all who cared to offer them.

Twenty-two men attended with an unspecified number of designs, both drawings and models,[2] but notification was received from 'Messrs. Smeaton and Gilbert' that their designs were not ready and the committee agreed 'in compliment to their respective characters' to put back the day of submission to 1 November.[3] On that day they selected, from over fifty submissions in all (some men having offered more than one design), fourteen to be retained for full consideration. Meeting again on 22 November to make a choice, they appear to have given preference to a design by Robert Mylne (fig. 59) until it was formally objected by one or more members that the elliptical arches in his design were 'deficient in strength and stability'. They therefore decided to consult a panel of eight 'gentlemen of the most approved knowledge in building geometry and mechanics for their opinion and advice'.[4] What was submitted to each of these men was Mylne's own drawing of the largest arch of his design and an alternative drawing of a semicircular arch of the same span, viz 100 ft, also drawn by Mylne at the request of the committee.[5] The fourteen remaining designs some-

how got reduced to eleven and these were left to 'lie upon the table' until the experts had replied.

It is possible that the only question still at issue was whether Mylne's design would be built as submitted or with the arches altered to semicircles; but if so neither the candidates nor the public were aware of it, for a controversy in writing about the merits of the designs started and went on for the whole three months until the committee met again to make a decision. It was started by a person with no reputation for 'building geometry and mechanics' but with formidable literary

59   Robert Mylne.

and intellectual powers – none other than Dr Samuel Johnson. He had apparently been persuaded by his friend John Gwynn,[6] who was one of the candidates, to exercise his literary art in favour of semicircular arches, which were used in Gwynn's design, as in eight others of the remaining eleven. Johnson mentioned neither Gwynn nor any other candidate except Mylne in his three letters to the *Gazetteer and new daily advertiser*, but he asserted that three designs were acknowledged to be much better than the other eight and that two of the three had semicircular arches while one (clearly Mylne's) had elliptical arches. It is impossible to say now, and it would have been difficult for the committee or the readers of the newspaper to know then, to which two designs as well as Mylne's he was referring. It appears that he did not sign his letters. His argument, written in layman's terms, was certainly persuasive:

The first excellence of a bridge built for commerce over a large river is strength; for a bridge which cannot stand, however beautiful, will boast its beauty but for a little while. The stronger arch is therefore to be preferred . . . All arches have a certain degree of weakness. No hollow building can be equally strong with a solid mass, of which every upper part presses perpendicularly upon the lower. Any weight laid upon the top of an arch has a tendency to force that top into the vacuity below; and the arch thus loaded on the top stands only because the stones that form it, being wider in the upper than the lower parts, that part that fills a wider space cannot fall through a space less wide; but the force which laid upon a flat would press directly downwards is dispersed each way in a lateral direction . . . In proportion as the stones are wider at the top than at the bottom they can less easily be forced downwards, and as their lateral surfaces tend more from the centre to each side, to so much more is the pressure directed laterally towards the piers, and so much less perpendicularly towards the vacuity . . . If the elliptical arch be equally strong with the semicircular, that is, if an arch, by approaching to a straight line, loses none of its stability, it will follow that all arcuation is useless, and that the bridge may at last, without any inconvenience, consist of stone laid in straight lines from pillar to pillar. But if a straight line will bear no weight, which is evident to the first view, it is plain likewise that an ellipsis will bear very little, and that as an arch is more curved its strength is increased.[7]

Johnson wrote two more letters and was answered at least twice by somebody who was conversant with Mylne's design and with Continental bridges.[8] Discussion about the supposed weakness of Santa Trinita Bridge in Florence – a bridge of very low arches, but not elliptical – was bandied about, and at least two designs were offered to the committee by plates printed in monthly magazines.[9] In January appeared a pamphlet entitled *Observations on bridge building and the several plans offered for a new bridge*, which was addressed to the gentlemen of the committee and described and criticised all the eleven designs 'upon the table'; it was signed 'Publicus'. The descriptions give many details of the designs, both as regards the functional and structural forms and the 'architecture' or ornament, and the criticism is both logical and amusing. This well-informed writer was even able to repeat

explanations given by the designers verbally to the committee, and the reason is that 'Publicus' was, beyond any reasonable doubt, Robert Mylne himself.

Mylne's emergence as the leading contender for the commission was unexpected, and indeed largely fortuitous. He had arrived in London from Rome less than seven days before the Common Council agreed to their committee's detailed proposals for raising subscriptions and starting the bridge, and he had planned to go home to Scotland shortly. He was actually on board a ship to leave before the imminence of the bridge proposal changed his mind and made his career.[10]

His father was an Edinburgh mason, a successful man who became Surveyor to the City of Edinburgh, but without great pretensions to being an architect and of modest means. Robert always made much of the fact, however, that his forebears had for six generations held the title of Master Mason to the Crown of Scotland. He was apprenticed to a wright (i.e. carpenter) whom he served for six years, after which he was admitted to the lodge in Edinburgh and worked for a year making wood carvings for the Duke of Atholl's castle at Blair Atholl.[11] Then, accompanied by his brother William, an apprentice mason, and with a joint allowance from their father of only £60 a year (and which was sometimes months in arrears), he travelled to Paris, Marseilles and finally Rome, where Robert remained for fully four years, studying architecture by every means available to him.[12] He also made contact with a number of wealthy English gentlemen who came there on their 'grand tours', teaching them architecture and drawing and hoping to obtain patronage from them when he returned to Britain. In September 1758 he entered the concourse of the Academy of St Luke and won first prize. As he was the first Briton to do so it was reported in newspapers throughout England and Scotland and he wisely decided to return to Britain the following year to seek employment.

It is known that he had studied the aqueducts of ancient Rome, but most of his studies in Italy were of buildings and their ornament. An intention to see Holland on the way home as 'a country so new and so much connected with the mechanical part of our trade' suggests that he was interested in 'hydraulic architecture' but, even if he was, he must have worked hard during the two months or less which it took him to produce his design for Blackfriars Bridge.

He was in no doubt, however, that he would have to do more than design well. He wrote to his brother in mid-August: 'It is necessary, that, besides an ingenious drawing, and impudent face, and good friends, I should have a fine shell and lodging that bespeaks affluence'; he requested money from his father to provide 'an entire new shell from top to toe', and wrote separately to his mother to ask her for half a dozen shirts. By 30 August he was 'very busy making interest with and showing all and even more merit than I have to the fat aldermen of the City. My design', he added, 'advances and is approved of by the principall man in the committee.'

The 'principall man' was John Paterson, an attorney who was currently the City Auditor, and later a Member of Parliament, and was the most prominent member of the bridge committee from 1756 until at least 1770. After the question about his arches had been referred to the experts Mylne kept up his efforts to persuade everyone with influence that his design was the best; and he wrote to his father on 24 January, 'My whole time is taken up about it ... In my defence, I have been obliged to turn author, in pamphlet and in newspapers. I have been obliged to speak in public, and reason with every species of men from astronomers down to porters.' It is easy to identify two of his writings as the 'Defence of elliptical arches and iron-rails',[13] published in answer to Johnson's first letter, and the pamphlet signed by 'Publicus'.

The pamphlet begins by setting out his very logical approach to the problem. He wished to avoid the use of long or steep approaches of what he called 'forced earth' (in modern terms, compacted fill) and so proposed that the height of the end arches should be only 12 to 15 ft above high-water and the height of 'forced earth' not more than 17 ft, basing these requirements on his observations at Westminster Bridge, the ends of which he thought to be a little higher than was necessary for boats and certainly too steep for carriages. His requirement for height of the largest arch above high-water was 26 ft minimum, a figure obtained from discussion with barge masters and watermen; it would only demand a gentle gradient of the roadway on the bridge itself from the end arches to the middle. Next, he would have the piers and their foundations as small and thin as possible, consistent with giving adequate support to the arches and spandrels. There is nothing to suggest that he even tried to use theory either in design of the arches and piers or to calculate the 'fall' of the water through the bridge. It was obvious that any bridge which obstructed the flow no more than did Westminster Bridge would cause no sensible 'fall' and so there was never any argument about the 'fall' under Blackfriars Bridge. In the dimensions of his arches and piers Mylne conformed with the rules of Palladio and Gautier.

The width of the river made it obvious that the bridge should be of nine or eleven arches and Mylne chose nine. He drew them elliptical and so provided wider openings, for any given rise and span, than he could have done with semicircles or segments. Such arches were almost, if not entirely, unprecedented in Britain, but Mylne had seen large bridges on the Continent with elliptical arches, and some with lower ellipses than he now proposed. Yet these gave him a bridge whose height met all the requirements he had set himself: the 'forced earth' was only $12\frac{1}{2}$ ft high, the height of the end arches over high-water 17 ft and the height of the middle arch 28 ft. The latter was a semi-ellipse of 100 ft span and 41 ft rise.

60   Design of a *ponte magnifico* by Piranesi, 1750.

There is little or no room for error in this reconstruction of his reasoning, in view of what he wrote in the pamphlet. He referred also to the need for a reasonable height, at least in midstream, to make the bridge grand, and he made the line of the top of his bridge in elevation a circular arc of very long radius, to give it 'a light form, taking it in a profile view' (see fig. 69). In this decision there was a vital compromise between aesthetic dogma and the functional requirements, because he was proposing to erect on each cutwater a pair of Ionic columns which would support an entablature forming the floor of a recess (or refuge) from the footway; and all architectural theory required columns – the antique supports of Greek temples – to carry a *horizontal* entablature. It is difficult nowadays to imagine the strength of this dogma amongst people who professed knowledge in architecture, but the bridge was disparaged for this fault by various critics all through its life of a hundred years.

There was a striking change of architecture between the designs submitted for Westminster Bridge in the 1730s and those presented to the Blackfriars Bridge committee in 1759. The source of ideas was no longer Palladio's plate of the bridge of Rimini and other ancient models, but the fabulous drawings of neo-classical 'triumphal bridges' and 'bridges of magnificence' made by members of the Franco-Roman school in the 1740s.[14] The members of this school applied all the ornament of surviving triumphal arches to their drawings of bridges. Grand colonnades and loggias lined the sides of the roadways, or more properly processional ways, which were carried on horizontal lines high over the water by rows of equal semicircular arches. At the centre of this school was G. B. Piranesi, whose own version of the *ponte magnifico* (fig. 60) was published in 1750 and whose four volumes of *Le antichita Romane* in 1756 included many accurate engravings of both the ruined aqueducts and the surviving bridges of Rome, showing precise (though sometimes incredible) details of the internal construction of the bridges and their enormous foundations.[15] And Piranesi was a warm friend of Mylne, an acquaintance at least of William Chambers, and soon to be tutor and friend of the son of George Dance; as well as opening his doors to virtually all the English who went to Rome at about that time.

Dance the elder and Chambers drew large ornaments on the tops of their designs for Blackfriars at midspan, John Gwynn drew columns over the cutwaters and sculpture all over the bridge,[16] and all three set their road levels too high and too near horizontal to permit the approaches to be of a reasonable gradient and cost. Only Mylne allowed himself to distort the image of the triumphal bridge enough to meet the practical requirements of traffic and the purse of the City of London, and because he did so a few of the Roman students' dreams were built with Portland stone in London. In a wealthier age Rennie would come a little nearer to their hopes in his Waterloo Bridge in 1817, building columns like Mylne's with a horizontal entablature and therefore all of equal height as the classic tradition required.

Fig. 61, which is thought to be the end of one version of John Gwynn's design, will serve to illustrate the grander designs submitted, and also the clever blend of ridicule and serious criticism which Mylne used throughout his pamphlet to discredit his opponents' designs. He wrote of Gwynn's drawing:

61   Part of design for Blackfriars Bridge, attributed to John Gwynn.

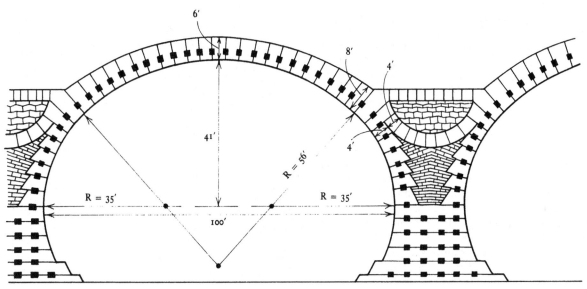

62 First design of arch for Blackfriars Bridge by Robert Mylne.

There is placed upon the two middle piers a group of small columns like organ pipes, to support either a recess or a clock. The form is taken from those of Westminstre-bridge, which look more like a centry-box, than any form produced by the squareness of stone. There are no more columns till you come to the piers next the shore, and these intermediate ones are covered by a thing like twisted glass, which supports, what are things of vast importance to the benefit of mankind, a globe or a pine-apple! The groups of columns next the land, being shorter than the middle ones by the fall of the bridge, and no intermediate gradation between, look like dwarfs of columns or over-grown banisters.

To return to the structural aspects of Mylne's design: he sought to understand the relative strengths of arches by imagining their physical behaviour, as Johnson did, but with much better knowledge of the common forms and techniques of masonry; and by the time he wrote his pamphlet, but probably also when he first submitted his design, he could offer a complicated process of construction of the arches, spandrels and piers which he claimed would reduce the horizontal thrust of the arches on the piers to nil. The 'ellipse' of each arch was to be built of three circular segments, one of long radius (56 ft in the largest arch) across the middle of the span and two of equal shorter radius (35 ft in the largest arch) at the ends. Mylne (and Simpson, see below) identified the 'haunches' as the points at which the long-radius arcs met those of short radius. In the pamphlet he explained:

the weakness is, that the middle part of the arch makes a lateral pressure against the haunches of it. These haunches of the arch [the designer] has secured in such a manner by ruble work, a counter arch, and an horizontal course of stones, that the spaces between the feet of the great arches are filled up for 35 feet high from the top of the pier, so that, if I understand it right, all from the haunches of the arch downward becomes a pier or abutment, to support a small part of the arch in the

middle as a segment of a circle. This middle part, if built like other arches would make a lateral pressure against these abutments, but to take that away he has placed cubical stones, which he calls joggles, in the joints of the arch; so that every stone tends to fall perpendicularly by its being carried along with the one above it, and not shoved aside as in other arches, which is the cause of the lateral pressure.

Fig. 62 is a reconstruction of these proposals, the dimensions being abstracted from the experts' replies to the bridge committee.

The logic of Mylne's explanation was less compelling than Johnson's, but that is at least partly because his understanding of the structural actions was more thorough. Johnson's mistakes were, firstly, to suppose that the strongest possible arch shape must be preferred to any other, even one which was *adequately* strong, and, secondly, to compare arches alone without reference to other parts of the structure – both being mistakes which Emerson had made in his mathematical treatment of arches (see chapter 5). Mylne made neither of Johnson's mistakes but his explanation still contained a serious conceptual error. The joggles could only work as he said they would if the voussoirs were all slightly separated from each other, in other words if the piers moved away from each other by a small distance; and the cause of such a movement was certain to be that horizontal thrust of the arch which he claimed to have 'taken away' by inserting the joggles. The thickening of the arch from the crown towards the haunch was often practised, but the splitting of the arch at the haunch into two members, the segment of 35-ft radius and the reversed arch, was original. The general concept was obviously to employ a 'skeleton' of stiff members of hewn stone and fill between them with rubble, and the drawing with its curving bands of white stone is distinctly reminiscent of the longitudinal sections of

Roman bridges published by Piranesi in 1756. The deep curve of the reversed arch suggests what Professor Robison wrote many years later and was broadly confirmed in a letter to him from Mylne,[17] that the all-round pressure of the two 35-ft radius segments and the reversed arch on the rubble work between them was expected to make the whole pier act as a single resistant mass. The rubble was to be of Kentish ragstone, cheap but flat and square. The reversed arch was not envisaged as carrying the horizontal thrust from haunch to haunch of the main arches (for which it was too deeply curved) but the horizontal course of hewn stone from haunch to haunch would certainly have performed this function and thus relieved the piers of nearly all horizontal thrust so long as the arches on both sides of a pier stood and were equally loaded.

In the pamphlet Mylne added a list of other features of his design, all of which seem to have been well thought out. The most original in British practice was the plan of the cutwaters, drawn 'nearly like the bow of a boat', which meant that they were pointed at the ends but rounded at the shoulder angles where they met the sides of the pier (see fig. 69).

If the third design which Johnson thought to be still in the running was John Smeaton's (fig. 63), then his mention of it was irrelevant to his argument because its arches were not semicircles; and if it was not Smeaton's then he was dismissing the man whom Mylne considered to be his most serious rival. He had good reason for thinking so. In the first place, Smeaton's design was much nearer to his own in the dimensions he thought important than were any of the other nine. Secondly, Mylne had been advised by a leading member of the committee (presumably John Paterson), when he first decided to compete for the commission, to make a joint design with Smeaton because he was then considered the chief contender.[18] This proposal was put to Smeaton but he must have rejected it. The design which Paterson had exhibited in Parliament in 1756 to

63   John Smeaton. Bust by Chantrey.

obtain the enabling Act had been made by Smeaton;[19] like his later one it had nine segmental arches and was described to Common Council as 'like that at Westminster (but in a plainer taste, and upon more frugal principles)',[20] but he estimated its cost at £120,000 and it was on the basis of this estimate that the committee calculated, and Parliament approved, the sum they needed to borrow. Their report to Common Council described Smeaton as 'one of the ablest engineers of this kingdom' and he had been recommended by the President of the Royal Society a few months earlier for the job of building the third lighthouse on the Eddystone rocks off Plymouth.[21] This was the commission which had established him as Britain's foremost

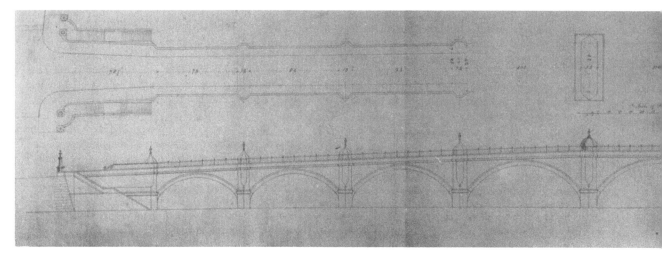

64   Smeaton's design for Blackfriars Bridge, 1759.

engineer and the completion of the lighthouse was announced during October 1759 just before he presented his design for Blackfriars Bridge. He had already, from that work alone, more experience of 'hydraulic architecture' than Chambers, Gwynn and Mylne together.

He was also of quite a different background from Mylne. The son of a Yorkshire attorney, he gave up a legal training against his father's wishes to apply himself to mathematics and mechanics, working as an instrument maker in London until he made an impression at the Royal Society by his improvements to scientific and mechanical instruments. In 1752 and 1753 he made extensive experiments on wind and water mills which led to another honour in 1759, the award of the Society's Copley Medal for a paper on the science of wind and water mills. Throughout his engineering career he lived for part of every year at his home in Yorkshire and reserved some of his time for making astronomical observations and carrying out other scientific research.[22] It must follow that the commission to build the bridge was less important to Smeaton than it was to Mylne, and if Smeaton canvassed the councillors of London he probably did so as an equal and certainly left no record of it. He responded to Mylne's pamphlet with the air, and the accuracy, of a man of science and, unlike almost all the other writers in the public controversy, he signed his piece.[23]

His reason for writing was to answer the criticisms of his design made by 'Publicus' but his pamphlet has added interest today because it reveals his basic attitudes and methods in bridge design better than any of his later reports. Of the design he was defending we have a full plan and elevation and also several preparatory drawings at larger scale of the arches, piers and foundations.[24] It had nine arches, all circular segments of 120° arc (but with sharply curved 'springer-stones' to turn the ends of the curves down to the vertical) and varying in span from 75 to 110 ft (fig. 64). They sprang

from about 10 ft above low-water, whereas Mylne's rather higher arches sprang from 3 ft above low-water.[25] The leading dimensions at the middle arch, compared with Mylne's, were:

	Smeaton	Mylne
Span	110 ft	100 ft
Rise	31¾ ft	41 ft
Thickness at crown	3½ ft	6 ft
Thickness of pier at springing	15 ft	20 ft
Thickness of pier at foundation	25 ft	25 ft
Height of soffit at crown above high-water	26 ft	28 ft

The outlines of the two arches are shown in fig. 65.

There was clearly little to choose between these designs in the convenience offered to traffic over and through the bridge; the differences were in the structural dimensions and the architectural treatment. Mylne had seized on the thinness of Smeaton's arches as evidence of weakness and on the narrowness of his piers to claim that none of the arches could stand without the mutual thrust of its neighbours. Smeaton's more difficult task was to defend the proportions he had chosen not from any textbook's rules but from his own measurements of standing bridges and to some extent from theoretical study. With regard to the latter he wrote:

As he does not know that the quantity of the lateral pressure of arches, under different circumstances, have ever been truly and clearly determined, he diligently applied himself to this work, before he determined the thickness of his piers; and having succeeded herein equal to his wishes, he has given such a size and form to his middle piers, as will resist several hundred tons more lateral pressure, than the middle arch will exert thereon, and this he declared before the committee.

It is a pity that no record of this investigation of arch thrust has survived, as it was probably the most thorough theoretical study of thrust made by an actual

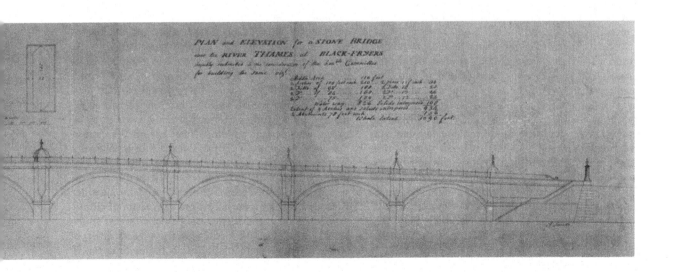

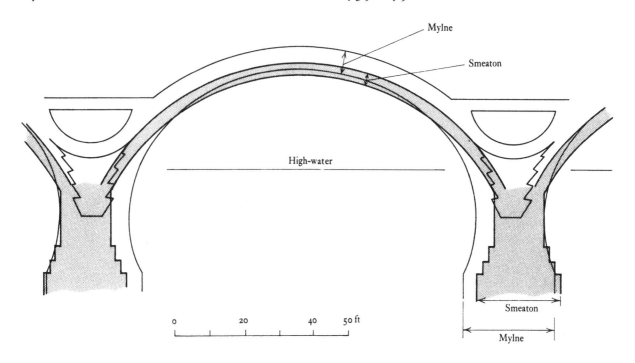

65   Sections of middle arches of Mylne's and Smeaton's designs.

designer between the time of Labelye and that of the Rennies. The committee, however, could not understand theoretical arguments and Smeaton was forced to rely on his reputation to persuade them that his design was safe. 'If the success that has hitherto attended his undertakings, is insufficient to produce any reliance upon his judgement, in a matter that he has made his study . . . he is not the proper person to be employed on this occasion.' With regard to the central issue of the controversy, the shape of the arches, he wrote in words rather reminiscent of Emerson's *Mechanics* (see chapter 5) that: 'Having chose such a figure for them as having no tendency to vary their original form, either by their own gravity, or the incumbent load, they admit of being made of the least thickness possible with the strength required.'

Of the architecture of Smeaton's design Mylne had written, 'It is the meanest and poorest of all the designs', and it is obvious enough that it bore less ornament than any of the others; but a modern eye might judge that it conformed very well to the description that Mylne had given in praise of his own design. 'Light, yet not too slender; and modest, without bordering on rusticity . . . The particular parts of this design are all dictated by utility.' If these were Mylne's aspirations, they were Smeaton's too, for he wrote in his reply that 'Mr Smeaton studied to avoid every ornament that did not naturally arise out of the subject, looking upon simplicity to be the greatest beauty; this he apprehends to have been the rule of the Greeks, utility and a fitness of parts being strictly attended to';

and he continued, with an implied criticism of Mylne's twin columns:

As the ancient orders are not applicable to bridges as such, ornaments borrowed therefrom have always been used very sparingly, both among the ancients and moderns; even the Gothick builders . . . when they set about to build bridges, strictly adhered to the simple and chaste . . . from whence arises, as Mr Smeaton apprehends, a kind of elegance, that stands in no need of ornament drawn from foreign subjects.

At the end of his pamphlet Smeaton challenged Mylne as a gentleman to disown the erroneous pamphlet of his supporter 'Publicus' or else reveal his identity; 'otherwise', he concluded, 'Mr Smeaton must look upon Mr Mylne to be himself the author'. No response to this challenge is on record.

Opinions founded on mathematics and mechanics were expressed by the experts consulted by the committee,[26] but also by Smeaton in the quotations given above and by William Emerson in a letter to a newspaper. Of the eight experts consulted four professed ignorance and indeed only one of the eight, namely John Muller, had even a tenuous claim to be called an expert in building. They included a clergyman, the Astronomer Royal, a teacher of medicine, a lawyer, two professors and two others unknown. The important opinions were expressed by the two professors, Muller and Thomas Simpson, and the outsider Emerson. Muller reported strongly in favour of elliptical arches on the evidence of 'demonstrations' published in his book on fortifications five years earlier[27] and which other mathematicians, including Simpson,[28] knew to contain mistakes. Perhaps the committee did not know. Emerson had made no serious mathematical mistakes in his writing on arches but, as shown in chapter 5, he had failed to account for the successful use of arches of widely varying shape and had left unanswered the practical questions of how thick to make piers and the independence or interdependence of adjacent arches. His letter in the *London chronicle* on the eve of the committee's decisive meeting was as brief and dogmatic as Muller's reply to the committee, and totally opposed to it. He said, 'Elliptical arches are good for nothing', and referred to the derivation in his *Mechanics* of the 'strongest arch' for a given span, rise and crown thickness. 'I am no judge what is agreeable to the eye', he added.[29]

In contrast with these curt pieces of advice was Simpson's reply to the committee, reasoned and relatively long. His biography is even more bizarre than Emerson's; for when turned out of home for his failure to work seriously at his father's trade, he had lodged with a friend whose mother was a widow, and after learning some mathematics and also some astrology from a pedlar and gaining some fame as a village fortune-teller he married the widow and she bore him two children, the first of them, it was said, at the age of fifty-five. When he became professor of mathematics at Woolwich Academy with a house provided, she enjoyed the house with her children and friends while ill-treating and half-starving her husband till his death at the early age of fifty, just a year after the controversy we are describing; and afterwards she obtained from the King, as a reward for her husband's 'great merits', a pension and house at Woolwich where she lived to the age of one hundred and two.[30]

Hunger and discomfort, however, did not blunt Simpson's sharp mind. His letter to the committee discussed the dimensions of the main arch and the spandrel work of Mylne's design, showing that the arch was thicker than necessary, and gave results of detailed calculations he had made for the size of piers necessary to resist the horizontal thrust of the elliptical arch and the alternative semicircular one which Mylne had drawn for the committee. There was virtually no difference for the two arches drawn but he noted that if the semicircular arch had tapered in thickness as the elliptical one did it would have been superior.[31] His calculations must have been thorough, and were probably more rigorous than Smeaton's. They may have been based on a correction of Muller's 'demonstrations', but probably used French authors as well. He was extending them later into a treatise on arches, but after his death they were given by his widow to a military officer who refused to let them be seen again.[32]

Simpson's letter should have assured the committee, if any mathematician could, that Mylne's design was sound. Whether it did we shall never know because their minutes were destroyed in the tollhouse of the bridge when it was attacked by rioters in 1780; but on 22 February 1760, they decided by a show of hands, but supposedly by a large majority, in favour of Mylne. A week later they appointed him surveyor for the building at £350 a year and he directed the work in every detail until it was finished in 1769.[33]

London's third bridge was now to be built, like the second, by a man who had never built a bridge before. It was a bold choice for the committee, but it is obvious that Mylne had been an impressive candidate. He now applied himself to the building of the bridge with the same resolution and diligence he had used to obtain the commission; this is very clear because the facts about the building come to us mostly from his own pen. Working drawings were made as construction went forward – a separate drawing for every course of stones in the piers – and a written order was issued to the relevant contractor with each drawing. Mylne kept copies of both the drawings and the orders in a folio volume and from the beginning of 1762 he also kept a diary of cryptic jottings; and from these two sources a narrative of his work can be gleaned. Seven folio engravings of the machines and methods used, published by one of his assistants, R. Baldwin, in 1787, clarify some of the technical details referred to in the diary.[34]

Several letters to his family during the first twelve months show that Mylne had to contend with resentment at the appointment of a Scotsman right from the start. The poet Charles Churchill took this attitude in his satirical poem *The ghost* in 1763, and a scurrilous drawing was engraved showing Mylne as 'the Northern Comet with his fiery tail, turned bridge-builder', directing the work on scaffolding while breaking wind at well-dressed passers-by who are supposed to be the designers he had defeated in the competition. On the other hand he had to restrain his father from sending Scots masons to get work on the bridge because he was unwilling to ask the contractors for the favour of employing his family's friends.[35] His letters to his father show a firm commitment to honest dealing and more than a trace of humility and youthful uncertainty behind the pride and hot temper which his contemporaries remembered. One of his workmen has often been quoted as saying 'Mr Mylne was a rare jintleman, but as hot as pepper and as proud as Lucifer.'[36] But in November of 1760 Mylne wrote to his father: 'Would you believe it, father, I have been drawn into the temptation of bribery varnished over with every symptom of politeness . . . when you have a leisure moment, pray for one who stands on the pinnacle of slippery fortune and the world's esteem.'[37]

The committee and their surveyor began their task of building cautiously. The Westminster Bridge Commissioners' records were consulted about the form of their contracts for building and their famous piling machine was borrowed. Then Mylne drew up full plans, specification and estimates of the quantities in the work of the various trades, and tradesmen submitted prices for each item. Four different contractors were appointed, with estimated total bills as follows:

Ballast contract – for digging foundation pits, disposing of the spoil, placing 'rubbish' or fill over the arches, and gravelling the road   £ 2,500

Carpenter's contract – for piling, caissons and centring   £15,300

Smith's contract – for cramps, straps and other fixings in carpenter work, chain-bars, etc.   £ 3,000

Mason's contract   £90,250

TOTAL   £111,050

This would be a cheap bridge and after completion it was claimed that the actual cost was less than the estimate, though a lot of extra money was required for property purchases and not a little for Mylne's fees as surveyor, the committee refusing to pay the latter until the matter had been referred to legal counsel.[38]

Mylne had ridiculed in his pamphlet the proposals of Smeaton and Dance to found the piers of the bridge in cofferdams but he cannot have been unaware of the inadequacies of the foundations laid in caissons at Westminster Bridge. The well-known story of the sinking of a pier was only one of the warnings; the appearance of the bridge would show to any architect that all the foundations had shifted and settled, for the Commissioners had had a survey made, while the Blackfriars controversy was going on, by four men of whom two were amongst the final candidates for Blackfriars (Dance and Phillips). They reported:

The works in some parts are rack'd and settled; the caps of the saliant angles, and the springing courses of the arches, are some of them out of a level, there appear chasms in the fascades, there are many spalls and fractures in several of the archstones . . . the ballustrades are in a serpentine form,[39]

all these being evidences of serious foundation movements.

In adopting Labelye's method of founding in caissons, therefore, Mylne must needs improve it, and before each pier was founded he had the river-bed bored to sample the ground and wrote down the findings with a recommendation for a grid of piles under the caisson. His proposal was approved by the bridge committee each time and the river-bed was then excavated by the ballast team to a depth at which he considered the gravel firm; this depth varied from 2 ft to 9 ft 6 in. The piles were then driven as deep as possible and all cut off at the same level just above the excavated bottom (fig. 66), which was 16 ft below low-water in the worst instance. The number of piles under a pier varied according to the firmness of the bottom from 62 to 104, which meant that they were spaced at from 5 to 7 ft in both directions. Their lengths in the ground must have varied also, but they are shown in Baldwin's prints as 12 ft long.

All this was done within 'guard works' which comprised sheet or plank piling driven only 1 ft 6 in. into the river-bed but strengthened by about forty square piles driven 10 ft and standing 25 ft high from the bed so as to mark the place of work and fend off river craft. The area enclosed by the guard works was a rectangle considerably larger than the base of the pier.

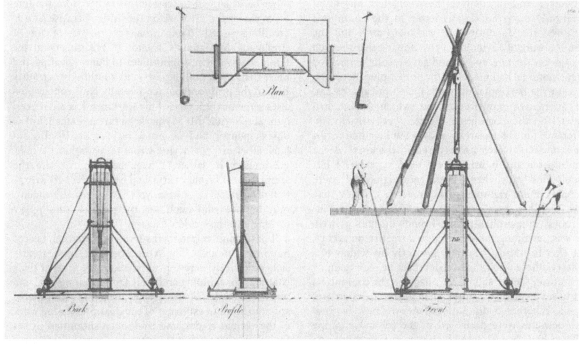

66   Cutting off pile-heads under water (from Baldwin, 1787).

67 Sections of caisson with one course of masonry (from Baldwin, 1787).

Collisions of barges with the guard works were not infrequent.

While the piles for a pier were being driven the caisson was made (fig. 67). Two 'stages' or jetties had first been built on piles in the river near the northern shore. One was close inshore and its deck probably higher than the level of most high tides. On this the carpenters assembled the bottom of each caisson. The caisson bottom was a rectangle of timber measuring up to 85 ft × 33 ft, consisting of a solid layer of 12 in. × 12 in. timbers 33 ft long laid side by side, with 3-in. planking running in the other direction under them and a chequered 'grating' of more 12 in. × 12 in. timbers over them. These three layers were connected with iron fixings including, it would seem, bolts, screws and straps. When the bottom was assembled it was floated at high tide to rest on the second, and lower, stage. There the sides were erected on its edges and fixed to make a joint as water-tight as possible. The sides were 27 ft high (compared with 16 ft for the Westminster Bridge caissons) and made of 12 in. × 12 in. timbers laid longitudinally with 3-in. vertical planking inside and

68 Floating the caisson by two barges (from Baldwin, 1787).

out. These were braced apart by the two floors shown in fig. 67, tied across by 'chain-bars' (probably iron bars with screwed ends to take nuts) at these two levels, and further braced at the corners by diagonal struts. There was a sluice in the side to be used when sinking the caisson and also a removable panel, comprising the bottom half of one end, which was taken off when the sides were floated away after founding a pier (see fig. 68).

The completed caisson contained 68 cwt of iron in its screws, straps and fixings, and 19,000 cubic feet of timber. It therefore weighed about 350 tons and Mylne calculated that its draught would be 4 ft 3 in. when empty; as the depth of water in the river at low tide was generally about 7 ft it would float over most parts of the site. However, after the first two uses the first course of stones for a pier was always laid and mortared in the caisson before it was floated off the lower stage and it then drew nearly 7 ft while floating out to be sunk at its proper position. While the bottom alone was being fabricated on the upper stage it had to be loaded with loose stones to prevent it floating off at high tide. Once the sides were on, the caisson would float as long as pumps were worked to throw out such water as seeped in; generally it seems to have been pumped dry continuously, while moored within the guard works, until the masons had built four courses of stone, making 10 ft height of the pier. Then the position was adjusted as accurately as possible and the sluice opened to sink it on to the bottom of the excavation and, as was hoped, the heads of the piles. As the stone was already in place, there could be no fixings between the timber bottom and the pile heads. The fixings between the caisson sides and bottom were undone and the open box of the sides floated off as shown in fig. 68, with the help of two flat barges and large triangle brackets made of timber balks and fixed to the box, the barges rising with the tide to lift the box high enough to clear the stonework with its half end panel removed. The box was re-used as the sides of the next caisson while the masons continued to build up the pier between tides. They could not have built more than one more course (of 2 ft 6 in. height) within the caisson without coming up against the braces, so the whole process had had to be nicely calculated.

Entries in Mylne's diary show that the caisson was a

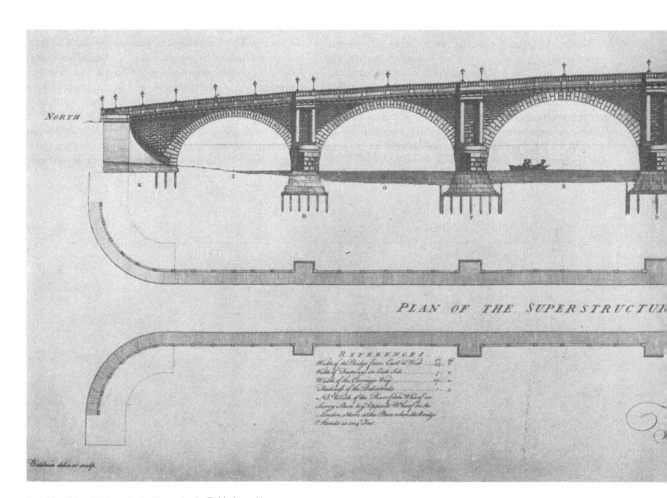

69   Blackfriars Bridge as built. Engraving by Baldwin, 1766.

difficult craft to control. In June 1762, for instance, the caisson for the north middle pier (the second to be built) was afloat within its guard works. On 8 June, part of the second course of masonry was set but then the mason work had to stop because the remainder of the stones for the course had not arrived from Portland. These stones had been specified individually on a drawing fully sixteen months earlier and such delays in the supply of stone were a common problem throughout the work. When the stones arrived the course was finished on 21 June and the caisson then touched ground, lying level from end to end but 5½ in. out of level from side to side. It could still be floated if pumped dry at high tide and it was not till 17 July that, with the third course apparently finished, Mylne grounded it finally, or so he thought. On checking he found it within ½ in. of the correct level at all points but out of line and position, so that the span of the arch was 3 in. too great at one end of the piers and 5¼ in. too great at the other. He corrected most of this by having the fourth course laid a little askew on the third, but the caisson surprised him by floating again on a 'great tide' on 24 July, carrying now five whole courses of stone weighing, as he calculated,

1,188 tons. He seized the opportunity to move it slightly and get the position just right. Another course was finished on 9 August – presumably working right against the bracing – and the sides of the caisson removed two days later.

While this was going on in the middle of the river there had been difficulty on the stages too. When the next caisson bottom had been finished on 20 July it fouled the stage instead of floating off and part of its ballast of stones had to be thrown off and later retrieved from the river-bed. When Mylne got the caisson bottom on to the lower stage he laid on it a 'great quantity of stones' to ballast it but it too surprised him by floating on the next tide and settling again inconveniently on the stage, so that it had to be adjusted on the next day's tide and loaded even more. Then, he wrote in his diary, 'it floated no more – seemingly little damage done'.

The box forming the caisson sides was used six times. On the second last occasion, for the third pier from the north end, the caisson was launched and towed out to the guard works on 25 October 1764, but at the end of the year, some of the stone for the fourth course not

The New Bridge at Black Friars London.

London, Printed for J. stayl . [?] opposite Great Turnstile Holborn having yet left Portland, the caisson was allowed to ground in a position which offered the least resistance to the movement of large ice-packs which were then in the river, and the three horses which had been in it working the chain pump were taken ashore. It lay there until 19 April 1765, but by then was leaking too much for three horses to pump it dry. After 'plaistering' of leaks and caulking of joints, removal of some stone, and several additions of pumps and men, it was floated briefly on 10 May by the efforts of four horses and eighty men working '6 square pumps, 3 common do., 1 square chain pump; the horse chain pump; 2 square pumps at east end'. Similar armies worked several times before the pier was finally fixed about 25 May. The sides were floated off on 7 June. In handling the box near the northern shore, however, on 15 June it was strained too much and a number of the chain-bars snapped and the sides were bent outwards, while the whole thing dropped on to the river-bed. It took some time to repair, but the next pier was founded by 5 September with the use of only six horses and eight 'pumpers'. It was 3 in. too far west, which Mylne apparently accepted, but was also $7\frac{1}{2}$ in. higher at the west than at the east end and he had this excess trimmed off by the masons.

The box was now dismantled, as Mylne had decided to use the long sides as bottoms for the two piers nearest the shore at either end. The water at these piers was shallow enough to suggest that a cofferdam of no great height might be pumped dry at each low tide. For the first of these two foundations, however, he adopted a method which required no cofferdam at all. It was a rather frightening method but it worked, though only just. The depth of water at low tide at this point was only 3 ft and dense gravel had been found only 2 ft deep in the river-bed; so if the timber bottom (1 ft 6 in. thick) could be sunk to its bed with two courses of stone (measuring 5 ft in height) already on it, it would stand 1 ft 6 in. above low-water and allow the masons to work on between tides. Mylne therefore had two courses built on the timber as it lay on the lower stage, and the day before a spring tide in February 1766 he floated over the stage at high tide a barge which was fitted with a number of rods or bars, probably hanging vertically from the gunwales. These rods were attached in some way to the timber and as the next tide rose the empty barge lifted the bottom with its two courses of stone – but only just, for the barge was down to the gunwale on one side and Mylne put on a 'wash boarder' to keep the water out. Thus the barge was towed across the river, but as it passed the last pier before its own place the turning tide carried it broadside against two guard piles. It heeled over another 6 in. and was only just saved from sinking by applying 'more wash boards and dung'. The water was pumped or baled out and a cable taken ashore to a crane which pulled the barge free and got it into place where, says the diary, the bottom 'took the ground very well'. At low tide they worked to disengage the barge but could not get all the nuts off, so

as the tide rose again the barge had to be 'skuttled' to prevent it lifting the bottom or breaking the fixings. It was successfully removed at the next low tide and the masonry of the pier resumed.

Details of the founding of the northern end pier are not so clear. It appears that the river-bed was above low-water level, but the excavation had to be dug almost 10 ft below low-water to reach a gravel stratum. Baldwin's print of the whole bridge (fig. 69) shows three timber bases in this pier, each with one or more courses of stone over it and under the next timber. The diary records another perilous journey of the second of these, suspended under the barge while water lapped over the gunwales, the timber and masonry sagging 1 ft in the middle and the barge sunk for safety at the end of the trip. This was on 16 March 1767, but in June Mylne was attempting to pump the water out of a cofferdam; sand came in through the gravel so quickly that he could not clear enough of the surface to lay a single stone. Another bottom was sunk in August and the pier was finished on 15 October. It had taken at least fifteen months' work.

When Mylne decided to use the caisson sides as bottoms for these piers he was abandoning the earlier intention he had copied from Westminster Bridge of using them as bottoms for the abutments. For his abutments he first drove a grid of piles, then dug between and around them to a good gravel stratum, cut off the piles and laid the first course of masonry between and around the pile heads. The abutments were built in small sections enclosed one at a time by a 'fence', or cofferdam of timber, and needed a gang of men with pumps and scoops to throw out the water and silt which kept seeping in.

In a very short time after the acceptance of his design Mylne had altered his proposals for the masonry of the piers from those shown in fig. 62. The stone joggles were left out and the widening at the base obtained by vertical steps at the edges of the first five courses. The whole width of the bottom course of stone was 8 ft more than the narrowest part of the pier and the timber bottom must have protruded 2 ft 6 in. more all round. A balk was fixed to it to form a kerb round the bottom course of stone and the masonry was wedged tightly within it. All courses of stone were from 2 ft to 2 ft 6 in. high and the layout of stones in each course was shown precisely on a large-scale drawing. The first of these drawings is dated 31 May 1760 and while there were some simplifications as the years passed the general principles were not changed. The external surfaces were always of Portland stone, well hewn and all wider in the bed than the height of the course. Tying lines of Portland were used along the middle of every course and at one, two or three points across it (fig. 70). There appears to have been little use of iron cramps, but wooden dovetails cut from blocks of oak 20 in. × 10 in. × 5 in. were fitted into prepared holes in the tops of the stones on each side of 'every cross joint', which is thought to mean the vertical joints between adjacent

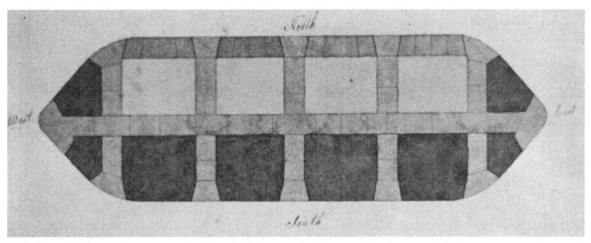

70   One course of masonry in a pier, drawn by Mylne. The white areas are rubble masonry, the dark grey areas are hewn Portland Stone with the positions of joints not specified, the pale grey are Portland Stone of specified shapes and sizes.

stones forming the outside face of a pier. The facing stones were also cut to taper outwards, which Mylne called 'arching' them, and the end stones of each tying line tapered inwards to provide abutment to the 'arches' (see fig. 70). The 'pockets' between the face stones and tying lines were filled, in alternate courses, with rubble masonry and with Portland stone of random size and shape. This must have eased the problems of supply of Portland at least a little. The 'arching' is similar in principle to methods of dovetailing in stone such as Smeaton had used in his lighthouse, but very much simpler. The concept of a skeleton of strong stonework encasing weaker material was noticed above in Mylne's first design for the spandrels (fig. 62).

The design of both the spandrel and arch shown in fig. 62 was also altered when the time came for their construction. All the joggles in the arches were omitted, and when Mylne found that the 'ragstone' from the Maidstone quarries would not be sufficiently regular in shape to provide the firm filling needed under the reversed arch, he replaced the reversed arch by a second horizontal course of hewn Purbeck stone similar to the higher course (of Portland) shown in fig. 62.[40] The upper horizontal course was keyed with driven wedges to make it an effective strut between the haunches of the main arches. The main arches were reduced to thicknesses close to those suggested by Thomas Simpson, namely 5 ft at midspan and 6 ft at the haunches for the middle arch and less for the shorter arches. Each arch course was permitted to be built of stones alternately the full thickness of the arch and half of the thickness, making headers and stretchers as in brickwork.

The faces of the spandrels above the main arches had been shown with the courses arranged radially in Mylne's winning design but by 1764, when he was ready to build the first spandrel, he had noticed serious fractures of the spandrel faces built radially at West-

minster Bridge. He therefore changed his method to courses laid parallel to the gently curved line of the parapet. To meet the square-cut ends of the arch stones, these courses were made to vary in height as shown in figs 69 and 72, an effective treatment of a question of appearance which troubled many designers.

Although Mylne drew an elevation of one of the ribs of each of the centres for the carpenters to build from, their finer details have to be inferred from the jottings of his observations when they were loaded by the arches. The main principles are quite clear, however. He believed that all the weight of the bridge should be brought gradually on to the piers and foundations and so he supported the centres entirely from the steps of the lower courses of the piers (fig. 71). The trussing of the centres was similar to that of the Westminster Bridge centres, all the diagonal members acting as struts when the arch stones were placed over their top ends. That the members could not act as ties is shown by some notes Mylne made in April 1764 when the middle arch was being built. Wherever there was a tendency for a member to extend it simply moved away from its end connections with other members. Clearly the connections would not resist any large tensile force.

This looseness of connections meant that the centres were none too rigid and as a result Mylne observed openings and closings of joints in the arch masonry as the loading on a centre advanced upwards from the springings to the crown of the arch. When the centre was struck the movements continued and he observed and recorded them all. The crown of the first arch to be finished dropped 2½ in. when the centre was struck, while the 'tails' of joints (i.e. ends at the extrados) at the haunches opened. Joints also opened right up the spandrels beside the twin pilasters (see fig. 72) as the crowns of the adjacent arches sank, but there was some movement in the opposite sense when the spandrel

71   Centre for Blackfriars Bridge (from Baldwin, 1787).

72   Work in progress at Blackfriars Bridge. Engraving by Rooker, 1766.

filling was placed over the haunches. Mylne contrived to control the opening of joints sometimes by placing fill over piers or haunches. He recognised quickly the only matter of concern, which was a tilt of one of the middle piers when the centring of the arch on one side of it was struck, and which he found to be caused by the ballast men having excavated 18 in. lower than the timber platform at one side of the foundation when clearing the side of the pier for the erection of the centring. He had a double row of piles driven close against that side of the foundation to prevent the ground from being pressed out into the cavity.

For the actual easing and striking of the centres he adapted the multiple-wedge system devised for Westminster Bridge but he planned it fully in advance, inserting wedges as the centres were erected and aligning the wedges with the span of the arches so that there was a wedge under each end of each rib of the centre (figs 71 and 72). A wedge was moved to ease the centre by striking the end with a sledge hammer or else by an arrangement of 'hoops' and 'screws', which probably means screw jacks. Mylne recalled the operation with pride in his letter to Robison in 1799.

There was no concussion . . .I could stand under the keystone and ascertain the awful subsiding of the middle of a one hundred foot arch, five feet thick in the middle, and stop or proceed with that motion as I pleased. The hand that writes you this trembled not . . . One day struck the centres, and after a proper interval for observation . . . another day cleared the arch of all the complicated timbers.[41]

As a description of the method this is not really clear but it was obviously superior to Labelye's method and the speed of working can be judged from the fact that a set of centring was eased and taken down from one arch and re-erected for another between 20 April and 11 July in 1767.

Blackfriars Bridge was opened to traffic on 19 November 1769 and continued in use until 1864. By then it seemed old and inconvenient in its gradients and its width, but that was due to a century's change in vehicles and traffic intensity. For its day it was well designed and very well built in all respects except one. The Portland stone used for all the exterior was not resistant to atmospheric decay when used in a river, and was especially susceptible to frost damage. Within twenty years serious repairs to the masonry were necessary[42] and over the century of its life a large proportion of the facework must have been replaced. It is likely that Westminster Bridge had suffered frost damage before Blackfriars was started but that the fact was not remarked because there was so much cracking due to foundation movements. With its famous name, its brightness and its excellent cleavage, Portland was a dangerously attractive stone for bridge-builders. As he learned so much from the building of Blackfriars Bridge, Robert Mylne probably also learned the weakness of Portland stone, for in January 1768 he jotted in his diary just the comparison with Westminster Bridge which we are making now: 'Frost broken up . . . parts of stone under high water mark scaled off. Do. at Westr. Bridge but not quite so many.'

8

JOHN SMEATON AND ROBERT MYLNE

My profession is as perfectly personal as that of a physician.

Letter of John Smeaton, 15 January 1783

When the magnificent is procured by the simple and the genuine, it pleases universally.

Robert Mylne, Observations on bridge-building (1760)

In following the progress of Robert Mylne's greatest work we have passed by half the bridge-building life of the man who might have been building Blackfriars Bridge instead of him, John Smeaton. To survey the bridge designs, reports and standing bridges of these two men is to draw a skeleton of the bridge history of about thirty years.

More is known of the details of Smeaton's work than of Mylne's because after his death in 1792 the members of the profession to which he had given the name 'civil engineers' paid him the unique compliment of publishing all the surviving manuscript reports of his professional life,[1] while the office copies of all the drawings which accompanied these reports were bound and eventually placed in the library of the Royal Society in London.[2] The calendar of his bridge works which is given in table 2 is very largely based on these two sources. Another surprising fact, and equally useful for our study, is that every one of the bridges shown by table 2 to have been built to his design is still standing,[3] and all but one still in use. Mylne played a leading part in arranging for the publication of Smeaton's papers and himself wrote the preface,[4] and these facts, together with their regular attendance at the Society of Civil Engineers from its foundation in 1771, show that they had a considerable respect for each other. The sharpness of their pamphlets in 1760 must have been largely forgotten, though they did not hesitate to act on opposite sides in several legal conflicts, notably that concerning Hexham Bridge in 1782–8. Facts concerning Mylne's bridges are harder to find and a calendar cannot be attempted; but a selective study of some ten of his larger bridges probably reflects his general output very well. Drawings of his designs are rare, but enough survive to provide the sample of dimensions and proportions given in table 3b, to compare with those of some of Smeaton's designs in table 3a.

We consider first the general forms and architecture of the two men's bridges, including the proportions given in the tables and the widths, gradients and decoration of their bridges. With regard to decoration, their statements in the pamphlets in 1760 should be recalled. Smeaton denied any competence in 'the ornamental parts of architecture';[5] he was an engineer, self-styled and self-taught. Mylne had learned architecture in at least a semi-formal way in Rome and thought of 'the ornamental parts' as an acquired skill. In criticism of Smeaton's design for Blackfriars he wrote: 'A knowledge of architecture is not only an exact calculation of the great forms in buildings, but likewise the bringing forward some of the materials which compose the great mass, and giving them such forms, as will serve some particular useful purposes.'[6] Thus ornament was to be fused somehow with useful form. This is a more sophisticated principle than the commitment to a style which governed the work of most architects of the eighteenth century. It resulted in plainer bridges, and in some cases plainer than Smeaton's, for Smeaton was not strictly true to the principle he wrote in 1760 that ornament should 'naturally arise out of the subject'.[7] The ornament of his bridges was restrained, but it was ornament for its own sake, consciously applied.

Smeaton had been interested in bridges at least from his earliest days in London, about 1742. He watched the founding of the piers at Westminster Bridge[8] and collected most of the accurate engravings of the timber bridges up the Thames (see chapter 3) and of Westminster Bridge. One of his earliest engineering papers is a memorial he addressed to the Westminster Bridge Commissioners in 1748 after the pier had settled.[9] Doubtless he studied Palladio, Gautier and Bélidor and also the theoretical writers Emerson, Muller and Martin Clare.[10] A small manuscript among his collected drawings shows that he added practical observations to his studies of books. It is a list of measurements and proportions of existing bridges,[11] copied here as table 4, and it shows that the rules of proportion given by Gautier and Bélidor were broken by many standing bridges known to Smeaton. The span-to-thickness ratios of the arches are not directly comparable to those of table 3 because of the existence of the 'counter arches', that is, second arch rings overlying the first, which are noted as present in six of the bridges listed. The predominance of segments about 120° or less, the existence of span-to-thickness ratios greater than 30 with and without counter arches, and of a span-to-rise ratio of 12 in an 83-ft length of the Santa Trinita arch, were probably the most important guide-lines Smeaton derived from the list. The last two entries were added after it was first written and it is all but certain that he wrote the rest of the list before he made his second design for Blackfriars, since he claimed that the arch thicknesses in that design were confirmed by many such examples.[12] Of the proportion of arch span to pier thickness he seems to have made few calculations and he claimed in his pamphlet to have used a theoretical method[13] to check the very high ratios which he used in his Blackfriars design. There was a similar ratio in the Hexham (1756)[14] design (see table 3a).

In these designs 120° segments were drawn for all the arches, but this of course meant that arches of different span were of different radius. For the sequence of large bridges which were built, namely Coldstream, Perth, Banff and Hexham, he adapted his preference for arches of about 120° so as to make all the arches of a bridge have the same radius, and this meant that the centre made for the smallest arch, which was 120°, could be used for all the others with the addition of only short extra timbers to support the first one or two courses of the arches above the springings. For the bridges in question this gave a middle arch of 134° as the table shows.

The constancy of form and decoration in these four large bridges contrasts sharply with the variety shown in his earlier designs. Each of the Blackfriars designs had nine arches and the parapets and roads sloped in two straight lines to an apex in midstream, with gradients of 1 in 30. There were no steeper ramps from the river-banks up to the bridge, and the widths between parapets were 40 and 42 ft. By comparison, the early Hexham design (1756) appears to have been made with ill-judged economy, having four arches of equal span (76 ft) and a horizontal road and parapet across the middle two arches and half of each end arch, but then a steep slope of 1 in 7½ down to the banks. The width between parapets was only 11½ ft. After that Smeaton always adopted 18 ft as the minimum width of road on a public bridge. The Coldstream trustees insisted on 22 ft,[15] which he used in his final design there and also at Perth, but the traffic at Perth was much heavier and even before the bridge was finished the traveller Thomas Pennant remarked on its lack of width as its 'only blemish'.[16] From early in the nineteenth century there was regular discussion of widening and it was

widened by ugly iron cantilevered footways in 1869. Banff Bridge, built 18 ft wide, was widened in 1881 by extra stone arches,[17] but the widths at Coldstream (22 ft) and Hexham (18 ft) were sufficient for the traffic until the early 1960s, when both were widened neatly by cantilevered concrete footways, and the stone parapets rebuilt.[18] The maximum gradient Smeaton allowed in the approaches to any of these bridges – and all his other designs after 1756 – was 1 in 12, and he argued for this as a tolerable gradient in reporting on the Edinburgh bridge in 1770.

73 Catterick Bridge, downstream side.

TABLE 2. *Calendar of Smeaton's bridge designs and reports*

	Large masonry bridges					Small masonry bridges		Timber bridges	
Date	Spandrel ornament or other — Design a	Design b	Design c	Construction	Report	Design	Construction	Design	Report
1748					Westminster (sunken pier)				
1754	Blackfriars (other)								
1756	Hexham (other)				Pontypridd?				
1759		Blackfriars							
1760	Glasgow Wentworth Woodhouse?	Stockton Wentworth Woodhouse? (3)							
1762		Coldstream	Coldstream		Bristol				
1763			Perth	Coldstream →	London				
1764									
1765				Perth →					
1766			Montrose (N. Esk)						
1767					London	Stonehouse Creek (Plymouth)			
1768					Dumbarton	Newark (brick) Perth?			
1769					Newcastle Edinburgh				
1770					Edinburgh Aberdeen Newcastle (2)		Perth?		
1771		Newcastle (1-arch)			Berwick		Newark?		
1772		Dunballoch	Banff Braan	Banff	Hull	Altgran			
1773							Stonehouse Creek		
1774					Carlton	Amesbury (Queensbury)			
1775					Carlton	Amesbury (ornamental 3-arch)	Amesbury (Queensbury)	Richmond?	
1776							Amesbury ? (ornamental 3-arch)		

Chronological chart (years 1777–1790) of bridge projects:

1777	Hexham		Walton-on-Thames
1778	Hexham		
1779			
1780	Walton (brick)	Cardington Amesbury (ornamental 1-arch)	
1781			
1782		Hexham	
1783		Hexham	Harraton Hull (with J. Gwyn)
1784		Hull	
1785	Montrose (S. Esk – end arches)	Coldstream	Montrose (S. Esk)
1786			
1787			
1788			
1789			Cardington (brick)
1790		Coldstream	

Robert Mylne's contact with above bridges

Pontypridd	Visit 1764.
Blackfriars	Design accepted 1760. Surveyor for construction 1760–70 and for many later repairs and improvements.
Glasgow	Some advice 1772 re bridge built to design of William Mylne 1768–72.
Stockton	Visit and report re design by Joseph Robson, 1762. Answered letter re abutment, 1767.
London	Gave advice to the workman who demolished the middle pier, 1762.
Coldstream	Gave design of a centre in 1764 (not adopted). Visited Coldstream for two days, August 1766.
Perth	Visited Perth September 1772.
Dumbarton	Scheme of repair given to Col. Skene 1768, 'paid by appointment of brother'.
Edinburgh	Named as security for brother William in contracts for construction (1765) and repair (1770). Took part in negotiations after failure, August–September 1769. Probably at 'final' inspection, September 1772. Large loans to brother, 1769–72; contributed £500 to brother's loss.
Banff	Sent pozzolana to Col. Skene, 1773.
Newcastle	Surveyor for rebuilding of Bishop's part. Report 1772, design 1774, directed construction 1775–80.
Hexham	Visit 1778. Acted for county justices in litigation 1783–8. Printed reports 1783.
Montrose (S. Esk)	Attended Parliament at passing of Bill, 1791.

TABLE 3a. *Structural proportions of Smeaton's designs*

		Largest arch			Piers of largest arch		Proportions at largest arch		
Date	Design	Degrees of arc	Span (ft)	Arch thick-ness (ft)	Width at springing (ft)	Width at foundation (ft)	Span/arch thickness	Span/width at springing	Span/width at foundation
1754	Blackfriars	120	120	5 or 3?	15	unknown	24 or 40	8.0	unknown
1756	Hexham	120	76	2.5	10	14	30	7.6	5.4
1759	Blackfriars	120	110	3.5	15	25	31	7.3	4.4
1760	Glasgow	90	76	2.5	13	24	30	5.8	3.2
1763	Coldstream	134	60.7	2.5	14	18	24	4.3	3.4
1764	Perth	134	75	3.0	17	21	25	4.4	3.6
1772	Banff	134	50	2.0	12	15	25	4.2	3.3
1777	Hexham	134	51	2.25	12.5	16.5	23	4.1	3.1
1780	Walton	140	60	1.67	7 (existing piers)	12	36	8.5	5.0

TABLE 3b. *Structural proportions of Mylne's designs*

		Largest arch				Piers of largest arch		Proportions at largest arch			
Date	Design	Degrees of arc	Span (ft)	Thickness at crown (ft)	Thickness at springing (ft)	Width at springing (ft)	Width at foundation (ft)	Span/thickness at crown	Span/thickness at springing	Span/width at springing	Span/width at foundation
Designs for which drawings of arch section exist											
1759	Blackfriars 1st design	—	100	6.0	8.0 (haunch)	20	25	16.7	12.5	5.0	4.0
1762	Blackfriars as built	—	100	5.0	8.8 (springing)	20	28	20.0	11.4	5.0	3.6
1759*	Blackfriars 1st design	85	78.5	6.0	8.0 (haunch)	20	25	13.1	9.8	3.9	3.1
1762*	Blackfriars as built	85	78.5	5.0	6.0 (haunch)	20 (minimum for pier)	28	15.7	13.1	3.9	3.1
1773	Aray	100	65	4.5	6.5	18	29	14.4	10.0	3.6	2.2
1774	Newcastle	113	50.7	2.5	4.75	18	26 (old pier)	20.3	10.7	2.8	1.9
1785	Dubh Loch (Inveraray)	106	60	3.25	3.67	(single arch)		18.0	16.4	(single arch)	
Design for which no arch section exists (arch thickness measured on elevation)											
1773	Tonbridge	120	22	1.5	1.5	4	12 (old pier)	14.7	14.7	5.5	1.8

Note. All arches are segments of circles except Blackfriars. Except at Dubh Loch, depth of arch stones on elevation is constant.
* In these lines the pier and spandrel is considered to be a massive abutment up to the haunches (i.e. tops of the short radius segments of the arches). The long radius (56 ft radius) segments are considered as arches springing from the haunches (see p. 68 and fig. 62).

TABLE 4. *Bridge measurements and calculations by Smeaton ('Designs', vol. IV, fol. 94)*

	Span feet	Rise feet	Thickness of the arch	Prop. of arch to span
Watsons Mill bridge near Sowerby bridge Yorkshire a segmental arch	37	9	1 ″ 4	27¾
Badger in Lancashire a segment	48	abt. 12	1 ″ 7	30⅓
Greenlaw near Coldstream Scotland	34	abt. 8½	1 ″ 1	31,4
Weldon Mill bridge in Northumberland 2 arches with a pier of 12 feet between	50	12	1 ″ 10	27⅓
Great A bridge in Yorkshire a segment of about 120° a very fine arch	60	– – –	1 ″ 8	36
Newton Bridge near Bishop Auckland in the Bishoprick of Durham 2 arches, the lesser 90 feet the great arch about 120° with two counter arches one of 22 inches the other of 20 a little crippled	100	– – –	2 ″ 0	50
Jacks Bridge or Tanfield Arch near Newcastle two counter arches together 3–6	102	37 ″ 9	1 ″ 11	53
Swillington Bridge between Wakefield and Weatherby in Yorkshire 2 arches segments of circles with counter arches the pier settled both perpendicular and out of upright which has crippled the arches about 120° 3 ft 8 inches in the crown through all	37½	– – –	0 ″ 9	50
Middle arch of Ouse bridge at York two segments that meet in a very obtuse angle it has two counter arches of 19½ and 14 inches the thickness through all is 6 feet	82	27⅓	2 ″ 1½	38
Coln bridge arch is a segment with a counter arch of rubble	43	14⅓	1–3	34⅔
The Trinity Bridge at Florence two anomalous curves of the elliptick kind which meet in an obtuse angle the piers are 25 feet English they have a counter arch of 13 inches	95 ″ 10	15 ″ 0½	1 ″ 6	64

A segment from the middle of 83 ft 3 inches span has only 1/12th only 6 feet 11¼ rise
Durham Bridge two large flat thick segments of about 90 feet each
Welsh arch 140 feet span and 35 feet rise [Pont y Pryd see Atlas v. 4 p. 90]*

* This note was added by another hand. The rest of the list is in Smeaton's hand. It is printed here with the rough and varied punctuation Smeaton used. 1 ″ 4 clearly means 1 ft 4 in., 'abt.' means 'about', and a comma is used as a decimal point.

In its masonry finishes and ornament his first design for Blackfriars was an almost slavish copy of Wren's scheme for London Bridge,[19] but it had a simple iron guard-rail, a feature he repeated in his second design; ever thereafter he designed solid stone parapets except for the ornamental bridge at Amesbury, which has a classical balustrade. The second design for Blackfriars, though generally austere, had two special aesthetic details. First, the extension of each cutwater up the spandrel was dressed back from its triangular plan to a trapezium which at the level of the road enclosed a recess or refuge from the footway, and the adjustment from triangle to trapezium was made not by a sloping plane, as was common, but by a curved surface with a decorative flourish at its bottom tip (see fig. 64). Although he repeated this detail in all the designs of type b in table 2, none of those designs was built. It appears to have been adopted by Joseph Robson of Sunderland when he re-designed the bridge at Stockton-on-Tees,[20] but Robson's bridge was demolished in 1887 and to see the detail today one must look at the downstream side of Catterick Bridge which was built when the bridge was widened under John Carr's direction in 1792[21] (fig. 73). The bridge from which Smeaton probably derived the idea also survives, namely Colne Bridge in West Yorkshire which occurs in his list of dimensions (table 4). The curved surface there connects the point of the cutwater with a thin pilaster on the spandrel, not a trapezoidal recess.

The second original detail in the Blackfriars design was a course of stones with sharply curved soffit at the springing of each arch, tangent both to the curve of the arch and the vertical face of the pier. Though he may have drawn it on the Blackfriars design in anticipation of a prejudice against segmental arches, Smeaton clearly liked it, for he drew it again in all his four designs for Lord Rockingham's estate at Wentworth Woodhouse near Sheffield[22] and also in those for Glasgow and Stockton and the first design for Coldstream. In the designs for Glasgow and Wentworth Woodhouse the arches were segments of very low profile, as little as 65° of arc in the Wentworth Woodhouse designs. These are the most original and most striking designs that Smeaton made. They were all of three arches with spans up to 60 ft. It is likely that the watercourse in the estate was well controlled and that there was no navigation, so he designed the bridge to leap the river with curves of soffit so long and low as to look like a twentieth-century bridge. There was more decoration than was his wont, but it was still restrained (fig. 74). Nothing like any of these designs was ever built in a nobleman's park in the eighteenth century and at Wentworth a regulation classical bridge of five arches was built from somebody else's design a few years later.[23]

The thought which seems to underlie these designs of low arches was made explicit in the comments Smeaton sent to Ferdinando Stratford in November 1762[24] about his design for a single-arch bridge at Bristol (see chapter 6).

I look upon it, that no limit to the span of arches, in proportion to their rise, has as yet been found; since the widest and flattest arches that have been attempted, upon right principles, have succeeded as well as the narrowest and highest, provided the abutments are good, and the stone and cement whereof they are composed are of a firm texture.

74 Design for a bridge at Wentworth Woodhouse.

But such boldness is entirely lacking in the designs he made himself from that time forward. He never drew an arch of more than 76 ft span after his second Blackfriars design, and 75 ft is the largest span he ever built.

His first design for Coldstream Bridge was made to support the application to Parliament of the trustees of Coldstream Bridge for a grant from 'the Supplies' (money voted to the King for the support of the Army and other services); but the trustees also called in evidence their appointed overseer Robert Reid who exhibited a design of his own.[25] When they had obtained the grant they set Reid to work quarrying stone and investigating the river-bed and adopted formally the dimensions of his design.[26] When they consulted Smeaton again a year later he adhered very closely to the arch spans intended by Reid but corrected a terrible gradient of 1 in 4 on the south approach (fig. 75) and improved the elevation, mostly by reducing the general height (fig. 76). Rather surprisingly, he adopted all the ornament Reid had drawn, comprising masonry rings on the spandrels with keystones on the ends of the vertical and horizontal diameters (the wall within the rings shown black), triple keystones on the main arches and modiglions under a plain cornice (see figs 76 and 77). Moreover, having drawn them once Smeaton made these elements his own and used them in nearly all his subsequent designs of large bridges. His practice in masonry finishes was settled at just the same time. The trustees of Coldstream Bridge had intended Reid to be the overseer of a contract with a mason from Sauchie named David Henderson, and during Smeaton's visit to the site in July 1763 Henderson asked 'if Mr Smeaton approved of making the piers rough or what they call

scrabled as he thought that would look both bolder and stronger'. Smeaton replied that

this being a matter of taste it must be determined by the trustees only, but as to his own opinion he would not only make the piers rough but the whole bridge, the projections only excepted which he would make very smooth work and this contrast will look prettier than if the whole bridge was smooth and at the same time save a deal of money.

Henderson replied that the smooth stones already hewn would build the whole of the piers and that 'it would look preposterous to have the piers smooth and the rest rough'. After some argument the piers were built with smooth stone and the spandrels and parapets with scrabled stone, but this was to suit the trustees' poverty, not their taste. Henderson left and the bridge was built by Reid with labour paid directly by the trustees. Smeaton recommended scrabled masonry (in England, 'scapled') for the piers and spandrels of the bridges of Perth, Banff and Hexham, with smooth-hewn arch rings, mouldings and quoins, and similar finishes were used on most large Scottish bridges for a considerable time. A notable example is the North Esk Bridge near Montrose which bears a huge inscription crediting its design to Smeaton, John Adam and Andrew Barrie, a local mason. In fact Smeaton's contribution was an earlier design for a different site and Adam's was only some advice,[27] but Barrie, as designer and contractor, transferred the finishes and many of the details of Smeaton's design to his own. Only the rings on the spandrels were omitted, except for two tiny elliptical ornaments, each with four keystones, on the two largest spandrels facing the town.

75   Design for Coldstream Bridge by Robert Reid, 1762.

A feature of the Coldstream design which Smeaton did not copy from Reid was the line of the parapet in elevation; he made it a continuous gentle curve. As mentioned in chapter 7, Mylne had done this in his Blackfriars design, and Smeaton had drawn it in one of his Wentworth Woodhouse designs. At Coldstream it fitted well over his five arches of constant radius but varying span and height. The rise, both of the arches and the parapet, was determined here, as at Perth, Banff and Hexham afterwards, only by considerations of river flow, road traffic and appearance, since there were no large river craft at any of these sites.

The economy of making all the arches of equal radius could only be realised if a single set of centring was used for each of the arches in turn. Robert Reid was unhappy with the slowness of construction caused by this, but Smeaton insisted. Reid also designed a trussed centre like those used at Westminster Bridge but Smeaton thought it unnecessarily complex because there was no need for a clear passage under the centring. Although Reid's estimates showed his own design to be slightly cheaper, the bridge committee ordered him to follow Smeaton's design. It required three rows of piles at the quarter-points of the arch span and two rows of posts at

76   Design for Coldstream Bridge as built (Smeaton, 1763).

the springings standing on the offsets of the bottoms of the piers; each row was capped by a 12-in. square beam[28] (fig. 78). This, Smeaton wrote, would 'distribute the supporters more equally under the burthen' than the Westminster Bridge type, but 'preserving at the same time such a geometrical construction throughout the whole, that if any one pile, or row of piles, should settle, the incumbent weight would be supported by the rest'. To provide this 'geometrical construction', or what would nowadays be called truss action, each rib was made a king-post truss – a structure very familiar to carpenters – and each load was resisted, where it came on to a rafter, by two raking struts, of which one was the rafter itself and the other went to the king-post or one of the posts over the piles at the quarter points of the span (fig. 78). The long bottom member of the truss was made an effective tie by iron straps tying the ends of the rafters and rim-pieces to it. These two strong connections, together with the straight single tie, made this rib much stiffer under load than those of the centres at Westminster and Blackfriars (figs 12 and 71), for in those centres many more joints at various angles would have had to resist tension if the structure was to be really stiff.

Smeaton's method of striking the centres seems to have been simply to knock out timber blocks 4 to 6 in. thick placed between the capping beams of the rows of piles and the bottom ties of the ribs at every point where they crossed – twenty-five blocks in all (fig. 78). This

77 Coldstream Bridge.

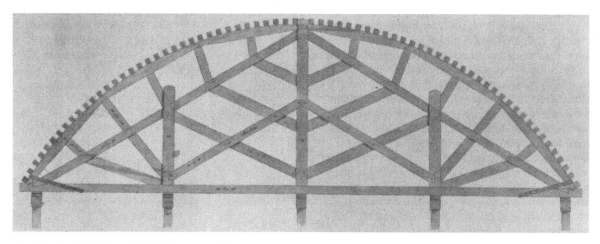

78   Centring for Coldstream Bridge (Smeaton, 1764).

was crude, but there is no record of difficulty, although the bridge committees' minutes at Coldstream and Perth[29] are very full.

The Perth centre was very similar in principle but larger in size and so required more strapped and bolted joints. For Banff and Hexham Smeaton simplified the design to what was little more than a few sets of radiating struts, one from each rim-piece of each rib to the nearest pile-capping beam, and he allowed the struts to be plain rough timber, round or square at will. Such 'design' seems more the province of the tradesman than of Britain's leading engineer, but Smeaton made drawings and gave careful instructions about these centres.

He has often been said to have built elliptical arches. In fact he used them only twice in large bridges, though he had considered them as early as 1753 for Blackfriars.[30] He made the end arches of Perth and Hexham Bridges elliptical simply to lower their crowns more than he could do with circular segments, and thus reduce the height of the approaches. He also made the three arches of his only classical design, the ornamental bridge in the park at Amesbury, very low ellipses (figs 79 and 84).

His designs for small masonry bridges show greater variety than those of large bridges. In the Amesbury designs the spandrels are of ashlar and he carefully drew every masonry joint on the elevations. The Cardington

79   Ornamental bridge at Amesbury.

and Newark bridges are of brick and entirely plain, but the latter is only small in the span of its arches, which is generally 15 ft. It consists of no less than seventy-four arches for flood relief through an embankment which carried the Great North Road over the flood plain of the Trent. Smeaton's arches, somewhat widened, carried the traffic of that road until it was diverted in 1971, two hundred years after they were built. They are not strictly as he designed them, for he drew segments of less than 90° and only 12 ft span, with every third or fifth pier built of double thickness to provide abutment to a group of arches if one of the next group should fail; the spans were increased by extending the arches to almost semicircular curves, with narrow piers all of the same thickness.

as the estates in which they stand, and this was doubtless his deliberate intention. The Warwick bridge is large and high, but very delicate; Dubh Loch is strong and rustic. Warwick is the only bridge built to his design which had a really difficult gradient, but that was probably consciously accepted, for the bridge's chief purpose must have been to decorate the park. It was built, as far as we know, to a design made in 1765,[32] one of his earliest after Blackfriars. Its grey sandstone arch with smooth V-joints, and every stone sized to form a regular pattern even on the soffit, the carved keystone projecting right across the soffit, and especially the balustrade beautifully carved from square pillars instead of the usual round, testify to the great care taken in its design. At Dubh Loch the stone is dark

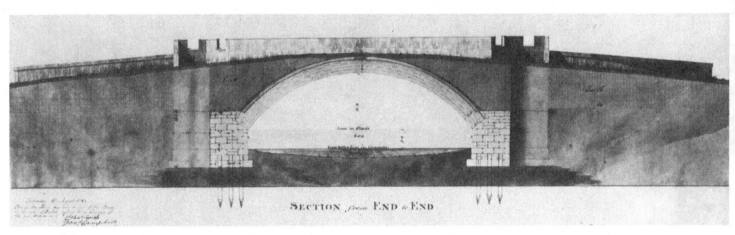

SECTION from END to END

80   Section of Dubh Loch Bridge (Mylne, 1785).

Although Robert Mylne was the first man to build elliptical arches in a large British bridge, there are no other elliptical arches in any of the bridges he is known to have designed. The arches he built after Blackfriars Bridge were segments of circles varying from 85° to 120° and therefore considerably flatter than any that Smeaton built, though they were of similar spans (see table 3). They were distinctly thicker, however, even at the crowns, and in every arch of which we know the section thickness there was an increase of thickness towards the haunch and/or springing. Of the principles by which he decided the thicknesses and the taper we have no surviving statement. In contrast with Smeaton he did not become more conservative with experience, and this may be because he was guided more by convention and less by science. That was true both initially and throughout his career.

On the façades of at least four bridges he showed an arch of constant thickness; only at Dubh Loch did he show a taper and it was not the same taper as in the section (fig. 80).[31] This is one instance of the freedom to design for appearance which is found in all his designs. The bridges at Dubh Loch (fig. 81) and Warwick Castle (fig. 82), though each of one large arch, are as different

grey, hewn at the joints but rough-faced, and castellated turrets at the junctions of the spandrels with the high wingwalls add to the sturdiness of its appearance. Its rustic air is carefully contrived. But there is nothing in either of these bridges that can be considered *added* ornament and the impression created is that functional form and ornament fuse together as he suggested they should in his pamphlet in 1760.

While Mylne was designing the Warwick bridge he was also at work on a much larger estate bridge. For the Duke of Portland at Welbeck Abbey in Nottinghamshire he designed a bridge of three arches, the two end ones 75 ft in span and the middle one 90 ft.[33] He took pains with this bridge and his important client, visiting the estate for the founding of each pier and sending a man named Peter Upsdell from London as resident supervisor. The local tradesmen, who were unaccustomed to such work, almost stopped at least once in fear of an accident, but Mylne made light of it. Late in 1767 all three arches were finished but before the centres were struck the middle arch underwent a serious movement and some of the stones cracked. Mylne assured the Duke that similar cracking had occurred at Blackfriars but ordered the centres to be left in place

81 Dubh Loch Bridge, Inveraray.

through the winter to allow the mortar to harden. On 13 February 1768, the middle arch collapsed. Mylne accepted responsibility, in a letter to the Duke, for allowing the use of too weak a stone, 'for on that alone', he said, 'depends the stability of great arches. I have been too adventurous . . . and I suffer much and heavily in my own mind for it.' He suggested that 5 ft extra should be added to each side of each pier and abutment and the arches rebuilt, all reduced by 10 ft in span but of the original height. How the Duke responded is not known. Workmen continued on the site until May, when the record in the Duke's estate papers stops. Mylne recorded in his diary neither the collapse nor any further dealings with Welbeck or the Duke.

In his two town bridges at Newcastle-upon-Tyne and Tonbridge, Mylne designed neat arch rings and cornices with the plainest of mouldings. The Tonbridge design[34] shows a solid stone parapet and the Newcastle one[35] no parapet at all (fig. 88), for a reason we shall return to. It appears that all the exterior stonework of both bridges was ashlar. In the Aray Bridge at about the same date he inserted some open balustrade at the express desire of the Duke of Argyll.[36] Ridley Hall bridge in Northumberland was designed in 1786[37] as a plain county bridge, but the county bridge at Romsey in Hampshire, built in 1782–4,[38] had carved medallions on the faces of the abutments and bold

projecting mouldings, because it was visible from Broadlands House, the seat of Lord Palmerston, and Palmerston paid for the embellishments. It was a single segmental arch with a roadway 18 ft wide, and had to be rebuilt in 1931 to improve the gradients and width.[39] The façades were rebuilt as before, however. The width of the roadway of the Newcastle Bridge was 24 ft, and inadequate from the start in such an important bridge, but that was not Mylne's fault.

He designed the temporary bridge at Newcastle, at least in outline, but he left the contractors to make their own design of trussed centres for the permanent bridge, using the starlings for support. At Romsey, Mylne designed both the temporary bridge and the centre even though, with two old piers in midstream for support, the centre was very simple.

We must also consider the methods used by Smeaton and Mylne for founding their bridges and their ways of dealing with the problem of scour. We start with Smeaton's engagement to save London Bridge from destruction by scour in February 1763.[40] He had still never had a bridge design adopted but his reputation as a river engineer was such that the Common Council of London sent for him at his home in Yorkshire, even though Mylne was working on Blackfriars Bridge less than a mile away and actually employed by the City.

The emergency was caused by the removal in the latter part of 1762 of the old pier under the middle of the new large arch (see chapter 6). The lighterman, named Parsons, who undertook the demolition was also employed at Blackfriars, so he discussed his task at London Bridge with Mylne and was warned that when he removed some piles of the starling and the pier foundation the inner rubble and timbers might quickly disintegrate. That was exactly what happened; as Mylne recalled many years later, 'the whole pier . . . dissolved away in the midst of that impetuous agent the *fall* under the bridge'. The current then attacked the gravel below, which had never been exposed before, and the new channel became deeper at every tide. When Smeaton arrived post haste from Yorkshire he found 'the current making hourly depredations upon the starlings, the south-west shoulder of the north pier undermined six feet, and the original piles, upon which the old works had been built, laid bare to the action of water, and several of them loosened'. This refers to the remaining piers and starlings which now supported the large arch. He took soundings of the bottom and immediately ordered quantities of the largest stones available to be dumped in the channel. Carts were hired the same day to bring the stones of several City gateways, recently demolished and sold but now bought back in haste by the Council. Catastrophe was thus averted and Smeaton then made proposals for forming a stable bottom. He drew a longitudinal section of the bottom to be formed under the new arch, a streamlined shape rising from deep water at the upstream end of the starlings to a sill 5 ft below low-water under the upstream face of the arch

and then descending at a slower gradient to the downstream end of the starlings (fig. 83). He wanted this shape made 'by throwing rough stones . . . the interstices of which in time will fill up with gravel and the fullage of the river, and become as compact and durable as a rock. The heaviest, roughest and largest stuff is best adapted for this purpose.' He also recommended that two arches be closed to the tide and the bottoms of others be raised, intending thereby to preserve the fall and save the City from having to compensate the waterworks company or pay for improving its machinery to supply as much water with a lower fall.

Not all that he recommended was done, if Mylne's account of the affair is to be believed, and when Smeaton made another survey in 1766 he found that the bottom of the large opening was no longer streamlined. But under the arch the depth was generally less than 8 ft from low-water and nowhere more than 11 ft, which he thought quite satisfactory. The increased velocity of the ebb and flow had, however, scoured great hollows in the river-bed up and downstream from the bridge, with a maximum depth of 38 ft downstream and dangerous depths of 26 and 25 ft respectively abreast of the ends of the starlings. He ordered only mild correctives this time, preferring to stabilise the slopes of the hollows than to attempt to fill them. He advised that 'the condition of the rubble bed should be frequently examined, and to have always in readiness a quantity of rubble'.

Many examples could be quoted of Smeaton's conviction that a bed of rock or rubble shaped by the fiercest flow of the natural stream, and with its

82 Bridge in Warwick Castle estate.

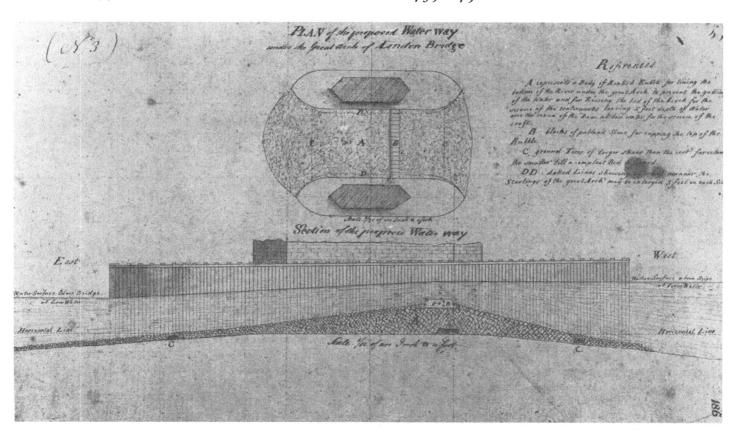

83   Design for the waterway through the large arch of London Bridge (Smeaton, 1763).

interstices filled with gravel and sand, was the best defence against scour. He held a poor opinion of the common alternative, the 'framed and sett' river-bed, and he never proposed it for a bridge in a large free-flowing river. He was averse to the use of timber in this, as in any place where it could decay, because it then allowed the stones to work loose.[41] For small bridges on small rivers, however, he always designed a continuous platform of stone or timber, anchored and surrounded by sheet piles and with bearing piles under the piers, and extending across the whole river; and then built his bridge on top[42] (fig. 84). The present appearance of the small bridges suggests that their foundations have proved more stable than those of the piers of his large bridges.

The first of his four large bridges was started in 1763 a few months after his rush to save London Bridge, and all four were built on gravel bottoms. The Coldstream Bridge foundations[43] illustrate his preference neatly. The rock lay at moderate depths below the bed of the Tweed and he hoped that cofferdams could be pumped dry while he excavated down to rock and laid the bottoms of the piers and abutments on it. It was a natural preference for one who had founded a stone lighthouse on rock but he achieved it only at the first abutment and first pier of this bridge. At the second pier the gravel proved so leaky that the day-and-night

pumping was becoming too costly, so he changed to a piled foundation (fig. 85). Bearing piles were driven and cut off 3 ft under the existing level of the bed, then capped by a grating of timbers with a maximum size of 12 in. × 7 in. Round the edge of the grating he drove another 'fence' of permanent sheet-piling, 5 in. thick and spiked to the side of the grating. Then he packed the spaces within the grating with rubble stones and gravel and on top of this laid the first hewn-stone course of the pier. A special expedient was to make all the stones round the edge of the first course headers, overhanging the sheet-pile 'fence' by 9 in. and cut with a nib which went 3 in. down the face of the piling and thus restrained it from being pushed out by the pressure of the gravel inside. There were ties of cramped stones through the pier at the level of low-water also. The piles being short, probably never more than 10 ft long, Smeaton advised that they be driven by small 'cat's-tail' rigs of the old-fashioned kind, with a wooden monkey of no more than 4 cu. ft (perhaps 200 lb) worked by a gang of ten or twelve men; and all the piles were driven to rock without shoes.

The permeability of the gravel took Smeaton by surprise and he recommended at one time use of an Archimedean screw for pumping, at another a sprinkling of 'corn mould earth' on the river-bed outside the dam. In one letter he commented that 'Sundays are a sad rubb with you to work under water', meaning presumably that the pumps had to stand idle on the Scottish sabbath. And the gravel proved as unstable as it was porous. To avoid disturbance the piles of the cofferdams were cut off instead of being drawn, and the spaces between them and the piers were filled with rough rubble stones which had to be watched and replenished for many years.

At the site of the fourth and last pier there was virtually no gravel overlying the rock and neither a cofferdam nor bearing piles could be driven, so Smeaton had to adopt the method which was always his last resort, though he knew it to be often the cheapest, namely that of a caisson (fig. 86). Compared with those Mylne was using at Blackfriars it was a light and simple thing, the bottom made of two layers of 3-in. fir and the sides of single 3-in. boards held together by framing members and cross braces. The height was only 9½ ft, to reach from the rock to 1 ft above the water. There were, of course, no tides to be allowed for.

The cofferdams at Coldstream were started before Smeaton took over the design, but they were clearly made with only one thickness of sheet-piling. He had

suggested such dams to the Westminster Bridge Commissioners in 1748 and the Blackfriars Bridge committee in 1759[44] and when he moved on to Perth he designed the cofferdams to be of sheet piles 5 in. thick notched into 9-in. square gauge piles at 9 ft spacing. The outline of the dams was oval in plan to give better resistance to the water pressure outside. All the piers were founded on piles, with details very similar to those of the second and third piers at Coldstream. But the gravel at Perth proved very compact and piling was difficult, so Smeaton was content with bearing piles only 6 ft long and he had to suggest that gravel and 'corn mould earth' – probably what is now called loam – should be mixed and banked up against the outside of the cofferdams if the sheet piles could not be driven more than 2 ft. This was to reduce percolation under the dam. Spring tides rose 8 ft at Perth but there was little rise at all at neaps. Rubble stones were used to fill the excavations and form small mounds round the bases of the piers.

The foundations at Perth proved very firm and scour has never threatened them seriously; but Coldstream Bridge's piers have been at risk since the day it was

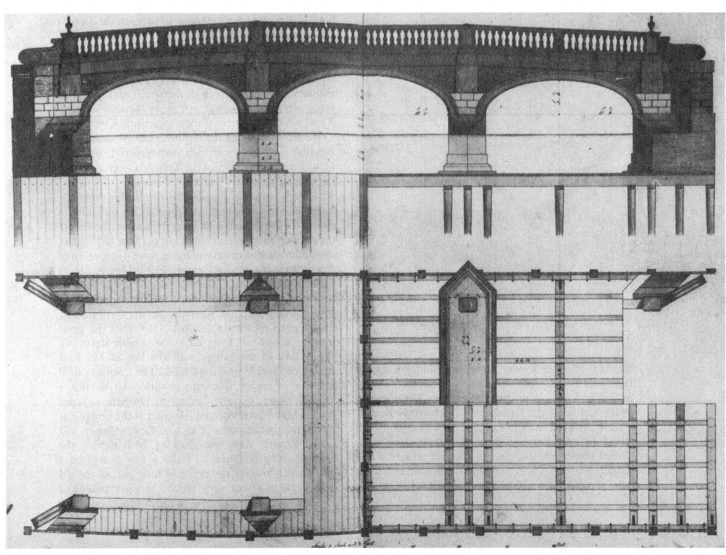

84   Design for ornamental bridge at Amesbury (Smeaton, 1776).

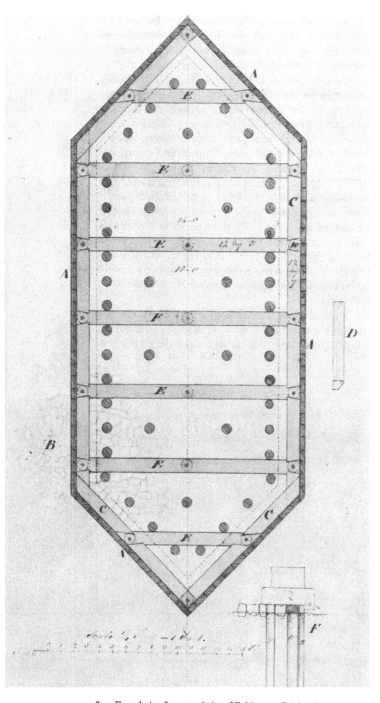

85 Foundation for second pier of Coldstream Bridge (Smeaton, 1764).

In the designs he made for Scottish bridges in 1772 Smeaton made a further economy by designing the sheet-pile 'fence' round the base of the pier to act as cofferdam as well (fig. 87). There were extensions of all the sides upwards, to be removed when the pier reached water level, and instead of the masonry overhanging the sheet-piling the piling stood against the outside face and had to be tied by long iron bolts into the joints of the masonry. The foundations of Banff Bridge, presumably built by this method, appear to have lasted well and still have no extra protection.

One of the first decisions recorded in the minutes of the Coldstream Bridge trustees was that the foundations of their bridge and the piers up to water level would be built not by contract but by day labour. Perth Bridge was built by tradesmen paid at agreed rates per unit of work[45] and Banff was built as a military bridge, probably by direct labour.[46] So none of the three suffered the risk that foundation work would be skimped by a contractor who had underestimated the task. In two written contracts for bridges by Robert Mylne all foundation work is specifically excluded, and such clauses were common but by no means universal (see the Edinburgh bridge story below). The exclusion in Mylne's two contracts means that the foundation details of those bridges are not exactly known, but in both instances he built, at least in part, on the piers of the former bridge. One was the small three-arch bridge at Tonbridge, where the design drawing shows two old piers, 12 ft thick, cut down to about 6 ft below water level, fully capped with new masonry about 1½ ft high, and new piers only 4 ft thick then built on top.[47] The other was the important bridge at Newcastle-upon-Tyne, which is worthy of more description.[48]

We have described the old bridge at Newcastle and mentioned its dual ownership in chapter 6. It may now be added that the valley of the Tyne has steep sides and the approaches to the bridge were narrow, crooked and very steep, coming downhill to the bridge from the towns of Newcastle and Gateshead. This difficulty of approaches, together with deep hollows in the bed of the river up and downstream, such as all the old bridges caused, made it necessary to move some distance up or downstream if a modern, convenient bridge was to be built. Smeaton discussed the choice of a new site in a report to the Common Council of Newcastle in May 1771 but their response was only to order small repairs. Events overtook them six months later when the 'great inundation' of the Tyne on 17 November destroyed almost half of the bridge with the loss of five lives among the people who lived on it. The Council called Smeaton to report, in company with John Wooler, a local engineer, but the Bishop of Durham engaged Robert Mylne to advise him. His part of the bridge was a total ruin, while only one arch of the town's part had fallen. Reports were published by both parties, the town showing readiness to build a new bridge on a better site, although the repair of their part of the old one would not cost very much. Mylne's published

finished. More and more rubble was dumped round them year by year while Smeaton resisted Robert Reid's proposition of a small dam downstream like the one which had saved the foundations of Aberfeldy Bridge when he worked there in his youth (see chapter 2). At length in 1785 Smeaton gave in and the dam, with several subsequent increases in size, has preserved the foundations ever since, though not without the added help of concrete 'starlings' round the piers (fig. 77).

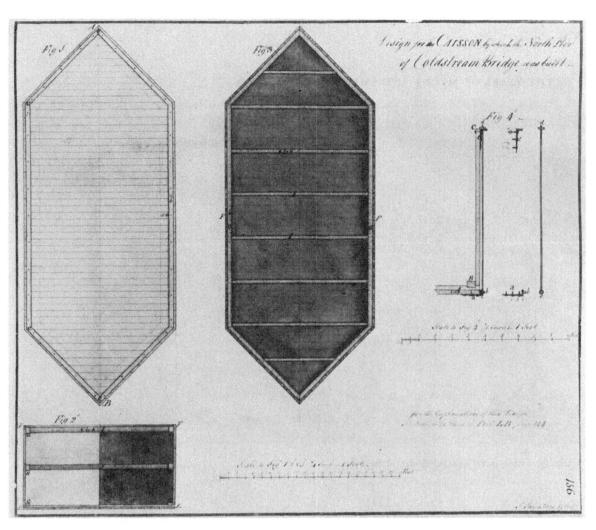

86   Caisson for fourth pier of Coldstream Bridge (Smeaton, 1765).

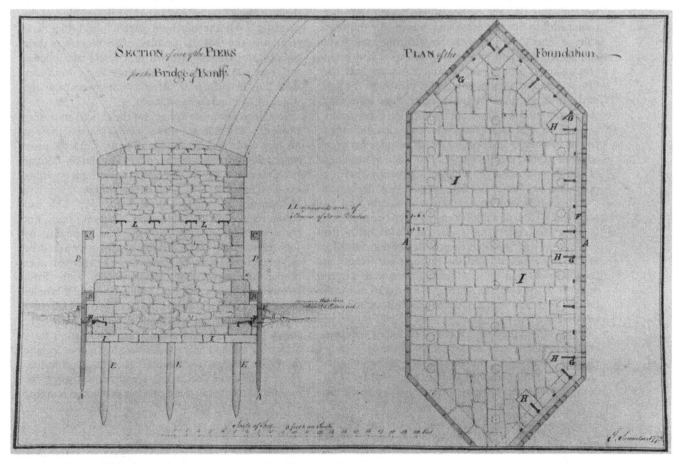

87   Foundation and pier of Banff Bridge (Smeaton, 1772).

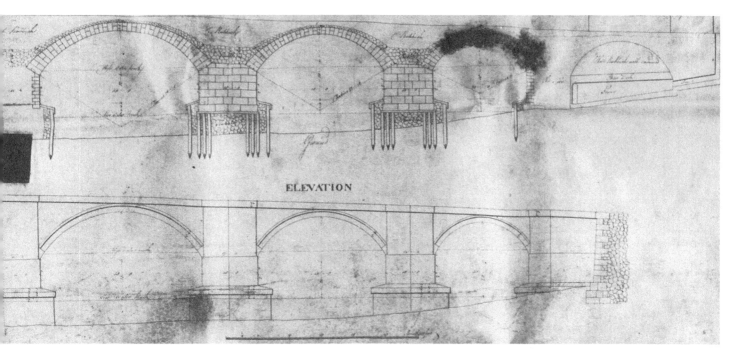

ELEVATION

88   Design for southern part of Newcastle Bridge (Mylne, 1774).

report was not addressed to his client but to the people of Gateshead and the lower part of Newcastle, as spokesman for his client. The crucial message, which he doubtless conveyed to the Council on 17 March 1772 when, says his diary, he 'delivered a message from the Bishop', was that the Bishop would accept no other responsibility but the one which was legally enforceable, to rebuild his part of the bridge of the same form and dimensions and on the same site as it had been. Discussion of a new site therefore ceased and the Council erected a temporary bridge parallel to the line of the Bishop's ruins. Mylne designed three arches for the Bishop and his drawing[49] shows thick new piers (fig. 88) of solid ashlar founded on wooden platforms laid at low-water level on the heads of piles between which is rubble, and the whole surrounded by starlings of rubble enclosed by more piles. Assuming that the piles were new, as the drawing suggests, it is likely that the ancient method was chosen to avoid the task of removing all the old rubble from the river-bed and then forming deep cofferdams to place new work below the bed. A drawing[50] made during the demolition, in 1885, of the pier which divided the Bishop's part from the town's shows that no new bearing piles were driven under it in 1775–80; the starling was reduced in width by driving a new facing of sheet piles, but its top still stood above low-water level.

The width of the bridge was 24 ft including parapets but no parapets were shown on the drawing, clearly because the Bishop had not yet agreed to abandon his ancient practice of leasing buildings on the bridge. The cutwaters were also carried up to the road to provide the same support for houses as before. In 1779 when the bridge was finished a plan was actually drawn showing the areas of twenty 'hanging shops proposed to be built over the river' all along each side of the Bishop's part of the bridge. Eventually this proposal was dropped but the Newcastle Council had to pay for the building of the parapets.[51] Of the northern part of the bridge it is enough to say that the Council, with Wooler as their engineer, decided too late that they would take it down and there was then no point in rebuilding it any wider than the Bishop's part. After twenty years' use the whole bridge had to be widened.

In at least one other design, for the entirely new Aray Bridge,[52] Mylne drew a pier of similar shape to those in his Newcastle and Tonbridge designs. Thus the most consistent difference between his foundation designs and Smeaton's was that Smeaton never built anything like starlings until his last large bridge. This was Hexham Bridge which he designed in 1777.[53] He had been a regular visitor to Hexham for twenty years, but the first bridge actually built there was designed by John Gott. After that had been totally demolished by the 'great inundation' of November 1771 several other people, including John Wooler, had made fitful attempts to design and build a bridge. When Wooler found quicksand 4 ft below the river-bed he expressed the view that only a 5-ft thick bed of masonry right across the river would make a safe foundation for a bridge – a method only used before by George Semple for Essex Bridge (see chapter 4). Though Smeaton chose a different site, nearly a mile downstream, his probing of the bed with an iron bar persuaded him that

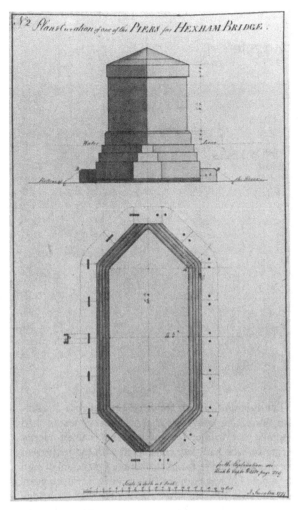

the strata were similar to what Wooler had found, a very weak material and possibly quicksand, covered by only a few feet of gravel. Whatever was under the gravel, he thought it 'so little compact or capable of bearing weight, that to drive piles in it would only weaken the stratum. The question therefore . . . was, whether there was a bed of gravel of sufficient thickness and compactness to bear the weight of a bridge, in case it was unwounded and unbroken?' He decided that the gravel would resist erosion and laid the foundations of four of the piers right on top of it without excavating at all. They were each to be surrounded (fig. 89) by a girdle of large hewn stones, each one ton in weight and cramped together, and a slope of concrete, called by the French name 'béton', from the girdle down to the river-bed. This could be regarded as a very small starling. So as not to disturb the gravel by piling, the bottoms of the piers were founded by a light caisson.

The river did not give him time to prove his design, for before the girdles had been placed a spate came down and scoured the gravel from under the upstream ends of all four piers. Smeaton's response was to increase the size of the 'starling', driving sheet-piling about 3 ft from the face of each pier, tying it round at its top with a waling and filling the space between it and the pier with rubble and sand, as well as packing up the voids under the piers by working in a diving bell and dumping a slope of rubble against the outside of the piling (fig. 90). Most of the piling was driven from 7 to 10 ft into the bed.

89   First design of foundations for Hexham Bridge (Smeaton, 1777).

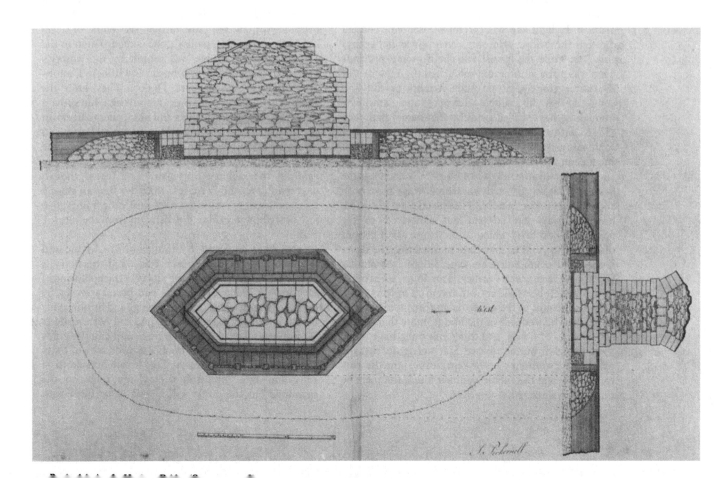

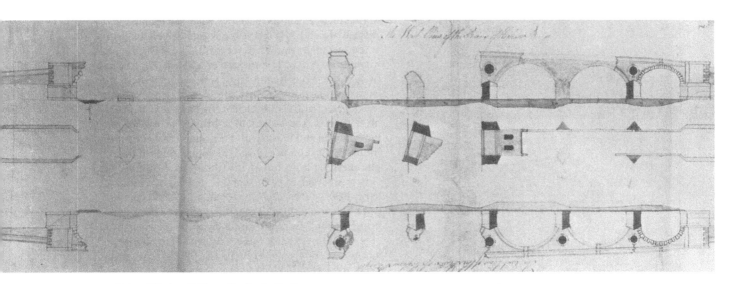

91   Ruins of Hexham Bridge, 1782, sketched by Smeaton.

In later floods there was evident scour of the spaces between the piers but no derangement of the piers or their surrounds till almost two years after the bridge had been finished. Then on 11 March 1782, a greater flood occurred after heavy snow followed by rain and a quick thaw, and it first undermined one pier and then brought down six arches all in the space of half an hour (fig. 91). Two piers founded on bearing piles with sheet-pile 'fences', as at Coldstream and Perth, were demolished as easily as those placed by caissons. The destruction, unlike that of Gott's bridge in 1771, took place in daylight and was carefully observed. The fall through the bridge before its collapse was between 4 and 5 ft, while the largest fall noted in any previous flood was 3 ft 9 in. Smeaton calculated the velocities of the water consequent on such heights of fall and deduced that his rubble defences and 'starlings' (though he never called them by that name) had been able to withstand a velocity of 930 ft per minute but could not withstand 1,100 ft per minute. His method of calculation was less exact than those devised by Labelye and others (see appendix 1) and the real velocity, if the height of fall was correctly measured, must have been about 1,350 ft per minute. The mistake he admitted, however, was that he had not conceived of the possibility of a flood which would cause a fall greater than 2 ft 6 in., and he may have recalled that the River Tweed had confounded his expectations in a similar way almost twenty years earlier. 'I am much surprised', he wrote to Robert Reid at Coldstream on 27 May 1765, 'at your account of the water rising three feet perpendicular on the west side of the bridge, more than on the east in time of floods, and think you must have made some mistake', but he advised, if it were true, that the passages between the piers be kept clear of impediments to the flow and the rubble mounds round the piers be carefully maintained.[54]

The litigation which followed the collapse at Hexham was long and complicated, for Smeaton had designed the bridge for a landowner named Henry Errington who had then contracted with the Justices of Northumberland to build the bridge for the small sum of £5,700. He was bound to maintain it for seven years but Smeaton insisted after the failure that no bridge could be built to stand on that site without much more expensive foundations than Errington had contracted for. The Justices called on Mylne who advised that the bridge could be rebuilt, but on the evidence of a too brief survey of the site. After six years of legal argument, Errington paid £4,000 compensation to the Justices and the bridge was rebuilt by the county's bridge surveyors William Johnson and Robert Thompson but at a cost of over £8,000. They built the superstructure exactly to Smeaton's design, but drove a forest of piles under the piers and more piles to tie down the timbers of the framing and setting which they laid across the whole river-bed, an excellent specimen of the craft of northern bridge-builders. A record drawing held by the County Surveyor,[55] of recent though uncertain date, shows no less than five lines of closely spaced piles right across the river under the frames, two just upstream and three downstream from the piers.

At an earlier date and a different bridge Mylne and Smeaton had played opposite roles, and this story is linked with an important innovation in the hollowing of spandrels.[56] Hollow spandrels had been proposed by Batty Langley for Westminster Bridge and by others for Bristol and Blackfriars Bridges, one had been made at Westminster over the 'sunken pier', and circular voids had solved the problem of the Pontypridd arch in 1756. In his designs for bridges at Coldstream and Perth in 1763 and 1764 Smeaton drew large rings on the spandrels but the circle within was to be filled with black rubble masonry – perhaps to look like a void – and the inside of the spandrels was to be filled with rubble masonry for 6 ft above the springings of the arches and with gravel from there to the road. Coldstream was built like that, and the gravel, it seems, of poor quality. At Perth all the arches were built, and the rubble filling placed, before any of the gravel was added. Only two of the arches at most were still incomplete at the end of 1769, and Smeaton had ordered in that year that the spandrels should be hollow. His decision cannot be certainly related to events in Edinburgh in August of that year but the dates are very suggestive.

Men of Edinburgh had fitfully planned for at least eighty years to extend the city to the north or south by bridging over the low-lying swamps which bordered the Castle Hill on both sides and building new spacious streets beyond them. There were serious plans of a North Bridge and New Town prepared in 1759 by William Mylne (fig. 92), Robert's younger brother, newly returned from his travels to France and Italy. An Act of Parliament in 1753 had authorised improvements of which the bridge to the north was one. Finally, a competition for designs was held in February 1765 and won by David Henderson, the mason who had worked for a time at Coldstream. There is good evidence that his design had rings on the spandrels and probably open cylindrical voids. It was favoured, over the other twenty-one designs submitted, by the gentleman architect Sir James Clerk who seems to have exerted a strong influence on the bridge committee. He thought the bridge should be different in style from 'a river bridge', because it did not cross a river. Nobody but Henderson offered to build to his design and he could not find adequate securities. And Sir James thought the style could be improved further; he wrote to Rome for designs in the style of ancient aqueducts and in haste made a sketch of one himself. The Town Council were willing to build this oddity if a contractor could be found, but it gradually became clear that the only acceptable contractor was William Mylne and he would build much cheaper to his own design. This was a bridge of three semicircular arches, each 72 ft span, and two small abutment arches, all on tallish piers and connecting long earth-filled approaches which were

92   William Mylne.

each hollowed by three rubble vaults at their high ends and carried the road to the crown of the hill to the south (at the line of the High Street) and on to the rising ground to the north (fig. 93). The architecture was unexceptionable, all the ornament being strictly classical, and all the construction conventional. Here lay a trap for William Mylne, for the height of his bridge was not conventional, and particularly the height of the earth-filled approaches. It was therefore too heavy a bridge for conventional proportions of spans to pier thickness and conventional thicknesses of retaining walls. Mylne's design complied with the common rules of proportion but in the overall dimensions it went beyond the common experience on which they were based. One of his two securities was his brother Robert, but if Robert saw the design he failed to see the danger.

93   Section of restored North Bridge, Edinburgh (Smeaton, 1770).

Sir James Clerk suggested referring the designs to competent architects in London, including Robert Mylne and Smeaton, but this was not done and William contracted to build the bridge, including its foundations, for a fixed lump sum.

He must have been aware of difficulties, for he thickened some walls and broadened the foundations, all without payment because of the form of the contract; and he virtually finished the bridge within the four-year contract period. Within that time both Smeaton and Robert Mylne had submitted schemes for the repair of a bridge at Dumbarton where a pier had suffered a very large tilt and settlement from the overloading of the very soft ground it stood on.[57] Robert's proposal has not survived but Smeaton's has, and the line sketch he made in May 1768 shows a plan of the spandrel over the reconstructed pier, the spandrel hollowed by longitudinal voids between walls which must run from back to back of the adjacent arches. A repair of this general form had been proposed by the architectural writer William Halfpenny for the sinking pier of Westminster Bridge in 1748,[58] but no practical use of it is known before 1768 and it is not clear that Smeaton's proposal was adhered to at Dumbarton, although he thought it was.[59] In the same year, however, William Mylne made a design for a seven-arch bridge over the Clyde at Glasgow at a site with a similar weak bottom and he hollowed each spandrel by a large cylindrical void to reduce the weight and perhaps also the pressure of fill on the spandrel walls.[60]

This may mean that he had learned the errors of his Edinburgh design from the experience of trying to build it, but he was learning too late; for on 3 August 1769, the great mass of fill in the abutment and approach at the south end of the bridge forced out the side wall, cracked the short arch and the abutment wall, and collapsed the three vaults in the approach; four people walking on the top were engulfed and killed by the subsiding earth.

Smeaton, being in Scotland, was asked to report and insisted on a lightening of the bridge, partly by a general reduction of the level of the roadway but also by higher arches and larger voids in the approaches and new arched voids in the spandrels between the large arches (see fig. 93). Both the latter proposals would reduce the pressure of filling material against the exterior walls of the bridge as well as relieving the foundations of weight. For the spandrel voids Mylne or his brother Robert, who came from London to advise and help him, made cylindrical arches fully 20 ft in diameter but which did not show on the outside walls. These arches, being extra to the requirements of the contract, were paid for by the Council, but William Mylne was broken, more in spirit than in pocket, by a protracted wrangle between him and his securities, the Council, and the 'feuars', or purchasers of property, in the new town north of the bridge. The feuars had to be satisfied about the safety of the bridge and one after another the walls of the approaches and the vaults inside them were all taken

down and rebuilt just to prove that they were sound. William continued the work in a responsible way, but constantly in debt, until early in 1773 when another scare arose about yet another wall, and he disappeared without leaving an address. He had gone with his dog to London, from whence he took ship to Charlestown in Carolina. For five long months in the spring of 1774 he lived with the dog in a backwoods shack in peace. He did not stay there, but he returned neither to Edinburgh nor to bridge-building. He had never in fact seemed the right sort of man. In the fifteen or twenty of his personal letters which survive there are scattered sentences which reveal a rather melancholy disposition, a desire for a simple life, and a lack of the ambition and the determination to succeed which Robert had in plenty. From Vicenza on the way home from his travels in 1757 he wrote with enthusiasm of the buildings of Palladio he had been viewing, but then went on, 'I have little hopes and very indifferent about making a fortune as in my present state of mind I can be contented with little and frequently reflect over the mending a hole in my old shirts or coat how bainful ambition is to the happiness of man.' And a little later he wrote, 'I have passed many a miserable day within these five years.' From New York in 1775 he wrote to his sister that 'Could a few hundreds be saved from the wreck of the bridge I would be planter, it is the way of life I prefer, one is there independent.' There were many hundreds due to him, but long before they were paid he had returned to Dublin to finish his life as engineer to the city's waterworks.

Robert Mylne's part in the Edinburgh bridge story was support for William's negotiations with the Council in 1769-70, advice at various times, and the loan of sums totalling at least £1,650, £500 of which he wrote off as a contribution to William's loss. He does not seem to have been convinced of a general necessity to make bridge spandrels hollow, though he may have made the first design for the Glasgow bridge and he made a cylindrical void over the only pier of the Aray Bridge, designed in 1773.[61] In his part of the Newcastle bridge he allowed the spandrels to be filled with rubble and 'rubbish' (see fig. 88), the latter term probably allowing the use of stone and mortar from demolished buildings and other alternatives to clean gravel.

Smeaton's reaction to the failure at Edinburgh brings us back to the bridge at Perth. His drawing of hollow spandrels for Perth Bridge (fig. 94) is only dated '1769' and might have preceded the Edinburgh collapse. Its importance is that it is a full design for the method of hollowing he had suggested at Dumbarton a year previously, the longitudinal voids shown as covered by pointed vaults of good rubble masonry and the whole structure tied from side to side over the tops of these vaults by iron chain-bars. Although he was critical of the standard of building in these spandrels when he inspected them about a year later,[62] they appear to have stood from then till now without serious derangement. By contrast, the smaller spandrels of Coldstream

94   Design for hollow spandrels of Perth Bridge (Smeaton, 1769).

Bridge, which were gravel-filled, had to be taken down and entirely rebuilt with longitudinal voids in 1828.[63] By then it was the accepted way of building spandrels between large arches. It accomplished at one stroke a reduction of load on the foundations, the elimination of outward pressure on the exterior walls of the bridge and a considerable mutual bracing of the adjacent arches against which the ends of the longitudinal walls were

built. Whether Smeaton derived it consciously from William Halfpenny's proposal for the repair of Westminster Bridge, invented it *de novo*, or had knowledge of medieval bridges in Persia or elsewhere, where similar constructions were certainly standing, it is impossible to say. His use of it in later bridges is interesting; for the bridge at Banff he designed smaller arches than at Coldstream or Perth, the longest span being 50 ft, and allowed these smaller spandrels to be gravel-filled, but with spandrels of similar size at Hexham he used longitudinal voids in at least some spandrels, presumably to lighten the load on the very poor ground.

Mylne and Smeaton were quite different in the way they organised their practices. Even before he had won his case against the City of London for fees for Blackfriars Bridge calculated at 5 per cent of the cost of construction, Mylne was working for other clients on those terms, and with travelling expenses to distant sites as an extra.[64] He was appointed 'surveyor' by each client and personally certified all payments to contractors and others, but in some instances at least there was also a resident supervisor, whether paid by Mylne or the client is not clear. Smeaton, in contrast, never seems to have charged a percentage fee. He submitted modest accounts for his initial designs and for his time and expenses on visits to sites and then left it to his clients to pay him a gratuity when their bridge was finished.[65] Several of his clients made no offer of a gratuity and though he resented this he was diffident about reminding them even of unpaid accounts for visits and expenses. As regards supervision, his visits were one or two per year while a bridge was under construction, as were Mylne's. For Coldstream Bridge the trustees had already employed Robert Reid as resident overseer before Smeaton took over the design. To Perth he sent

John Gwyn,[66] a tradesman whose work he had admired on contracts for the Aire and Calder Navigation, and Gwyn was paid a salary by the Perth Bridge Commissioners to oversee the contractors. All money transactions at Coldstream and Perth appear to have been controlled by the overseers and Smeaton did no routine checking of accounts. The overseers were responsible for carrying out his instructions and reporting both to him and to the bridge committees. At Hexham an engineer brought from London, Jonathan Pickernell by name, was employed at the site to oversee the work, though Errington's land agent seems to have handled all the money.

Pickernell went on to be engineer of Whitby Harbour, and Gwyn to be resident on other works of Smeaton's. Reid, who was not a young man even when Coldstream Bridge was started, remained there and became by 1770 a bailie, or magistrate, of the town. In 1767 he had contracted with the bridge trustees to build a tollhouse beside the north end of the bridge and, as the ground was almost 20 ft below the level of the road, he built a two-storey house for himself against the wingwall and placed the tollhouse as a third storey on top. The trustees were annoyed until Smeaton observed that the walls of Reid's house gave useful support to the wingwall of the bridge.[67] In 1790 Reid was still at Coldstream supervising repairs; he had been working at bridges for over fifty-five years, on and off.

9

BRIDGES BY ARCHITECTS

There is no reason why the labour of necessity or usefulness should not be embellished by taste.

Edward Cresy, *Encyclopedia of civil engineering* (1847)

This chapter is mostly about bridges designed by architects. Though Smeaton was often called an 'architect' of bridges by others, he never called himself anything but an 'engineer'. Several other engineers, such as John Wooler, also designed bridges, but architects claimed the great majority of commissions for important bridges until about 1790. The meaning of the term was not what it is today, however, and some of the men considered here acted more often as mason contractors than as designers or supervising surveyors.

Within three years of the Blackfriars Bridge competition three notable bridge designs were made in the south-east of Ireland. They resulted from the damage caused by a torrential flood in the River Nore on 2 October 1763 which destroyed four bridges over the river as well as several others on its tributaries.[1] The bridge nearest to the mouth of the Nore was that of Inistiogue and it was badly damaged but not destroyed. The design of its repair, as well as those of two new bridges to replace those ruined in the town of Kilkenny, was entrusted to George Smith, who had been director since 1761 of the work of making the Nore navigable between Inistiogue and Kilkenny.[2] A man of the same name had been employed under George Semple throughout the building of Essex Bridge in Dublin[3] (see chapter 4) and since the Semple family were active builders in the south-east it is very likely that this was the same George Smith.

The bridge at Inistiogue adjoined the estate of the Tighe family and they probably added to the £900 of government money which is recorded as having been spent on the construction. Smith's 'architecture' was applied to the downstream side of the bridge only and is directly derived from Mylne's Blackfriars design. It is actually more true to the triumphal bridge model, for its nine arches are all equal and semicircular, and the line of its parapet is therefore horizontal. The spandrels are of good dark-coloured rubble but decorated by pairs of Ionic pilasters in a pale and sharp-edged granite (fig. 95). The moulded arch rings are also of granite, and this

elevation appears to have needed little maintenance since 1762.

In Kilkenny Smith used a different, but just as positive, style, namely that of the bridge of Rimini (fig. 2). The larger of his two bridges, called Green's Bridge, has a horizontal roadway over most of its length, reached by symmetrical approach ramps; it has five arches, of which the middle three are equal, and a pedimented aedicule on every spandrel. The arches, however, are elliptical, and the arch rings sharply moulded but banded with square blocks (fig. 96). It is almost as true a copy of Rimini as was ever built in Britain or Ireland. The second bridge in Kilkenny, St John's Bridge, was of three segmental arches with a similar parapet profile and aedicules on the spandrels.[4] It was replaced by a concrete bridge in 1910.

These three bridges of Smith's may have been built after another, but quite unique, design in the same part of Ireland, namely the bridge at Graiguenamanagh over the Barrow, a sister river of the Nore (fig. 97). This replaced a timber bridge near the Castle of Tinnahinch and may have been financed by the owner of the castle, since there is no record of a parliamentary grant as was necessary for almost every large bridge in Ireland at the time. The date of construction was certainly before 1777[5] and the design and/or construction are credited in one pamphlet of 1895[6] to George Semple. The authority for this is not given and since Semple makes no mention of the bridge in his book published in 1776[7] it seems unlikely that he was either the designer or builder. The design and construction appear related, however, like Semple's Essex Bridge, to the designs and construction of Westminster Bridge. The elevations take the Rimini spandrel aedicule and change it to progressively smaller devices as the spandrels reduce in size towards the ends of the bridge, as John Price had done in his design for Westminster (fig. 3),[8] but the arches are all semicircles, while Price's were segments, and the arch rings are rusticated in a way suited to the use of the local slaty stone. The internal construction of the spandrels was revealed during repairs a few years

96 Green's Bridge, Kilkenny.

ago[9] and was similar to Labelye's work at Westminster, longitudinal and cross walls of rubble dividing the space into compartments filled with gravel, though the walls were apparently of mortared stone, unlike Labelye's dry rubble. Although it is hard to credit this design to George Semple, then, it is equally hard to dissociate it entirely from his knowledge of Westminster Bridge; and this suggests that his former assistant George Smith, or his brother John or another member of the Semple family, may have been responsible for it.

Together these four bridges established a local tradition which can be traced in the spandrel decoration of a number of other bridges on the two rivers and some others nearby, notably the bridges at Ennisnag, Athy, Brownsbarn and Maganey, the last being decorated with plain rings like a simplified Smeaton design, and dating from 1783 or later.[10]

Another reference to Mylne's design for Blackfriars could be seen in Dublin a year or two after Smith began the repair of Inistiogue Bridge. Queen's Bridge across the Liffey was built in the years 1764–8 under the direction of Colonel Charles Vallancey who, like Labelye and Desaguliers, was the son of a French protestant. He was appointed 'engineer in ordinary' in

97 Graiguenamanagh Bridge.

Ireland in 1762.[11] The arches of his Queen's Bridge are shown on an original drawing[12] (fig. 98) with all the voussoirs joggled, not by inserted blocks as in Mylne's first design for Blackfriars (fig. 62) but by interlocks cut on the bedding surfaces of the stones themselves. The foundations of the piers were designed to be laid in caissons and, according to the drawings, without any piles underneath. Although Semple had so firmly rejected caissons when building Essex Bridge half a mile downstream only ten years before, the foundations of Queen's Bridge have survived as long as Semple's 'thorough foundation' (see chapter 4); but perhaps not quite so easily, for the heads of some piles are to be seen now near the piers. There is no recent confirmation that the arches were built with joggled stones but the exterior corresponds exactly to the original drawings.

removed). The one-arch bridge is a conscious imitation of a medieval bridge, less than 12 ft wide between the parapets and with high and steep approaches. The surprising thing is that this arch is ribbed in the medieval manner (fig. 99); and Cavendish Bridge,[14] which was built at about the same time across the Trent near Shardlow, to Paine's design, also had five ribs under each of its three segmental arches. The foundations of Cavendish Bridge, which may have been the first of the three, were laid in caissons on a gravel river-bed, the platforms being shown in Paine's engraving only 2 ft below the bed, but protected by sheet-piling driven round the edges of the platforms. For the other two bridges he laid the foundations in cofferdams formed 'by sinking tiers of large squared stones, the interstices of which were rammed with a body of clay'.

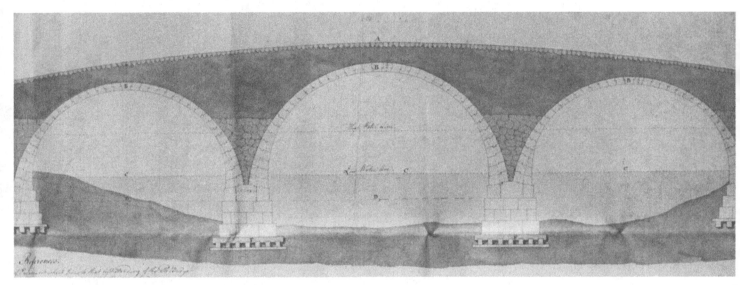

98   Section of Queen's Bridge, Dublin (c. 1764).

In England some notable bridge designs were made by James Paine, an architect with a very busy practice. The large number of Palladian houses he designed occasions some surprise that he never built a strictly Palladian bridge. He liked urns and statues, which he placed on the parapets and cutwaters and in niches on the spandrels or abutments, and his bridges never lacked ornament, but he seems only once to have included pedimented aedicules, and those not standard Palladian ones. They occur on a 'Roman bridge' built in Weston Park, Staffordshire, about 1765–70.[13]

In the years around 1760 Paine was working mostly in Yorkshire and Derbyshire and he built two bridges over the River Derwent in the park at Chatsworth, the larger one of three arches and the smaller of one arch. The former is of smooth and discreet masonry, to set off sculptured figures standing on the cutwaters and in niches at the abutments (some of the figures now

Paine's published engraving of the three-arch bridge shows the piers laid on rock at depths of 10 ft and 13 ft by this unusual method. Removal of such weighty cofferdams after use must have been difficult work. Cavendish Bridge was destroyed by a flood in 1947 but the two Derwent bridges still stand.

Paine moved to London when his practice grew and he became a landowner and Justice of the Peace in both the counties of Middlesex and Surrey; he was High Sheriff of Surrey in 1783.[15] These circumstances must be at least a partial reason why he was engaged to design four large stone bridges across the Thames between the two counties between 1774 and 1784. The first was at Richmond (fig. 100) and was projected by the owner of a ferry, as the wooden bridges at Hampton Court, Walton, Kew and Battersea had been. The others were at Chertsey, to replace an old piled timber bridge, and at Walton and Kew to replace the timber arched bridges

99   Beeley or Edensor Bridge at Chatsworth, Derbyshire.

100   Richmond Bridge. The new side built in the 1930s.

built in 1750 and 1759 respectively (see chapter 3).
Paine is reputed to have been a miser but he was
generous at least in the matter of Chertsey Bridge,
charging the county (of Middlesex) only for his
travelling and other out-of-pocket expenses and
nothing for his design or his time in supervision.[16]

In all these designs[17] he attended to the function of
the bridges, though he seems to have made the
gradients a little too steep and this was probably the
result of an architect's concern that a bridge should rise
high over the river and look 'grand'. He laid the pier
foundations in caissons and because the river at these
sites was shallower than at Westminster and Blackfriars
his caissons were lighter and less expensive than those
used by Labelye and Robert Mylne. Sheet-piling was
used round the perimeters of the piers, probably in all
the bridges, and bearing piles were driven in some
places under the bottoms of the caissons. It appears
from the specification of the caissons for Chertsey
Bridge that each caisson was sunk and fixed to the piles
before any masonry was laid, the pier being then built
without further floating of the caisson. The timber of
the caisson was yellow fir, except corner brackets and
pins, etc., which were of oak; and the same was true of
Kew Bridge. In every spandrel of all the bridges there
was a cylindrical void surrounded by an arch of brick,
and brick was also used as hearting for some piers and
abutments. The facing stone was from Purbeck or
Swanage after the first bridge, Richmond, where the
more expensive Portland stone was used for the
exterior; and the specification for Chertsey included
chain-bars to tie in the moulded architraves of long
curved stones which Paine applied as the visible rings
on the faces of the arches (fig. 101). In writing the
specifications he may have been assisted by Kenton
Couse, the Bridge Surveyor of the County of Surrey,
who certainly helped in the supervision of the work
both at Richmond and Chertsey, but there is no doubt
that all the designs were by Paine. He clearly made each
bridge as fine a composition as expense would allow,
but some refinements were lost between design and
execution. For Richmond Bridge, the first of the four,
he designed five arches, the middle one of 60 ft span,
with the road and parapet rising to a point at midstream. There were pavilions over the abutments with
large cavities under them. Over the cutwaters, which
were pointed, was a complex step-back to a curved-
faced pilaster which rose to form a shallow recess from
the footway. Large rectangular recesses were made at
the abutments when the pavilions were omitted,
presumably to reduce the cost. Each arch springing was
emphasised by three stepped rectangular blocks,
expressing rugged strength at this point (fig. 101).

Richmond Bridge stands (fig. 100) with elevations
and roadway much as Paine designed them, but the
gradients are very inconvenient nowadays. It was
widened by moving and rebuilding the upstream façade
in the 1930s. Chertsey Bridge has not been widened but
needed extensive repairs in 1821, 1842, and 1894. It is

101   Richmond Bridge. Detail on old side.

similar to Richmond Bridge in structural form, but
smaller, the largest arch of five being only 42 ft in span.
The bridge at Kew was of seven main arches, the largest
66 ft in span. It was too high, like Richmond, but was
plainer in elevation, with a solid parapet and only a flat
pilaster up the face of each spandrel above the cutwater.
Both in it and in the Walton bridge the parapet line in
elevation was a continuous curve. The façades Paine
designed for Chertsey and Walton Bridges had a very
unusual spandrel treatment. A recess, probably semicircular in plan, was supported over each pier by a
trumpet-shaped projection, of smooth pale stone,
starting from the point of a pyramid of masonry which
capped the cutwater. Such trumpets are still to be seen
on the little brick bridge at Godalming (fig. 102) also in
Surrey and built in 1782, but they were removed from
Chertsey Bridge in one of the nineteenth-century
repairs and Paine's Walton Bridge collapsed during a
flood in 1859, the pier which was undermined by scour
being the only one he built. The other two piers of that
bridge were those built by William Etheridge for the
earlier timber bridge which had failed through decay of
the superstructure (see chapter 3). Kew Bridge was
taken down in 1898 to make way for a wider structure
and so the only one of Paine's bridges on the Thames
remaining unaltered in appearance is Richmond, the
one he built first.

102   Godalming Bridge.

most important book, *London and Westminster improved*. It was a plea for concerted planning of improvements with a truly prophetic list of proposed alterations in the street plan of the cities. Only five years later he was actually employed in the building of four large bridges, and there were appended to two of these bridges extensive town improvements of the kind advocated in the book.

Gwynn's first appointment to design a bridge followed a refusal by Smeaton to accept the job because he was too busy, and the abandonment of a scheme designed by Mylne for widening and repair of the existing bridge.[20] It was the English Bridge over the Severn at Shrewsbury, Gwynn's home town, and he submitted two designs in September 1767, taking with him from London Richard Buddle, a master mason, who was to offer a price for the construction. A local architect named Thomas Farnolls Pritchard had been appointed surveyor to direct the widening according to Mylne's design, and William Hayward, another local man and future bridge designer, was acting as his assistant. Pritchard continued as surveyor after one of Gwynn's designs, of seven semicircular arches, had been adopted, but in November 1768 he was dismissed and Gwynn himself became surveyor. Many buildings on and near the bridge had already been demolished, a temporary bridge erected and timber bought for the

Only three of the final contenders in the Blackfriars Bridge competition became important bridge designers in the years after 1760. Smeaton was the engineer of the three, Mylne an architect; the third was John Gwynn (fig. 103) who could not be truthfully placed in either category before the year 1766.[18] Not a single building can be named as built to his design before 1766 although he had been writing on art and architecture since 1734; but his writings also include such titles as *An enquiry after virtue* (a poem) and *The qualifications and duties of a surveyor*. He was always concerned for the establishment of an academy of arts and worked much in partnership with Samuel Wale, a very skilled draughtsman. His first biographer, who knew him in his late years in Shrewsbury, described him as 'lively, quick and sarcastic; of quaint appearance, and odd manners . . . He was, I believe, poor, but high-spirited, and of unimpeachable integrity.'[19] In 1766 he published his

103   John Gwynn, painted by Zoffany.

foundations, all under Hayward's direction. A delay in contracting for the bridge was caused by insufficiency of the funds raised by subscription and Buddle eventually contracted in May 1769 to build the east abutment and first three arches only. He was exceedingly dilatory and often absent and he had to be discharged in 1771 and the third arch finished by direct labour. Already in April of that year Gwynn had estimated the cost of the remaining work to complete the whole bridge at £5,925 and then offered to undertake it himself for that sum. After a complete stoppage of work for months, a contract was made with him in May 1772 and the bridge was opened to traffic near the end of 1774. But he and the trustees were very slow in finishing their tasks. Small items of work were still being done at the end of 1777, both Hayward and Gwynn having now left the town; and Gwynn's account was not finally settled until at least 1784.

His second bridge design, also crossing the Severn and almost equally large, was offered to the Justices of Shropshire in July 1768 in answer to an advertisement for plans and estimates. The bridge was to be built at Atcham and Gwynn was again accompanied by Buddle who offered to build it for £5,000.[21] A contract was made immediately but Buddle was even slower about this bridge and only £1,240 was spent in four years, so Gwynn again became contractor in October 1772,

agreeing to complete the bridge by 25 March 1776. In the meantime he had been engaged by the trustees for the new bridge at Worcester in 1769, and to the task of designing the bridge there were added, over a period of about fourteen years, the laying out of extensive new roads on the west bank of the river, improvement of the streets in the town near the bridge and raising and repair of the quays,[22] amounting to a total expense stated in 1782 to be almost £30,000.[23] In May 1771 he was further appointed surveyor to the commissioners of the Oxford Paving Act, which provided for important changes in the street plan and approaches to the town as well as the rebuilding of Magdalen Bridge.[24]

What is noticeable about Gwynn's four bridges is their soundness of construction and the grandness of their architecture; but the scale of the 'town planning' he undertook at Worcester and Oxford is equally impressive. The success of his foundation methods is rather surprising, for the piers of English Bridge were founded without any piles on rectangular solid platforms made of two thicknesses of oak balks pinned together, with another 8-inch balk as a kerb on the edge. The first course of stones was a rectangle fitting tight within the kerb, the point of the cutwater being formed only in the second and higher courses[25] (see fig. 104). Some of the platforms were only 'just below the gravel bed of the river' when they were taken up in 1925–6,

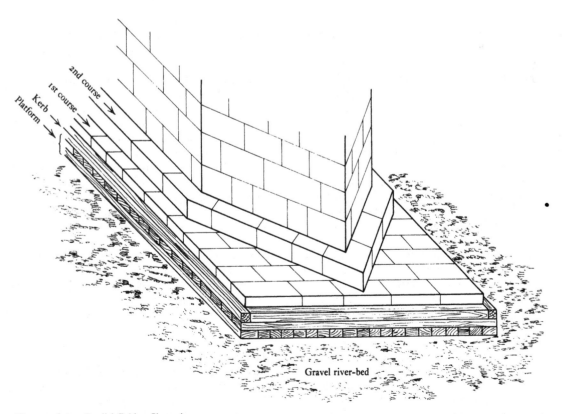

Kerb

Platform

1st course

2nd course

Gravel river-bed

104   Masonry of piers, English Bridge, Shrewsbury.

105   Atcham Bridge.

but they were in perfect condition and the superstructure showed no sign of foundation movement. Similar foundations were specified for the bridge at Atcham, with the addition of piles under the platforms if they were found necessary, and the engraving of the finished Worcester Bridge (fig. 106) shows shallow foundations without piles. The survival of these foundations must indicate that the bed of the Severn was more stable at their sites than at a number of other places where bridges were repeatedly damaged and undermined by floods (see chapters 11 and 12).

In his first two bridges, at Shrewsbury and Atcham (fig. 105), Gwynn made the parapets rise to an apex over the crown of the middle arch, and there he placed a large inscribed plaque surmounted by a pediment. The triple keystones in all the arches had sculptured fronts and they also protruded from the soffits across the whole width of the arches. The cutwaters were capped by moulded pyramids, the two largest ones at Shrewsbury carved into effigies of dolphins. The arches were all semicircular, with deep-cut V-joints both on the façades and the soffits. In the Shrewsbury bridge Gwynn even adjusted the dimensions of the ornament to improve the perspective of the bridge as viewed obliquely from the river banks. The depth of the V-joints was made to vary from $2\frac{1}{2}$ in. at the ends to $4\frac{1}{2}$ in. at the middle and there was an equal increase in the amount by which the spandrels were set back from the

face of the arches, while the projection of the cornice increased from 12 to 24 in. and the diameters of the stone balls which stood on top of the parapets increased from 15 in. at the ends to 24 in. at the middle. This must be the most fastidious construction ever applied to a masonry bridge in Britain.

The Atcham bridge, 18 ft wide, is now only used for foot traffic and its gradient is no longer a cause for concern. English Bridge was found too steep and, in the view of at least one of the committee, too narrow, immediately after its completion. Gwynn proposed a way of easing the gradients which involved lifting and re-laying the pavements, and Hayward carried this out in 1775. The road at the highest point was lowered by 15 in. (At just the same time the model of the design, which had stood as a pattern of the ornament for the masons throughout the time of construction, was sold.) Further lowering and widening were discussed repeatedly from 1872 to 1925, when the bridge was at last entirely reconstructed, but with very little change in the form or the details of the façades, and all the arch soffits of stone – a rare type of construction in the 1920s. The widening was from $23\frac{1}{2}$ ft between the parapets to 48 ft and the gradients were reduced from 1 in $17\frac{1}{2}$ to 1 in 35.

An early design for the Worcester Bridge[26] showed an apex over the middle arch but the final scheme (fig. 106) included the raising of the level of the streets at both ends and so the bridge was built with the parapet

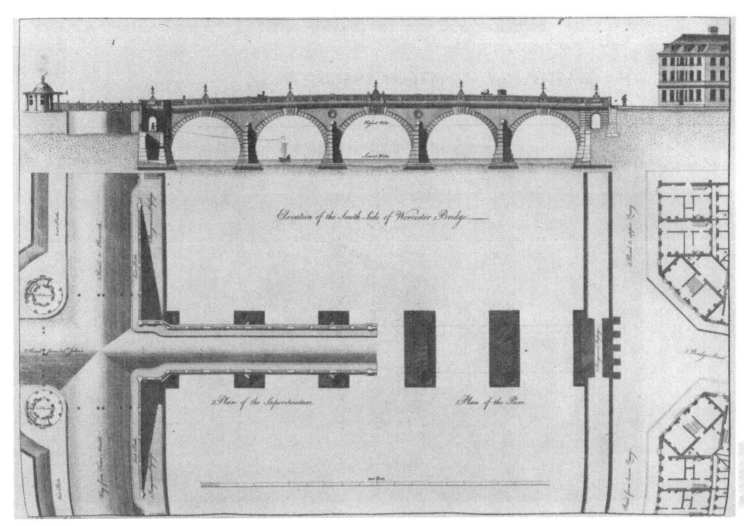

106   Worcester Bridge (from Nash, 1782).

almost horizontal. It was almost as ornate as English Bridge and boasted two circular tollhouses at the western end with colonnades round them and classical urns on the tops of their domed roofs – an extra monument to Gwynn's taste which was said to have cost the trustees £800.[27] In the Oxford bridge (fig. 107), the whole parapet and road had to be horizontal because the bridge consisted of two ranges of three short arches at the ends of a long causeway built to raise the road above low-lying ground between the two arms of the river. The arches, like all of Gwynn's, are semicircular, there are deep V-joints and sculptured keystones, and the spandrels and the walls of the causeway bear a series of large block-banded pilasters. Since the entablature is horizontal the use of pilasters cannot have offended architectural purists, of whom Gwynn was certainly one. The façades were preserved when Magdalen Bridge was widened from 28 ft[28] to 48 ft in 1882. The façades of Worcester Bridge were rebuilt with their decoration simplified when it was widened on both sides in 1931–2.[29]

Of Gwynn's four bridges only one, namely Atcham Bridge, was a county bridge. The other three were built by boroughs who obtained rights of toll by special Acts of Parliament and borrowed the capital they required on the credit of the expected tolls. By this means and because their bridges carried heavy traffic, towns were generally able to spend more on their new bridges than

108   Maidenhead Bridge.

counties. Two notable bridges were built by the Oxford mason John Townesend for boroughs on the Thames. The first was Maidenhead Bridge,[30] designed by Sir Robert Taylor and apparently his only design of a public bridge after his improvements to London Bridge with George Dance (see chapter 6). It was built between 1772 and 1777, of small semicircular arches with bold classical detailing (fig. 108). The second was Henley Bridge,[31] designed by William Hayward in 1781 and built after his death in 1782. The arches there are elliptical and the cutwaters rounded, their hemispherical caps pleated like sea-shells. Both were built of generous width and with gentle gradients and so remain useful today as well as elegant. In their structural proportions they are quite unadventurous, and an equally unadventurous but vastly cheaper bridge was built not far away at Sonning. It was built of brick in 1773 and the capital, which was probably only £1,150, was raised privately by several local landowners.[32]

For a few of their largest reconstructions the Justices of counties also obtained Acts and special rights of toll, but their constant maintenance and frequent improvement of the county bridges, especially by widening them, had to be financed out of local taxes and rates. In the first half of the eighteenth century they commonly accepted both designs and tenders for construction from masons or self-styled 'architects' but they almost invariably appointed a surveyor or overseer to check

that the work was done according to the mason's proposal or a written contract. The overseers were originally some of the Justices themselves but later on tradesmen were employed as overseers. From appointing the same man to oversee repeated contracts, the Justices eventually came to employ one or more permanent surveyors with annual salaries.

In the West Riding of Yorkshire two surveyors were paid small salaries from 1743 and they were the men who made the survey of all the county bridges in 1752 (see chapter 2 and appendix 4). One of them was Robert Carr, a mason and quarry owner, and many of the drawings in the survey were made by his son John who succeeded him as one of the surveyors in 1761. In 1777 the post was given to John Gott who for thirty years previously had been contractor, in sundry partnerships with his father, brother, son and nephew, for the regular annual maintenance of the county's bridges.[33] John Carr (fig. 109) had resigned after his appointment in 1772 as sole bridge surveyor of the North Riding with an annual salary of £100.[34] Because he held that office the number of bridges built to his design far exceeds that of any other well-known architect of the eighteenth century. He designed and directed the widening of sixteen bridges and the first building or total reconstruction of twenty-three others, and towards the end of his tenure of office (1803) he initiated a full survey of the North Riding bridges[35] similar to his father's for the

West Riding in 1752. The improvement in the standard of county bridges in fifty years is clearly shown by the statistics given in appendix 4, but the sizes of the North Riding bridges and the importance of the roads they served varied so widely that it is not possible to discern any constant principles about spans of arches, widths or gradients of carriageways, etc., which could be attributed to Carr. It is clear from one note in the survey that he thought elliptical arches weak and never built them; all his arches were segments of whatever height and span was appropriate to the bridge in hand, the largest span being 82 ft. The widths of his bridges varied from 15 to 32 ft, but were usually more than 20 ft, a width hardly known in the West Riding in 1752. On the elevations he almost always designed at least a neat arch ring, plain cornice and coping and some pilasters. Rustication was sometimes added and in a very few instances other classical motifs. His three most decorative façades were those of Aysgarth Bridge (fig. 110) and Greta Bridge, both single arches of more than 70 ft span with niches at the abutments, medallions on the spandrels and smoothly curved parapet lines, and Ferry Bridge, a three-arch structure designed as a special commission for the West Riding Justices, with similar niches at the abutments and trapezoidal pilasters over

109   John Carr, painted by W. Beechery.

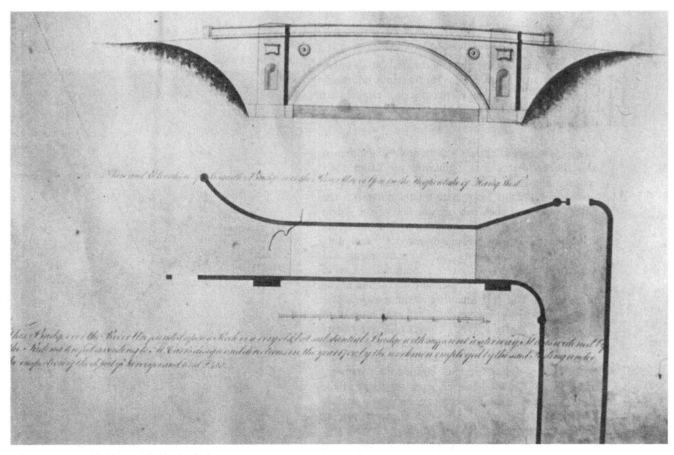

110   Survey drawing of Aysgarth Bridge (c. 1805).

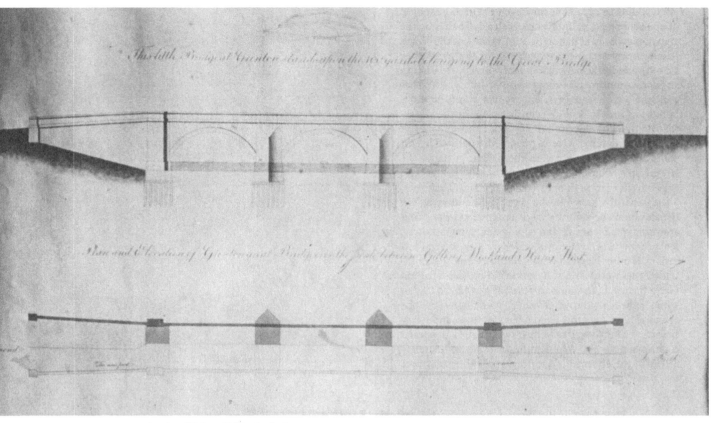

111   Survey drawing of Grinton Bridge (*c*. 1805).

the piers. Clearly his taste was for classical details but in ordinary county bridges (fig. 111) there was not much scope for style. There was plenty of style in the long list of country houses and public buildings he designed in the course of a very long career.

The architecture of bridges was the study of gentlemen and the architects they employed to design their country houses and estates.[36] Paine and Carr both made some designs for private estates but much the largest output of such designs was that of the Adam brothers, John, Robert and James. The surviving collection of their bridge drawings,[37] including sketches, landscape views and accurate plans and elevations, numbers well over a hundred. None of them, however, shows the internal construction or the foundations. The great majority of the drawings was certainly made by Robert (fig. 112), who studied in Rome from 1754 to 1758.[38] A series of about twenty sketches made after his return[39] shows that he was influenced by the ideas of triumphal bridges current there; he was a close friend of Piranesi (see chapter 7 and fig. 60). But he departed from the ideas of the Roman school in often designing segmental arches, sometimes of very low rise, and in never placing columns on the spandrels as Piranesi and Mylne (fig. 69) did. Adam's commonest spandrel decoration was apsidal niches with urns or vases in them; on plainer

112   Robert Adam. Medallion by Tassie.

bridges he simply used medallions (or 'paterae'). He often drew the spandrels with radial joints and he chose moulded arch rings, cornices and balustrades with wide variety but with obvious care. In many designs he added sculptured figures in the niches or on the parapets, and also the fluted and bas-relief panels, cords and swags with which in building design he created the 'Adam style'. It seems certain that none of his very ornate bridge designs was ever built, nor either of the ruined bridges he designed for Bowood and Syon House, but there are good examples of his simpler designs standing at Kedleston Hall in Derbyshire[40] (a three-arch bridge over a waterfall with horizontal parapet from end to end), at Audley End in Essex (three arches in a plainer style) and at Dalkeith House in Midlothian (a large semicircular arch of 70 ft span over a deep dell (fig. 113), for which the final design was made by James after Robert's death). Some of the designs have aedicules in the Palladian manner and several times Robert sketched covered bridges similar to the Palladian one which the Earl of Pembroke had built in his garden at Wilton in 1735–7. Pembroke had had the help of Roger Morris as clerk of works,[41] and Morris designed the Garron Bridge near Inveraray which was built in 1748–9 and supervised by Robert's elder brother John (fig. 114).[42] Morris was architect of the Gothic mansion of Inveraray Castle of which William Adam, John and Robert's father, was given the

114 John Adam. Medallion by Tassie.

task of local superintendence, and after William's death a strange castellated bridge was built at the mouth of the River Aray, probably to John's design. This may have been designed to match the Castle, but there are also a few Gothic designs amongst Robert's many classical ones.

Presumably Gothic designs were anathema to most clients and the two best bridges built in the early years of the brothers' practice are of loosely classical form. The first is a three-arch bridge in the grounds of Dumfries House at Cumnock in Ayrshire which was built by John between 1760 and 1762.[43] It may be assumed that he also made the design, for which an estimate was presented to the Earl of Dumfries on 11 February 1760, a few days before the acceptance of Mylne's design for Blackfriars Bridge. It is a pretentious bridge with regularly coursed masonry in the spandrels, wings and parapets, and moulded arch ring on the middle arch over which is an open stone balustrade (fig. 115). Four tall obelisks at the ends of this balustrade suggest the architecture of William Adam's Aberfeldy Bridge, but the most unusual thing at the date of construction is the form of the arches which are all ellipses, the middle one 50 ft in span. The second bridge was built in 1760-1 at Inveraray and is called the Garden Bridge, or Frew's Bridge, from the name of its builder. The final design was signed by John Adam. It consists of a single arch of 60 ft span and again of elliptical form, a form found in *none* of Robert's drawings, surmounted by spandrels of radial courses with masonry rings on them, both features which *are* found in Robert's designs, and a horizontal road

protected by a classical balustrade. There are vertical buttresses or turrets, of semicircular plan, over the abutments. The steep ramps up to the arch are stone-faced and penetrated by tunnels entered through doorways covered by pediments. The elliptical arches in these two bridges must be among the very first built in Britain, forestalling in construction, if not in design, those of Blackfriars Bridge.

In nearly all of Robert's designs the parapets sweep down at some point, much as his father's parapets in Aberfeldy Bridge turn down at the ends of the horizontal line over the middle arch. In Robert's designs of one arch, and some of three arches with horizontal parapet, this was the form of the parapets over the wingwalls beyond the abutments, but in the ornate design for Syon House at Isleworth, Middlesex (fig. 116), and a small design for Croome, Worcestershire, he drew the downward sweep very much as it is at Aberfeldy, just beyond the ends of the middle arch. Neither of these was built but the 'Aberfeldy profile' occurred in two large public bridges for which the Adams made designs, the bridges at Ayr (built in 1786-9) and Glasgow (1793-5). Paradoxically, the profile does not appear on any of the Adams' surviving drawings for either bridge, though there is something like it on one of the Ayr designs and on an engraving entitled 'the bridge and quays, agreeable to Mr. Adam's plan, proposed to be built at Ayr'. The date of this engraving is not known, and another manuscript design for Ayr has a continuously curved road profile with the parapet made up to horizontal over the middle arch (of three). The spandrels are drawn with radiating courses

115   Bridge at Dumfries House.

116   Design for a bridge at Syon House by R. Adam.

117   The 'new' bridge at Ayr, photographed c. 1860 (from Morris, 1912).

and medallions. The Glasgow design, which must have been made by James because it is dated 1793 after Robert's death, is similar but with five arches. Ayr Bridge was built (fig. 117) with five arches and the elevations altered, but Robert Burns attributed the design to 'Adams' in his poem *The brigs of Ayr*[44] soon after the construction began in 1786. Robert Adam was paid 30 guineas for the design several years later but that is the only mention of him in the minutes of the bridge trustees.[45] As the mason who built the bridge was a regular designer (see below) it seems likely that he was responsible for the design which was finally built. No mention of the Adams can be found in the minutes of Glasgow Corporation or the Chamberlain's ledgers, so the likeliest origin of the design for the Glasgow bridge is that it was a close copy of the bridge at Ayr.[46] It collapsed before being completed and was never rebuilt.

At a time when old bridges with houses on them were being taken down and replaced, the Adam brothers produced several designs for new bridges with houses. The first 'design' is a sketch by John of a 'design for uniting the Old and New Town of Edinburgh',[47] probably drawn in the 1750s. It is only an elevation and consists of a continuous line of buildings with several arches leading through. He may have intended a roadway along the top of the buildings. In 1768 Robert designed Pulteney Bridge at Bath as a street over the river with an elegant series of shops along each side of the road. It was part of a large town extension planned for Sir William Pulteney as a speculative development of his property, but only the bridge and half of its shops were built. The foundations (fig. 118), according to Thomas Telford, needed several attempts at underpinning before he directed a successful repair at some time before 1812.[48] A third 'bridge with houses' was designed for the Edinburgh Town Council about 1785. It was the South Bridge, continuing the line of the North Bridge (see chapter 8) across another deep but dry valley. The buildings were to be built beside the bridge, not on it, and because the bridge itself was of generous width they would not embarrass the roadway. The bridge being hidden, Robert Adam's taste was applied to the buildings. He was paid for his design but the bridge and buildings were built by others. The overall plan was not altered but the façades flanking the road were drastically simplified.[49]

The new bridge at Ayr (fig. 117) was built by a man of much humbler origin, but who, from the style of his bridges, must certainly be called an architect. He was Alexander Stevens of Prestonhall, Midlothian, and he contracted to build it in 1786 and finished it early in 1789.[50] There were five arches, the largest of 51 ft span, the parapet line was of the 'Aberfeldy profile', there was a length of open balustrade over each arch, a neat cornice, pilasters up the middle spandrels decorated with shields, and cast-lead sculptures in niches on the next spandrels. It was not the first time that Stevens had built the 'Aberfeldy profile', and it was also his habit to build bridges with plenty of ornament. Some of his commissions came from landed gentlemen, like the Adams', but his skill and training were those of a craftsman. He was probably first trained as a mason but by 1775 was also the largest tenant farmer on Lord Adam Gordon's estate at Prestonhall with a holding of thirty-eight acres and a substantial house.[51] He clearly

118   Pulteney Bridge, Bath, with one pier exposed down to the foundation. The modern cofferdam has sheet-piling, waling and struts like those used in the eighteenth century, but all of steel; and because steel is so much stronger than timber the struts are more widely spaced and there is much more free space within the dam. Modern pumps being powerful and the water shallow, only a single thickness of sheet piles is used and there is no clay seal.

owed some of his bridge contracts to Lord Adam's influence and his landlord was his cautioner in at least two large contracts. Stevens was almost certainly the builder, and probably also designer, of Hyndford Bridge over the Clyde above Lanark about 1773.[52] This was a five-arch bridge with interesting features, namely, cutwaters of the streamlined shape used in France, formed of two arcs of circles tangent to the sides of the pier and meeting at a point (see figs 119 and 121), and some architectural finishes, including moulded cornices, semicircular buttresses and recesses over the piers, and conical spirelets on the ends of the approach walls. For an early single-arch bridge at Blackadder House, Berwickshire,[53] he drew the Aberfeldy profile, and his second large bridge, Drygrange Bridge over the Tweed near Melrose, may have had the same profile when first built. It was his boldest design, of three high arches, the middle one 100 ft in span and only 2 ft 6 in. thick at the crown, and the side ones 55 ft in span.[54] The spandrels were hollow with two internal longitudinal walls to support the road, which was 16 ft wide between the parapets. The piers were required by the contract to be founded on rock and the cutwaters were shaped like those at Hyndford. These features show that Stevens was conversant with the most up-to-date structural techniques. His price was remarkably low, only £2,100, but he decorated the faces of the triangular buttresses which surmount the cutwaters with quatrefoil medallions and made a large circular panel on each spandrel with a classical urn of yellow sandstone standing before a black background (fig. 119). The rest of the bridge was of mixed red and yellow sandstone and he placed conical spirelets on the ends of the parapets. The bridge only ceased to carry main-road traffic in 1975.

In the autumn of 1784 he was at work on a bridge over the Teviot at Ancrum,[55] seven or eight miles further south on the same turnpike route, and there his use of the Aberfeldy profile is certain, for the bridge has never been altered. It is arguably his finest piece of architecture, for the decoration is more subdued than elsewhere. It has three segmental arches of 51 and 56 ft span and rather low rise, rusticated arch rings and a smooth parapet supported on corbels all in a pale pink sandstone (fig. 120). It has been bypassed for many years.

The next two designs attributed to Stevens rank with Aberfeldy and Drygrange as the most decorated public bridges in Scotland. They are those of the Teviot bridge just upstream of its confluence with the Tweed at Kelso, and Bridge of Dun over the South Esk near Montrose. The Teviot bridge was built by William Elliott of Kelso about 1795[56] but from its similarity to Bridge of Dun it may be guessed that the design was by Stevens. It is known that Stevens provided a first design for it in 1784, with an offer to build it for £1,100. He revised this offer to £1,230 in 1788 and it was still under consideration in 1792. The bridge as built by Elliott is of three high arches, the parapet following the Aber-

119   Drygrange Bridge.

feldy line, with modiglions under the cornice and well-detailed newels at the ends of the wingwalls. The cutwaters are of the same shape as those at Hyndford. From the top of each cutwater rise two columns, supporting the floor of a rectangular recess from the roadway. This mingles the architecture of Aberfeldy with that of Blackfriars Bridge, and in one respect the Teviot bridge is more true to classical models than Blackfriars, having a horizontal entablature over the columns. Stevens's contract to build Bridge of Dun is dated June 1785, and the price was £3,128.[57] The structural details and the Aberfeldy shape are similar to the other bridges but he placed three stone columns,

120   Ancrum Bridge.

each of three coupled shafts, on each cutwater to support the stone floor of a square recess (fig. 121). On the spandrels within the 'tabernacles' formed by these columns are deeply carved crosses, and inset panels on the faces of the long approaches (fig. 122) are carved with similar crosses and topped by rows of moulded corbels supporting the parapets. Such fulsome ornament shows that Stevens was devoted to decorative masonry, but he also contrived to make it cheap. This was the second bridge built by committees whose chairman was Alexander Christie, a former Provost of Montrose, and both committees resolved never to charge a toll on their bridges. As a result, all the capital had to be raised by private donations, including those of interested towns and counties, and grants of moderate amount from two public bodies, the Convention of Royal Burghs in Scotland and the Commissioners for the Forfeited Estates. The earlier bridge, over the North Esk (see chapter 8), could only be finished after making a further general collection at the doors of all the churches in Scotland,[58] but Bridge of Dun was built without such extra finance.

All these bridges, from Hyndford to Bridge of Dun, are still standing, though the last shows evidence of a large settlement of one pier, in spite of being founded on piles. The bridge at Ayr described above was Stevens's next design, and in his poem Burns put into the mouth of the Auld Brig of Ayr a prophecy that the new one would not stand more than two or three winters. In fact it stood until it was taken down in 1878.

Perhaps Stevens would have finished his life as a farmer if his patron had been a more successful landlord. Lord Adam Gordon was always in debt and after several earlier attempts he managed to sell the Prestonhall estate in 1789. It is a reasonable guess that Stevens then became a full-time bridge-builder. Only a few years later there was a firm known as Alexander Stevens and Son who in 1794-6 built a timber bridge across the mouth of the tidal basin of the South Esk at Montrose,[59] a crossing thought too difficult when Bridge of Dun was projected in 1784. Stevens designed Sarah's Bridge, a single arch of bastard elliptical shape and 104 ft span, which was built over the Liffey at Islandbridge near Dublin in 1791-3.[60] It bears no resemblance to his earlier bridges except those of competent design and construction. It is of generous width, almost 40 ft, and fenced with an iron handrail which is probably that erected when it was first built.

He was consulted about the project for a bridge over the Spey at Fochabers about 1793 and proposed a three-arch bridge at a cost of £14,000.[61] His last bridge, though his largest, was built to another's design and is therefore described in a later chapter; but the correspondence between John Rennie, the designer, and his clients, the Lancaster Canal Company, affords a few intimate glances at Stevens which should be taken here. Rennie advised the Company to accept Stevens's offer for building the aqueduct over the Lune in 1793, the largest stone aqueduct in Britain, although it seemed that he could find only a few thousand pounds' security,[62] and the Company agreed and signed the contract in July of that year.[63] In January 1795 Rennie wrote to their secretary that 'Stevens has been for this last fortnight seeing all the fine buildings in and about London',[64] presumably enjoying his first visit to the capital, but while there certainly discussing with Rennie the next stage of the work at Lancaster. When only a year later Stevens died at Lancaster, Rennie wrote, 'I most sincerely lament the death of poor Stevens – his loss will be felt by us all – In him the Company have lost the best of contractors and I a friend on whose advice I could always rely.'[65] The aqueduct was finished by his son (see chapter 10).

The last architect to be noticed in this chapter is Thomas Harrison. He also sprang from tradesman stock, being the son of a Yorkshire carpenter. His skill in drawing and mechanics attracted the attention of Sir Laurence Dundas, who sent him to Italy in 1769 to learn architecture.[66] On his return in 1776 he made a design for a triumphal bridge over the Thames at

121 Bridge of Dun.

122 Bridge of Dun.

123   Skerton Bridge, Lancaster.

Somerset House, as others were wont to do, but his first chance to build a bridge seems not to have come until he won a competition for design of a new bridge at Lancaster in 1782.[67] When he was appointed to direct the construction he proposed to the committee five alternative designs, with a book of notes to instruct them about bridges.[68] He gave figures for the widths of the new bridge at Newcastle-upon-Tyne (21 ft 6 in. within the parapets and 'much too little') and some Yorkshire bridges, and a discussion of the strength of different arch forms which led to the statement: 'Elliptical arches are the lightest and perhaps the most beautiful; and, when properly constructed, as strong as any other.' Not surprisingly, the design which was built and is known as Skerton Bridge is of five elliptical arches. Each is of 68 ft span and 19 ft rise, the parapet is horizontal from end to end, and Palladian aedicules over all the piers and at the abutments form the entrances to arched tunnels going right through the bridge (fig. 123). This is the closest copy of the bridge of Rimini ever built in Britain. It was very well founded on piles, is of good width, and does credit to Harrison. His next bridge was of similar but simpler style, over the Derwent at Derby, and was built in 1789–94.[69] He also designed major improvements to a bridge at Kendal in 1791[70] and to Swarkestone Bridge over the Trent in 1801,[71] and all this work was successful. We shall find him less reliable in chapter 13, at a time when he took to designing very large arches.

In 1791 there was a fearful challenge to current ideas about bridge design. The speculative builder and architect John Nash (fig. 135), who had passed through success to bankruptcy in London in 1783, and was making good in Wales,[72] offered a design to a bridge committee at Newport, Monmouthshire, in competition with three others. The committee accepted his design and work commenced; by November 1791 £700 had been spent.[73] The design was a single masonry arch of 285 ft span, enormously high, being a segment of a circle of 310 ft diameter.[74] It surpassed by no less than 100 ft the spans of the largest stone arches in the world!

Suddenly the scheme was dropped and Nash was paid off. One topographical writer in the following year said that 'the county gentlemen rejected a proposed single-arch bridge'[75] and we know nothing of their reasons. But the dreams of enormous arches, which had been thrust aside by committees for fifty years, had become a real project on a real site for just a few months. It was time for great changes in bridge-building.

10

NAVIGABLE AQUEDUCTS

... the ingenious Mr Brindley ... has erected a navigable canal in the air ... as high as the tops of the trees.

Letter to the *St James's chronicle*, 30 September 1763

The idea that a bridge might be built not only to cross a stream but to carry one was familiar to men of the eighteenth century, who studied the aqueducts of ancient Rome as great specimens of the mason's craft. When large aqueducts came to be built in Britain, however, they were for inland navigation, not water supply, and there was sometimes the possible alternative of taking the waterway down the valley side by a series of locks and ascending by a similar series on the other side. If the Duke of Bridgewater had done this to get his canal across the valley of the Irwell he would have had to find a source of water south of the river to fill the southern part of the canal. Instead he decided to take it over the river by a bridge, or navigable aqueduct, of three large arches at Barton; and the sharp debate that took place between his supporters and opposers in Parliament and between him and his friends outside it is evidence of the unconventionality of the proposal.[1]

The Duke himself appears to have been fully convinced. He had taken the trouble to study the greatest artificial waterway in Europe, the Canal of Languedoc (or Canal du Midi), in the course of his 'grand tour' and must have seen two or three large navigable aqueducts there.[2] In Lancashire he was living beside the building of his canal and had spent long hours planning it with two practical 'engineers', John Gilbert and James Brindley (fig. 124). Gilbert was his land agent and administrator of the canal project, but had served an apprenticeship in the manufactory of Matthew Boulton the elder in Birmingham. Brindley's success as a millwright and builder of steam engines had earned him the nickname 'Schemer'. By 1758 he was employed in surveying a line for the Trent and Mersey Canal, and he was responsible in later years for drilling important tunnels both for canals and for mines;[3] but in this he may have learned much from John Gilbert who first proposed and began to cut navigable 'soughs' into the Duke's mines at Worsley before Brindley joined him and the Duke there.[4] Brindley arrived in July

1759[5] and the decision to build Barton Aqueduct, if Parliament could be persuaded to allow it, was taken before the end of that year. It would allow the canal to run on one level from the coalfaces inside the hill at Worsley to the quays at Manchester without a single lock. The aqueduct had to be high enough to permit river craft to pass under it.

We shall probably never know which of the three men first thought of building it, but it has almost always

124 James Brindley.

been assumed, and never denied, that Brindley directed the construction. He probably designed it as well, since he seems to have had more practical experience of masonry work than Gilbert or the Duke. It is said that the Duke was persuaded by Brindley himself to ask the opinion of another eminent engineer and that this man, having viewed the site and the design, replied, 'I have often heard of castles in the air but never before saw where any of them were to be erected.'[6] The Duke, however, ignored even this sarcasm and his Bill went forward aided by most unconventional evidence given by Brindley. In the committee room of the House of Commons he made some 'puddle' from clay, sand and water and formed it into a trough which held water without any leakage, and this showed how the water of the canal could be contained on the top of the aqueduct.[7] The committee reported favourably and the House passed the Duke's Bill into law.[8] The epoch-making canal, with a long, high embankment across the Irwell valley and the first navigable aqueduct in the country, went ahead.

The aqueduct had three arches, the middle one 63 ft span, the side ones 34 ft (fig. 125).[9] They were segmental arches of low rise, the middle one rising only 12 ft, or one-fifth of the span. It was two brick lengths, or $1\frac{1}{2}$ ft, in thickness. The piers were of stone and so were the spandrels by 1784, but the latter may have been first built of brick.[10] The clear height of the middle arch above the river was 26 ft and the overall width of the aqueduct was 36 ft, much more than that of many later aqueducts. There was a waterway 18 ft wide and a tow-path on each side 9 ft wide, with no parapets or guard-rails. The spandrels and the upper walls which retained the water were plane and vertical, but there was some buttressing at the ends, achieved by widening the abutments and approaches to 45 ft.

It was filled and opened to traffic on 17 July 1761,[11] only fifteen months after the enabling Act had been passed, and it is said that Brindley was so nervous that he stayed away from the opening ceremony.[12] This statement was first printed in 1820 by the Rev. F. H. Egerton, Eighth Earl of Bridgewater, and he claimed to be quoting from notes he made immediately after conversations with his uncle the Duke. From the same notes he recounted some trouble during the construction of the aqueduct. There was apparently a distortion of one of the arches, probably in the autumn of 1760, which was corrected by placing more weight on the crown and lightening the load on the haunches. After this adjustment of load it was covered with straw, perhaps to protect it from frost, and allowed to stand until late in the next spring by when, we may assume, the mortar had set well and the arch had become stable. Egerton said the arch curve remained 'irregular' (possibly the pointed shape shown in fig. 125) and he gave credit for the remedial work to John Gilbert. After

125   Barton Aqueduct.

126 Rugeley Aqueduct.

this hazardous start the aqueduct gave good service for over a century.

Within months of the completion of Barton Aqueduct the Duke obtained another Act of Parliament[13] for a long branch of his canal to the Mersey estuary near Runcorn. It had to cross the Mersey first by an aqueduct at Stretford, and for this crossing Brindley designed a single arch of 66 ft span and 16 ft 4 in. rise.[14] It still stands, probably unaltered except for patching of the brickwork and masonry and a sinking by at least 9 in. of the crown of the arch. The arch is $2\frac{1}{2}$ brick lengths in thickness (1 ft 10 in.), half a brick thicker than the middle arch at Barton but still unusually thin for an eighteenth-century bridge. The spandrels are of sandstone and a batter is achieved by the simple method of stepping some of the courses back from the course next below. The width is generous, over 20 ft of waterway and an 8-ft tow-path at each side. The wingwalls are splayed but now badly cracked. The passage through the arch was actually too small for the river in full flood and an extra flood channel had to be provided in the 1830s.[15]

In building another bridge at Dunham Town on this branch of the canal, Brindley is said by Samuel Smiles to have pumped the water from the excavations by a steam engine of his own invention, called a 'Sawney'.[16] This is not mentioned by any earlier writer and Smiles does not reveal his source of information. The atmospheric engines of the day were much too heavy to be used on construction sites and Smeaton is credited with

the first design of a suitable engine in 1765.[17] If Brindley did use the 'Sawney' at Dunham Town, it was probably in the same year.

The instantaneous fame of the Bridgewater Canal, and especially of Barton Aqueduct, brought Brindley commissions for surveying and directing canals all over the country. In the ten years before his death in 1772 he was the engineer of many aqueducts and a recently discovered manuscript tells much of his methods in designing and constructing them. A situation which recurred often was that of a canal crossing a river only 15 to 30 ft below its level. Early examples were the Great Haywood aqueduct[18] which carries the Staffordshire and Worcestershire Canal over the Trent, the crossing of the Sowe by the same canal, and the crossings of the Dove[19] and the Trent (near Rugeley, fig. 126) by the Trent and Mersey Canal. It appears to have been Brindley's method to widen the river by cutting into the banks,[20] thus compensating for the obstruction of flow by the piers. He then built a string of low, short arches of brick, but with one or two courses of stone at the springings. There was also some stone in the spandrels, probably for the sake of appearance. The spandrel walls were built with a regular batter. Perhaps the most surprising feature is the lack of either rounded or pointed cutwaters. The arches of bridges and aqueducts on the Staffordshire and Worcestershire Canal were made of a single radius,[21] probably to make the centres interchangeable. Brindley toured the works of this canal once every three weeks in the first year of its

127   Kelvin Aqueduct.

construction and at longer intervals later, giving verbal orders which were written down, and sometimes drawn, by the canal company's supervisors. He ordered clay to be laid on the extrados of every arch of an aqueduct in three thin layers to make a thickness of 4 in. in all (there was presumably a much thicker horizontal lining of puddle at the bottom of the canal, but no specific order for this is recorded). The piers were founded on piles only if firm ground could not be found. On poor ground a penned floor or 'inverted arch' in the waterway was ordered several times. At Great Haywood 'pile planks' were driven both up and downstream, probably a continuous sheet-pile wall right across the river. He also ordered for Great Haywood 'the top of the arches there to be laid over with three-inch planks and the planks floored with flat bricks set in mortar', and this may have been a regular practice, for such timbers were found during the demolition in recent years of the Little Lever Aqueduct on the Manchester, Bolton and Bury Canal,[22] which was engineered by Brindley's most successful pupil, Robert Whitworth, long after his master's death.

After Brindley's successes there were few problems left to be solved in the construction of masonry aqueducts. Some designers, however, departed from the forms he used to the extent of curving the spandrels, upper sidewalls and wingwalls horizontally as well as vertically, the better to resist the outward pressure of the spandrel filling and, more especially, the pressure of water in the canal at the top of the structure. In multi-arched aqueducts curved sidewalls and spandrels required very large buttresses over the piers or cutwaters to resist their outward thrust, and these buttresses gave to the aqueducts a really massive appearance and an expensive bulk of masonry. A prime example is Whitworth's Kelvin Aqueduct at Glasgow, carrying the Forth and Clyde Canal over the Kelvin and started in 1787 (fig. 127). Whitworth adopted the boldest of rustication with deep V-cuts for all the horizontal joints and very large stones. It was the largest British aqueduct at its date of construction: 445 ft long, 56 ft wide at the spring of the arches, of four arches 50 ft span,[23] 62 ft high from the surface of the river to that of the canal, and the walls above the arches were strongly curved in plan. We owe to notes made by the young John Rennie in 1789,[24] when the aqueduct was just about finished, the knowledge that the layer of puddle over the bridge at the bottom of the canal was 3 ft thick, and his measurements of the structure conform both to Whitworth's proposals recorded in the canal company's books and the dimensions as it stands today. It shows signs of having leaked and been repaired, and of some localised decay of the freestone.

Rennie himself designed many aqueducts, and in some of his earliest and biggest he used alternating areas of rough-faced and smooth-faced stone to provide the robust texture which he, like Whitworth, thought they should have. In the single-arch aqueduct by which the Lancaster Canal crosses the Wyre north of Preston, the spandrels have a regular vertical sequence of two courses rough followed by one smooth, and the ring of the arch has a similar sequence of two rough voussoirs followed by one smooth. The wings are full quadrant curves in plan and vertically curved as well – both regular features of Rennie's larger aqueducts – and topped by a full classical entablature. This and the Lune Aqueduct (fig. 128), both designed only three or four years after his visit to the Kelvin, show that Rennie recognised at the very start of his bridge-building career that the horizontal line of its top made an aqueduct, and especially a high aqueduct, an appropriate subject for classical decoration. Over the Lune he built semicircular arches, which were most fitting for a classical elevation, and at the Wyre where the arch had to be

lower he did the next best thing by making it a semi-ellipse.

The structure over the Lune[25] is the largest all-masonry aqueduct ever built in Britain. It has five arches each of 70 ft span. It was contracted for in 1793 and finished in 1798, costing £48,000[26] although Rennie's first estimate was only £16,647[27] (probably excluding foundations). The structural design looks flawless in retrospect, and it is astonishing that Rennie, with no experience of designing such large structures, should have produced so mature a design. The spandrels were designed to have three longitudinal voids covered by pointed arches and the masonry was then to be made up to a level surface over which a 3-ft bed of puddle would be laid on the whole bridge. When this was well bedded and compacted a concave canal bottom of stone was to be laid on it and the side walls of the canal and the exterior walls then built up, with puddle packed between them. For the construction of the arches the centres, designed by Rennie, were to span from pier to pier, though a later drawing[28] shows a centre with a support at midspan; they were to be set higher than the designed level of the soffits at the crown by an amount sufficient to allow for the dropping of the crown when the centres were struck. An unusual provision was for four bars of 'rough iron', 4 in. × 1 in. in section and of length equal to the width of the bridge, to be set in four of the joints between courses of each arch, secured by 'dovetail bolts' and run with lead. The piers were to be tied in several of their horizontal joints

128   Lune Aqueduct.

129   Dundas Aqueduct.

by similar bars, but that was not so unusual. Before the bridge was built he added a suggestion that similar bars be laid across the top of the spandrels under the puddle clay at $12\frac{1}{4}$ ft spacing for the whole length of the bridge, and fixed at their ends to two longitudinal bars in the bottom joints of the parapet walls.[29] This he considered to be a much cheaper alternative to the arching of sidewalls horizontally to resist the outward pressure of the water. But it took him over a year to persuade the canal company that it was a reasonable extra expense, although within that time he had heard of William Jessop's use of iron bars in just the same positions to repair an aqueduct on the Cromford Canal which had failed by the water forcing out the side walls and splitting the arch. At the same time he introduced reversed arches between the haunches of the main arches, and he repeated both these and the iron bars in other large bridges in later years.

The design of the foundation works at the Lune was equally thorough. Borings were made to depths as much as 37 ft below the water surface, the water being no more than 10 ft deep.[30] The excavations were deep and the depth of founding, the type of piles and the details of the timber platforms, etc., were decided by the engineer or his representative on site. The procedure was to dredge out the gravel in the river bottom by a 'drag', as deep as possible. Then a cofferdam was driven, of two timber walls, the inside one composed of grooved sheet piles 6 in. thick and about 20 ft long and the outside one of square piles spaced about 12 ft apart and sheeted with boards laid on edge from above the water surface down to the river-bed, the space between the two walls packed with clay and puddled. The water was then pumped out by an unusual, if not unique, method. A description by a man who was probably an eye-witness reads:

A fire engine [i.e. steam engine] was erected in the line of the piers a little behind where the south east abutment is built this worked an iron balance wheel to the axis of which were fixed cranks at each end one communicated with the engine beam the other worked an horizontal piece of wood that communicated with the pumps when they were fixed in the coffer dams.[31]

Within each cofferdam the ground was excavated to a firm stratum and the rectangle of the foundation platform marked out. Sheet piles were driven round the edge, but apparently no bearing piles, and then a three-layer platform was made, the lowest layer being whole balks laid tightly side by side along the length, the second layer half balks laid transversely and the third an open framework of whole balks jointed at their crossings. The void spaces in the third layer were filled up with masonry and the first course of stones of the pier proper laid on this surface. Each course in the piers was about 2 ft thick.

Alexander Stevens contracted to build the super-structure in July 1793 and was present all through 1794 while the foundations were laid by direct labour. The standard working week was six days of ten hours, at two shillings per day for a labourer and two shillings and twopence to two shillings and sixpence for tradesmen; but many of the tradesmen were paid at piecework rates for grooving and shoeing piles, sawing and other tasks.[32] The steam engine for draining the excavations was bought and operated by the canal company. The contract work began in 1795 and was finished by Alexander Stevens junior in 1798.

Since the Lune, his greatest aqueduct, stands today with no visible distortion and only the smallest traces of decay, it is surprising to find that Rennie's next two aqueducts in order of magnitude, both of which were designed and built at about the same time as the Lune, have weathered badly and one of them is greatly distorted and now disused. They were both built to carry the Kennet and Avon Canal over the Avon and their interest now is in their architecture. The one at Avoncliffe has the alternate rough and smooth courses used in the Lune and Wyre structures, and both of the Avon aqueducts have curved wings and elegant classical entablatures. On the second of them, the Dundas Aqueduct (fig. 129) at Lympley Stoke, the decoration is really massive and the flat top of the cornice protrudes a full 4 ft from the face of the bridge. There for the first time Rennie put twin pilasters on the face of the spandrels over the piers, a decoration of the same 'style' as the twin columns which Mylne had introduced at Blackfriars in 1760 and for which Stevens had shown a fancy in his Scottish bridges. Rennie, however, in building them at Lympley Stoke, was using them as Piranesi and others had first drawn them (fig. 60), on a bridge with a continuous and perfectly horizontal entablature high in the air, and flanking a semicircular arch. Neither Mylne nor Stevens had had such a perfect subject for classical embellishment, and indeed Rennie never had such a perfect one again. But it is most revealing that, while he devoted such care to the classical styling of these two aqueducts, he was also urging the proprietors to swallow *their* pride and have them built of brick because the local stone from a succession of quarries had been found susceptible to frost damage.[33] However skilful his use of the classical idiom, Rennie was putting stability and durability first. Both the Avon bridges needed considerable repair of the stonework even before they were finished.

11

THE FIRST IRON BRIDGES

The whole of our knowledge and reasoning with respect to cast iron depends in a great measure upon judgement or conjecture with a little assistance from mathematical and mechanical principals. If this is the case a man of genius and of a mechanical turn of mind may be as capable of forming a right judgement as the professional architect.

Anonymous notes on 'Thomas Paine's patent iron bridge' (c. 1792)

There is one bridge design by Robert Mylne which was not considered in chapter 8. It was found a few years ago at Inveraray Castle and is probably the earliest surviving design of an iron bridge. There are two drawings, both in Mylne's hand and both dated 1774.[1] They show a bridge of two arches to be built on the pier and abutments of the old 'town' bridge at Inveraray. This old bridge had been superseded as a public bridge by the building of the Aray Bridge at the mouth of the river, first to John Adam's design and again in 1773–5 to Mylne's (see chapter 8). For the 'town' bridge Mylne designed two very light arches (fig. 130) of 43 ft span and apparently each of two iron ribs carrying a timber floor of cross-beams, longitudinal joists and boards. In his diary at the end of 1777 there is an entry 'Castle or metal bridge. Cost £2500', 1777 being the year when the old town bridge was demolished. The design (fig. 130) shows a gateway across the bridge, probably because it was to be the entry to the estate from the old town, and light guard-rails; both are of iron and in Chinese ornamental style.

Though Mylne lived well into the era of iron bridges this is his only known design in iron and apparently it was not built. It was very like the published designs of Abraham Swan[2] and others for arches of timber with Chinese decoration in iron, and most designers thought of iron as a substitute for timber, not for stone. It would have to be used in thin bars or plates and so must be framed together as timber was to make structures that were stable. Progression from a timber arch design to an iron one is explicit in a sequence of designs made by Thomas F. Pritchard of Shrewsbury in 1773–5. Elevations of his three designs were published by his grandson John White in 1832 (fig. 131).[3] The first is a timber arch of 136 ft span and 20 ft rise designed for a crossing of the Severn at the new canal port of Stourport in 1773; the second a masonry design of the same span made in 1774, to be constructed on a cast iron centre; and the third a 'Design for a cast iron bridge between Madeley and Broseley' dated October 1775. In the first design the spanning structure seems to be concentrated in two deep ribs or frames of braced

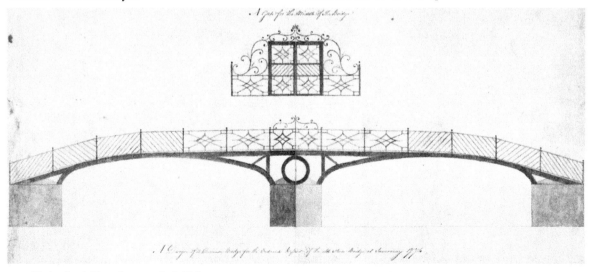

130   Design for a bridge at Inveraray by R. Mylne, 1774.

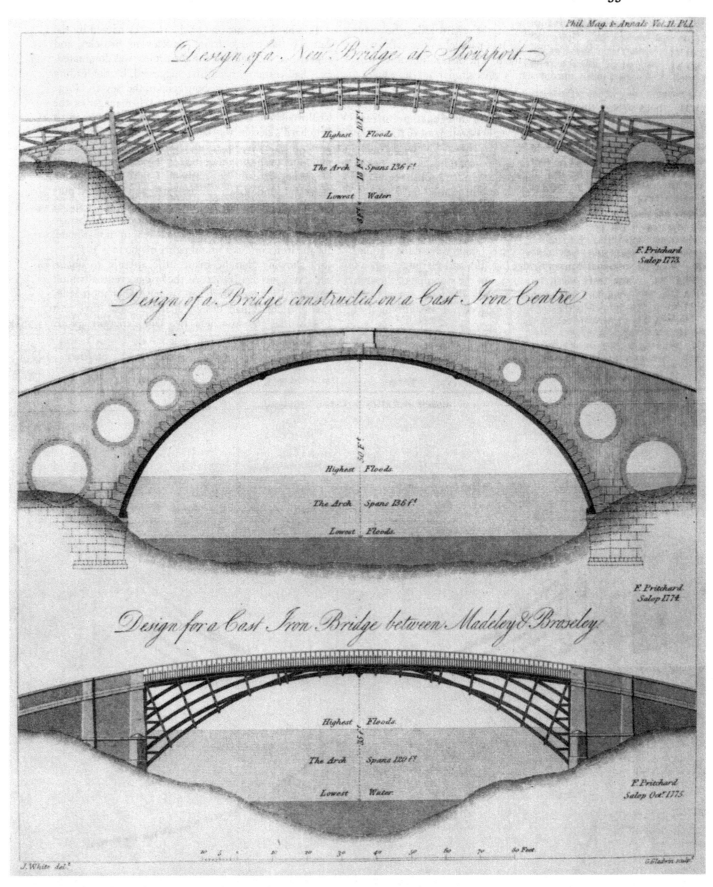

Phil. Mag. & Annals Vol.II. Pl.I.

Design of a New Bridge at Stourport.

Highest Floods

The Arch Spans 136 Ft.

Lowest Water

F. Pritchard.
Salop 1773.

Design of a Bridge constructed on a Cast Iron Centre

Highest Floods

The Arch Spans 136 Ft.

Lowest Floods.

F. Pritchard.
Salop 1774.

Design for a Cast Iron Bridge between Madeley & Braseley.

Highest Floods.

The Arch Spans 120 Ft.

Lowest Water.

F. Pritchard.
Salop Oct.r 1775.

J. White delt.

G. Gladwin sculp.t

131 Designs by T. F. Pritchard, 1773–5 (from J. White, 1832).

timbers in the planes of the guard-rails, the main members lying more or less parallel to the soffit with radial pieces crossing them. The disposition of the iron bars in the third design is quite similar but the whole of the structure is below the road, so it is possible that in this design Pritchard intended to have more than two ribs. The span is 120 ft and rise about 29 ft, putting the crown of the arch 35 ft over low-water and 16 ft over the highest floods. The arch rib shown on the elevation consists of one bar forming a complete segment from abutment to abutment and four other broken segments, three of them concentric with the first and above it, rising from the abutment walls till they intersect the line of the road, the fourth rising more steeply from below the complete segment and crossing it. In this crossing of members there is a slight echo of the framing of Schaffhausen Bridge (fig. 36), and a stronger echo of the timber design of 1773, but the stopping of the arch members where they intersect the road is not to be found in either of those sources and was untypical of good timber framing.

Pritchard became one of the trustees named by the Act for the bridge from Madeley to Broseley and subscribed 100 guineas, but the project was dominated, as it had certainly been first suggested, by the leading ironmasters in the neighbourhood of the Severn Gorge and particularly Abraham Darby III, manager of the Coalbrookdale Company.[4] Darby agreed, as soon as Pritchard's design was submitted to the committee of subscribers in 1775, to undertake the whole project, including even the obtaining of the Act, for the amount of money already subscribed, £3,150. A surviving estimate[5] (undated) made by Darby and Pritchard puts the cost of construction alone at £3,200, which shows that Darby was willing to lose money on the bridge right from the outset. The largest item in the estimate, for the ironwork, is a round figure of 300 tons at £7 per ton, showing that he made little attempt to obtain accurate costs in advance. As the bridge eventually built contained 378 tons of iron and much higher approaches than Pritchard had designed, it may well have cost Darby the £6,000 which two commentators quote.

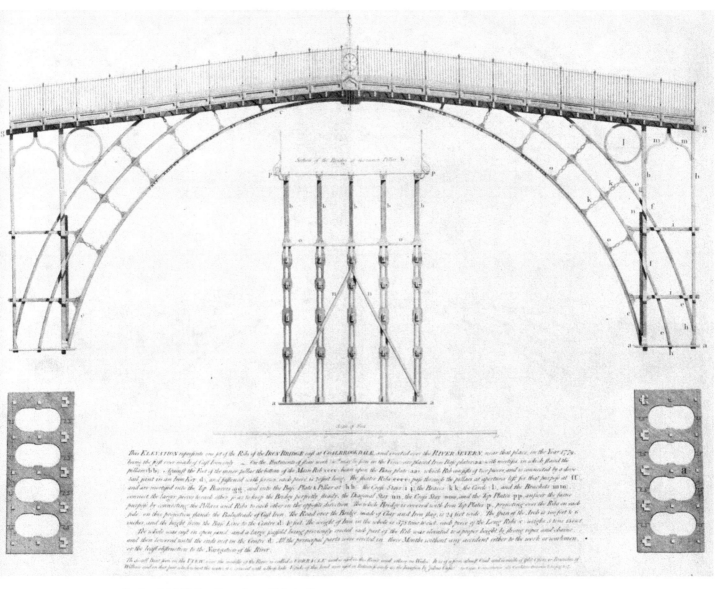

122   Coalbrookdale Bridge. Engraving published 1782.

Changes in Pritchard's design were considered from October 1776 onwards and resulted in an increase in width from 18 to 24 ft, a greater height over the water, reduction of the span and large changes in the layout of the ironwork. Authoritative details of the construction were given on an engraving issued by the Coalbrookdale Company in 1782 (fig. 132):

On the abutments of stone work are placed iron base-plates (aa) with mortises, in which stand the pillars (bb). Against the foot of the inner pillar the bottom of the main rib (cccc) bears upon the base-plate (aa), which rib consists of two pieces, and is connected by a dovetail joint in an iron key (d), and fastened with screws: each piece is 70 feet long. The shorter ribs (eeee) pass through the pillars at apertures left for that purpose (ff), and are mortised into the top bearers (gg) and into the base-plate and pillar (hh). The cross-stays (ii), the braces (kk), the circle (l) and the brackets (mm), connect the larger pieces to each other, so as to keep the bridge perfectly steady, and the diagonal stay (nn), the cross stay (oo), and the top plates (pp), answer the same purpose by connecting the pillars and ribs to each other in the opposite [i.e. transverse] direction. The whole bridge is covered with iron top-plates (p), projecting over the ribs on each side: on this projection stands the balustrade of cast iron. The road over the bridge, made of clay and iron slag, is 24 feet wide. The span of the arch is 100 feet and 6 inches, and the height from the base line to the centre (d) 40 feet. The weight of iron in the whole is 378 tons 10 cwt: each piece of the long ribs (c) weighs 5 tons 15 cwt. The whole was cast in open sand and a large scaffold being previously erected, each part of the rib was elevated to a proper height by strong ropes and chains, and then lowered until the ends met in the centre (d). All the principal parts were erected in three months without any accident either to the work or workmen, or the least obstruction to the navigation of the river.

Virtually all the connections between the members were made by dovetailed ends or by passing one member through a hole cast in the other, and bosses or brackets were cast on to allow mutual bearing, many of the joints being tightened with iron wedges (fig. 132);[6] there were some screws but no bolted connections. It has never been established who made this design but the details show that the minds of ironfounders were dominant; it is likely that Darby and his foreman pattern-maker Thomas Gregory, and Daniel Onions who, according to Telford, directed 'the practical operations', all had some part in it.[7] Neither the form given to the arch nor the framing around and over it reflect any of the current thinking of architects, engineers, and mathematicians. It had little influence on later designs but, becoming quickly famous, it was an important stimulus to the use of iron.

The only serious error in the design was not in the ironwork but in the approaches and abutments which, being high and built on the steep and unstable sides of the gorge, were forced towards the river. An arch of flatter curve and sprung from points high on the abutments, as many later iron arches were, would have resisted such movement of the abutments, but the high arch at Coalbrookdale exerted little horizontal thrust

133 Thomas Paine, engraved after Romney.

and the masonry of the abutments and approaches had to be repaired frequently from 1784 onwards.[8]

The second projector of iron bridges in Britain was neither a bridge-builder nor an ironfounder but a political writer. The second patron was a wealthy landowner and Member of Parliament. The projector, Thomas Paine (fig. 133), was in America playing an important part in the War of Independence when the Coalbrookdale bridge was designed and built; the patron, Rowland Burdon of Castle Eden in Cumberland, was making his 'grand tour' of Europe and spending much time in the company of John Soane,[9] a young architect who had won the King's Travelling Studentship at the Royal Academy Schools in 1778. In 1776 he had won the Academy's Gold Medal with a design for a triumphal bridge.[10] Soane surveyed bridges carefully, including the wooden Schaffhausen and Wettingen Bridges (see chapter 3) which were amongst the longest spans in the world.[11] The friendship he made with Burdon was to continue for more than fifty years.

When the American war was finished Tom Paine turned his mind to designing bridges. In making models of his designs he had the help of a Leicester 'mechanic' named John Hall who had emigrated to Pennsylvania in 1785. 'The European method of bridge architecture, by piers and arches, is not adapted to the condition of many of the rivers in America on account of the ice in the winter', wrote Paine in 1786 as he submitted two bridge models to the State Assembly of Pennsylvania.[12] He was looking for ways of spanning wide rivers with timber or iron structures needing no supports in the water, that is, by single arches of very long span and very low rise. At the suggestion of Benjamin Franklin, who was President of the Assembly, he shipped one of his models (a model in wrought iron and wood of an iron bridge of 400 ft span) to Paris in 1787 for the opinion of the Academy of Sciences, and took it to London later in the year to show to Sir Joseph Banks, the President of the Royal Society. He obtained encouraging reports from men of science both in Paris and London and, as his hopes of raising money for a large bridge at Philadelphia were fading, he stayed in Europe trying to have his scheme adopted for bridges over the Seine, the Thames, and the Liffey in Dublin. In September 1788 he took out a patent in London and sought the interest of some large ironfounder. He first intended to approach John Wilkinson, who was the chief competitor of the Coalbrookdale Company in Shropshire, but when two of the Walker family of Rotherham came to London to see his model he quickly made terms with them. Although he intended to dispose of his patent as soon as the invention was proved viable, he went twice to Rotherham before May 1789 to be present in person when a trial arch rib of 90 ft span and 5 ft rise was cast and erected. He set out the curve as a catenary but found it was practically indistinguishable from a circular segment for such a flat arch. He made the depth of the arch at the crown 2 ft 9 in. ($\frac{1}{32}$ of the span) and erected the rib on a centre, working from the crown downwards; he noted that before the extreme ends of the ironwork had been placed the whole had flexed upwards to rise off the centre at all points except the ends. The spaces between the ends of the rib and two adjacent walls were packed with timber to provide abutments and the rib, which itself weighed 3 tons, was loaded with a further 6 tons of pig-iron, after which the abutments yielded horizontally $\frac{1}{4}$ in. at each end and the crown of the arch dropped by 2 in. The load was left on for some weeks and the rise and fall of the arch measured through the daily cycle of temperature change, the movement up and down being $2\frac{3}{10}$ in. at the crown. Paine was confirming one of the advantages which the Paris Academy's report on his proposals had praised, namely that the design of such an iron bridge could be tested by erecting and loading only a small part of it. In letters to several people he repeated with pride the statement of a local gentleman, Mr Foljambe, that: 'In point of elegance and beauty it far exceeds my expectations, and it is certainly beyond anything I ever saw.'

Since no drawing of the rib survives we have to deduce its construction from written descriptions and a drawing of a Paine-type arch which John Soane made in November 1791, apparently a copy of another drawing by somebody else. It is likely that this other drawing was sent to him by Rowland Burdon (fig. 134).[13] Paine said that he had taken his idea from a spider's web and that a rib of his arch looked like part of a web. There were a number of concentric curved bars, 3 in. × $\frac{1}{4}$ in. in

134   Paine's iron bridge adapted for Sunderland.

cross-section, of which the lowest formed the 'soffit' of the rib and the upper ones stopped where they intersected the line of the road. The bars were initially straight and were simply bent to the required curve, and they could be 'composed of pieces of any length joined together' – though how they were joined we do not know. They were spaced apart by 'blocks, tubes or pins' placed radial to the curves. It appears that 'blocks' were used in the rib at Rotherham, each block having a bolt-hole through it from end to end and a bolt and nut to clamp the bars against its ends. This method of assembly made the rib easy to transport in small pieces and it could also be taken down and re-sited if necessary, advantages which Paine, like other advocates of iron arches after him, was keen to advertise. The shape of the main curved bars suggests that they were probably of wrought and not cast iron.

There was considerable interest in Paine's model in London and the trial rib at Rotherham. In July 1789, he proposed to the Walkers that they should make a whole bridge and send it to London where he would erect it for exhibition and then offer it for sale, or alternatively exhibit it again in Paris in the hope of obtaining an order for a bridge over the Seine. The Walkers readily agreed and all went well during trial erections of the bridge at the works, but it did not arrive in London until May 1790. Edmund Burke was then on the point of publishing his *Reflections on the French Revolution*; and

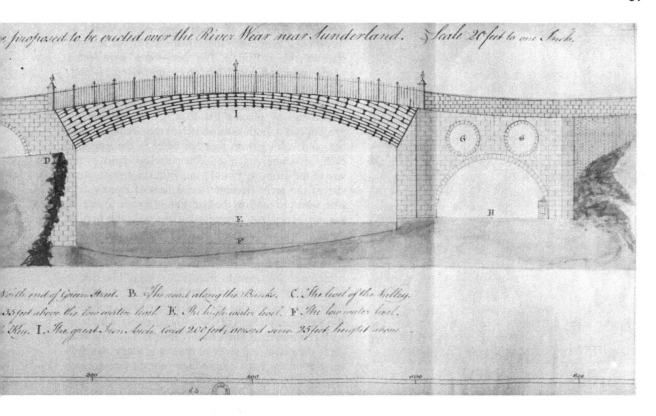

Paine felt bound to write a reply. But Burke delayed his publication for six months and Paine was spared from having to choose, as he thought he might have to, between writing *The rights of man*, on which his fame has rested ever since, and erecting the bridge towards which he had worked for five years. It was erected on a bowling green next to the Yorkshire Stingo public house at Paddington during August 1790 by three carpenters and two labourers under Paine's personal direction. This was because the man he had engaged to direct it was injured in a fall on the first day. 'I am always discovering some new faculty in myself either good or bad', he wrote to Thomas Walker, 'and I find I can look after workmen much better than I thought I could'.[14] The bridge was of five ribs and was 22 ft wide, 110 ft span and 5 ft rise. The arch was of catenary shape and it weighed 36 tons, of which 22¾ tons were wrought iron and the remainder cast iron,[15] proportions which confirm the supposition that the main curved bars were of wrought iron. There was a wooden floor, designed by Paine himself, and only wooden abutments. These yielded and, though it stood for over a year and was seen by large crowds and visited by members of the Royal Society – and doubtless every engineer and architect in London – it was taken down in October 1791 and the iron sent back to Rotherham.[16] Burke's book had been published in November 1790 and the first part of Paine's *The rights of man* in March 1791. In November

of that year he wrote to his former helper John Hall that he was engrossed in his 'political bridge' and he took no further interest in bridges of iron until many years later. But only a few weeks before Paine wrote this letter to Hall, John Soane had been making his copy drawing in London. The title he put on his copy, and so presumably the title of the original drawing, was 'A slight sketch of Thos. Paine's patent bridge proposed to be executed over the River Wear near Sunderland'. Fig. 134 is Soane's copy. There is no indication of who made the original, but it seems certain that Paine knew nothing of it. There are five concentric bars at the crown and twelve at the springings; the span is 200 ft and the rise 25 ft.

The committee formed to promote the building of a bridge at Sunderland had failed to raise any more than £4,000 by voluntary subscriptions, and when Rowland Burdon promised to supply all the remaining cost they committed the whole direction of the project to him.[17] He sought advice about the design from Robert Mylne, from the mathematician Charles Hutton, from Soane and possibly from Smeaton, and at some point John Nash (fig. 135) considered himself the architect and made a design. The use of iron was suggested by several different people and Paine's patent and his bridge at Paddington were clearly what provoked this. Burdon also consulted Messrs Walker at Rotherham. One of the inquiries to Smeaton concerned an arch of 400 ft span

135   John Nash. Wax bust, *c.* 1820.

proposed by 'Mr Nash',[18] but it was not named as the Sunderland project. Smeaton declined involvement. Late in 1791 Mylne sent to Burdon a list of the longest known bridges, including Schaffhausen and Nash's arch 'now building' at Newport, Monmouthshire, named as 285 ft in span.[19] At some time, then or later, he prophesied that Burdon's bridge at Sunderland would fall.[20] Hutton, in June 1792, reported on the question of whether to build one arch or two, on the use of timber or stone, and on the choice between three possible sites.[21] Surprisingly, he made no mention of iron though he favoured a single very large arch. We know more of the influence of Soane's advice than of its technical content, for in Burdon's very last letter to Soane more than forty years later he referred to his famous bridge and added, 'You, my friend! were the only person who whispered to me to "go on".'[22] According to Burdon's own account,[23] a stone arch of 200 ft span was rejected on the advice of some of these experts and he then turned to iron, examining the bridges of Coalbrookdale and of Paine, but 'he disapproved of their principles', adhering rather to 'the ancient construction of bridges, by the subdivision of the parts of the arch, in the manner of key-stones'. He took Thomas Wilson, who was probably a native of Sunderland, to Rotherham to oversee the making and erection of a single rib at Walkers' works and then contracted with them for the supply of the whole arch, six ribs spaced 5 ft apart, of 236 ft span and 34 ft rise. Wilson was put in charge of the work on the site, the foundation stone was laid on 24 September 1793 and the bridge completed in 1796 (figs 136 and 137).

Burdon's principle of using iron arch stones, or 'key-stones' as he called them, made the bridge as easily transportable as Paine's. The 'key-stones' were each a single casting 5 ft deep from top to bottom and 2 ft long, of three parallel bars in the curve of the arch connected by two in the radial direction. They were connected end-to-end by placing flat bars of wrought iron in grooves cast in both sides of each of the circumferential bars and bolting them together twice in the length of each 'key-stone' (fig. 138). The correspondence of the size of the grooves, 3 in. $\times \frac{3}{4}$ in., with the cross-section size of the main circumferential bars of Paine's arch ribs, seems to confirm the assertion in Rees's *Cyclopaedia* and several later accounts that 'the malleable iron' of Paine's bridge 'was afterwards worked up in the construction of the bridge at Sunderland';[24] and thereby to confirm also that the curved bars in Paine's bridge were of wrought iron (see above). The ribs at Sunderland were spaced apart by horizontal tubes at right angles to the ribs and with flanged ends bolted to the 'key-stones' (fig. 138). The spandrels were built up with cast iron rings and a timber floor laid over these to carry the gravel and stone of the road. There was thus almost nothing but a timber floor to resist lateral sway and with the considerable thrust in the arch tending to buckle the ribs they swayed sideways, presumably by bending at some of their many joints. A late report says that this happened 'soon after the centres were removed' to the extent of '12 or 18 inches', but that this was largely corrected by insertion of diagonal bracing and the use of wedges and ties.[25] Rees's *Cyclopaedia* in 1805 described this bracing as 'diagonal iron bars which are laid on the tops of the ribs, and extended to the abutments, to keep the ribs from twisting'. There was a report in 1806 that the bridge had failed,[26] probably referring to defects which had been repaired at a large expense in 1805.[27] But in 1812 Hutton could still write that the arch had 'settled several inches, as well as twisted from a straight direction, and the whole vibrating and shaking in a remarkable manner'.[28]

In spite of these troubles it was, in the words of a later master, Robert Stephenson, a bold experiment.[29] For a span of 236 ft and width of 32 ft, only 260 tons of iron were used, compared with 378 tons at Coalbrookdale to span 100 ft with a width of 24 ft. The ironwork was erected in the remarkably short period of ten days, on a scaffold rather than a centre (fig. 137), and thus solved the problem which had been a subject of lengthy discussion and a number of designs, namely how to erect the bridge without blocking the very busy traffic of keels and ships on the river.[30]

Contemporary with the start of the Sunderland project was a design by John Rennie. This was for export to the island of Nevis in the West Indies and so the need for transport in small pieces was real. The description of the design by Charles Hutton[31] says that each of the six ribs was to be 'of 3 rings, which were to be connected together by radii', which sounds a little like a description of Paine's construction (fig. 134). It

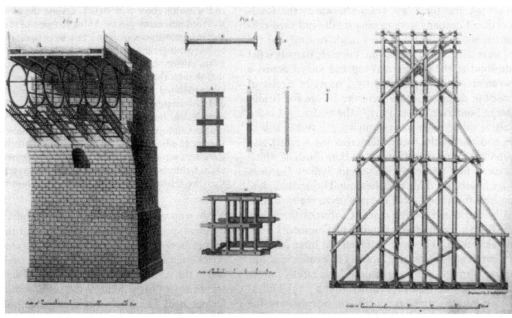

136   Abutment and scaffold for Sunderland Bridge (from *Edinburgh encyclopaedia*).

137   Scaffold and arch of Sunderland Bridge.

was to be 110 ft span and 13¾ ft rise, the spandrels filled with iron rings and the road to be laid on iron plates. This way of constructing the road, already used at Coalbrookdale, was to be adopted by almost all builders of iron bridges within the next few years, including Thomas Wilson, and it certainly helped to prevent lateral sway. In a second design for Nevis in 1794, Rennie changed the construction of the ribs so that the

radial bars were extended up from the top of the arch to the road and the span was reduced to 80 ft.[32] This general form corresponds to the short-span cast iron bridges which became a standard Telford design after 1811, when the initial fashion for rings in the spandrels had passed. Telford's small bridges were cast in only two pieces for each rib; whether Rennie, in his 1794 design, intended small or large castings is not known.

In 1795 and for a few years afterwards the Coalbrookdale Company were casting small- and medium-span arches with each half rib a single casting, but the ribs were different from Rennie's design, namely a flat plate solid at the bottom and top and joined across a tapered crescent-shaped void by a series of circles of increasing diameter. Such were the bridge for Bridgwater in Somerset, cast in 1795,[33] the bridge for Cound in Shropshire which was cast in 1797[34] and is still in use, and one of the two which cross the Kennet and Avon Canal in Sydney Gardens in Bath, both of which still carry foot traffic.[35] The second Sydney Gardens bridge has flat solid ribs and the Cound bridge has plate spandrels from the tops of the ribs up to the road line.

A bridge which was not cast at Coalbrookdale[36] was built for Sir Edward Winnington of Stanford Court, Worcestershire. This was designed by John Nash to span about 98 ft across the Teme. The profile was, like Burdon's bridge, a low segment of long radius but its construction was remarkably different (fig. 139). Unless the carrying structure was actually considered to be the X-braced handrails, it would seem that the arch ribs, apparently four in number, were only about a foot deep.[37] There is a graphic report of its collapse:

Sept. 26, 1795. This afternoon about four o'clock, the new iron bridge over the River Teme at Stanford, County of Worcester, suddenly gave way completely across the center of the arch, and the whole of this elegant structure was instantly immersed in the flood. In the fall the bars were all disjointed, and some of them that struck against the abutments were shivered into many pieces. At the moment of the crash which was instantaneous a man and boy were upon the bridge, the former with great presence of mind leaped into the river and swam safe on shore; and it is a circumstance truly surprising that though the boy went down with the fragments, he was also extricated unhurt. The bridge had been made passable, and only wanted the finishing of the side rails towards its completion, but no carriages had yet passed over it. The people employed had not left their work above an hour and were at an adjoining public house receiving their wages when the alarm was given . . . The misfortune is generally imputed to the lightness of the ironwork which was several tons lighter than the celebrated bridge at Colebrook-Dale.[38]

Nash, who was used to failures and not often depressed by them, registered a patent in 1797 for the construction of iron bridges, with the arches formed of hollow boxes open at the top and bolted together side by side,[39] and in the same year he built Sir Edward another bridge according to this patent on the same abutments as the former one. The boxes could be, and probably were, filled with earth or masonry, and the method provided real 'arch stones' of iron across the whole width as well as the span of the arch, thus avoiding all problems of lateral instability. This second bridge at Stanford was cast at Coalbrookdale and had spandrels filled with circles and a decorative handrail. It stood firm until 1905, but it appears that no more bridges were built in Britain under Nash's patent.[40]

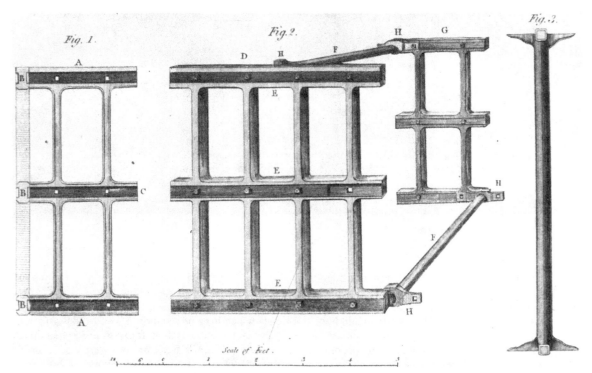

138   Framing of iron in Sunderland Bridge (from Patent Specification).

139   Bridge at Stanford which collapsed, 1795.

The floods in the Severn in February 1795 brought into the centre of iron bridge-building a civil engineer of the highest ability, namely Thomas Telford. In his hands the development took new directions and although his work overlaps some of the period which has been described already it is best left for later chapters. But almost exactly at the same time as Telford's earliest designs of iron bridges and aqueducts were made in the weeks following the great Severn flood, there was a proposal by the ingenious American Robert Fulton for a large iron aqueduct at Marple on the Peak Forest Canal;[41] and by a strange turn of the wheel Fulton was living in Paris and in regular contact with Tom Paine when Paine next took up his studies of iron bridges. Fulton made a drawing for at least one of Paine's inventions and must have discussed iron bridges with him,[42] but neither of them ever had a bridge built for practical use. Fulton visited Telford in

1795 and Telford considered that designs which Fulton published in a book in 1796 were copied from his own.[43] Paradoxically, Telford's interest in Paine was only in his political theories, at least so far as is known, and this landed him in serious trouble with his chief patron, Sir William Pulteney, just after the publication of *The rights of man*. Telford used Pulteney's postal frank to send a copy of the book to his home town of Langholm and the reading of it caused so much excitement that a few 'revolutionaries' were sent to gaol; Pulteney was very annoyed with his protégé.[44] Telford must have known all about Paine's bridge but he had not actually seen it[45] and in his article on bridges in 1812 he did not even mention it. He did criticise severely the design of the high abutments of the Coalbrookdale bridge and the use of the short 'key-stones' at Sunderland, which caused such serious instability.[46]

PART 3

The years of Rennie and Telford
1790–1835

12

RENNIE AND TELFORD – EARLY YEARS

I have hitherto made it a rule of my conduct to give such dimensions to works as to insure a certainty of success.

Letter of John Rennie to Samuel Gregson, 14 March 1795

Knowledge is my most ardent pursuit.

Letter of Thomas Telford to Andrew Little, 1 February 1786

If a decade were to be named in which major bridge-building slipped out of the hands of architects and into those of engineers, it would have to be 1790–1800. In those years John Rennie and Thomas Telford established themselves as the masters of bridge design, overtaking the older architects Mylne, Stevens, Harrison and Dance and younger aspirants like Soane and Nash.

Rennie (fig. 140) was the youngest of all these men.[1] He came of a prosperous farming family in East Lothian and was educated first at the parish school of East Linton and then at Dunbar High School, where in a public examination he showed 'amazing powers of genius' in mathematics and natural philosophy. In afternoons, vacations and at least a few days of truancy, he learned the skills of a millwright from a very famous craftsman, Andrew Meikle, whose workshop stood on the Rennies' land.

After leaving school he undertook contracts for mill work himself but in the winter months of 1780–3 he attended Edinburgh University to study natural philosophy under John Robison and chemistry under Joseph Black, as well as languages and music. Robison became his life-long friend and recommended him to James Watt in 1784 when Watt and his partner Boulton needed a man to design and erect the machinery for the Albion Mills in London. These being the largest flour mills in Britain and driven by the novel power of steam, Rennie was placed in the public eye and he used this limelight to establish a thriving business, first in mill work and very soon after in large canals and bridge design. By 1791, when he was thirty, he had designed his first iron bridge (see chapter 11) and by 1798 he had built the largest masonry aqueduct in the country, as related in chapter 10. He was master of all the principles of bridge design, and we shall trace the growth of these sure principles in various notes and reports he made between 1784 and 1799.

Robison must have instructed him in the equilibration of arches but Rennie looked beyond the theory for broad design rules. Soon after he left the university, on a journey into England, he wrote in his notebook:

Lowther Bridge.
It has often been my oppinione that the proportion of the load on the different partes of a circular arch were very unequal. I have had it fully verified for this bridge consists of two arches about 44 and 50 feet wide and about 16 feet high from the spring of the arch. One of these arches however has lost its

140　John Rennie. Bust by Chantrey.

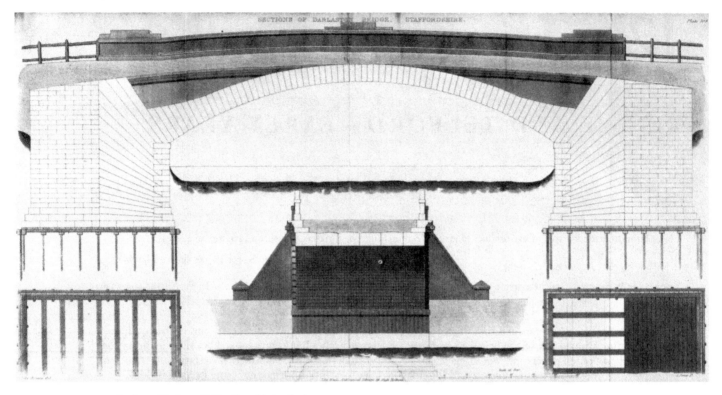

141   Sections of Darlaston Bridge (from Weale, 1843).

142   Detail at abutment of London Bridge during demolition in 1970. External voussoirs and spandrels seen at right, internal sloping courses at left.

form for it has falling [sic] in at the reins and raised at the top on the one and much the same on the other [side?] so that the arch instead of being a segment of circle has become a kind of parabola.[2]

One way of reducing the weight of spandrels over the reins, or haunches, was to make the arch elliptical, and the next day at Lancaster he saw Thomas Harrison's design for Skerton Bridge, of five elliptical arches. It became Rennie's habit to make his arches elliptical unless they could be very flat segments with correspondingly low spandrels. He also made the spandrels hollow. His notebook on the journey in 1784 includes full descriptions of the cofferdam and foundations of Skerton Bridge and of the Bridgewater Canal, including Barton Aqueduct, but his consuming interest at the time was in machinery of any and every kind.

In chapter 10 we have described his use of tie-bars in masonry, his choice of masonry finishes, his use of a steam engine for pumping from cofferdams, and the priority he gave to durability over appearance. He also gave special attention to the springings of arches and the mutual abutment of adjacent arches. In a report on the failure of the new Hutchesonstown Bridge in Glasgow in 1795, he wrote:

The failure of the bridge has principally arisen from the badness of the masonry, the weakness of the abutments and the improper mode in which the arch springs from them; in flat arches the springing should not be done abruptly, but ought to have commenced in a gradual manner from near the foundation.[3]

The manner of construction he favoured is illustrated by his design for Darlaston Bridge (fig. 141) and was seen during the demolition of his London Bridge in 1970 (fig. 142). Between adjacent arches of multi-arch bridges he always made an inverted arch of hewn stone to carry the thrust from one main arch to the next (see fig. 180). Fig. 143 shows two pencil sketches he made when he was designing Kelso Bridge in 1798, with rough lines of thrust drawn within the masonry of the main and inverted arches. These notes and sketches reveal an intuitive way of thinking about the necessary shape of arches, less precise but more realistic than calculations for equilibration.

Kelso was his first major road bridge and his initial report to the trustees[4] explained both his designs and his standards of construction. He advised strongly against reducing the cost by using rubble masonry or shallow foundations, and insisted that the foundations be laid by day labour, not by contract. The bedrock under the river was exposed in cofferdams and excavated to depths of at least 7 ft before the foundations of the piers were laid, this being 14 to 15 ft below low-water level. This represents a huge improvement on the foundations of Coldstream Bridge, built on a similar bottom about twelve miles downstream in 1763–6 (see chapter 8). For pumping from the cofferdams Rennie took power from a water wheel driven by the fall of a mill race on the south bank of the river,

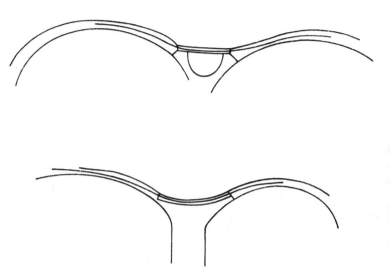

143   Freehand sketches of arches, inverted arches and lines of thrust (from Rennie's notebook, 1798).

presumably connecting the wheel to the pumps by a very long reciprocating rod, as he had done with the steam engine at the Lune Aqueduct (see chapter 10). For the form of the bridge he chose a level road and parapet with a high approach embankment at one end, because both of the sites considered had high ground at one end and a low river-bank at the other. He clearly would not consider an unsymmetrical bridge. The waterway of the old bridge, with six arches totalling 318 ft, had proved insufficient and he proposed to increase it by 42 ft if building five arches and by 30 ft if building three larger arches. The arches of both designs were elliptical, 'not only because those sort of arches give much more waterway but because such arches are better suited to the load they have to sustain'. Inside the spandrels were inverted arches and longitudinal voids above them. The cutwaters were semicircular. The external masonry was to be all ashlar with a wide cornice and pairs of three-quarter Doric columns engaged to the face of the spandrels (fig. 144). This was excellent architectural composition and Rennie never seems to have doubted his judgement in architecture in spite of his lack of training in it. While investigating the sites for the bridge at Kelso he had examined the Teviot bridge half a mile away (see chapter 9) and made scathing private notes. It was

but an unstable fabrick and is not likely to last for any length of time . . . The saliant angles are much too long and the bridge stands in a turn – the water running parallel to the roadway and at right angles to the piers – which piers are much too high and of course the whole bridge . . . it might have been at least six feet lower. The ornament on the sides is abominable the columns on the piers are Dorick – there is a kind of Dorick cornice with triglyphs but it scarcely has any proportions the rest of the cornice is simply little blocks but the top does not project 2 inches – in short it is the ugliest thing I ever saw.[5]

144   Kelso Bridge.

His prophecy was wrong, for the bridge still stands, but his aesthetic criticism was fair in the days of strict classicism. The confidence of his judgement was habitual. This successful man is described by his son as being 'naturally of a quick irritable disposition' which he suppressed so as to display only 'cool, steady and determined behaviour'; he had a Scotsman's auburn hair, and he stood 6 ft 4 in. tall.[6]

Thomas Telford (fig. 145) was also a Scot and a large man, but probably not as tall as Rennie. In business he had, according to one who knew him well, 'an apparent severity of manner',[7] but as a youth he was known as 'laughing Tam' and amongst his friends he was always a warm and often jovial companion. He was the only child of a shepherd who died when he was just a few months old, so he grew up really poor, but like Rennie he obtained a good elementary education at a parish school. Working in the vacations, not with a tradesman like Rennie but as a shepherd's boy, he nevertheless met the sons of influential families at the school. He was clearly gregarious and developed a deep affection for the friends of his youth as well as the hills and hamlets of his native Eskdale. He had a natural appetite for learning and especially for poetry but being poor he could not choose his career as Rennie did. Being bound apprentice to a mason he gave his days to learning the trade and in his spare time he added to his reading of literature the study of stone carving, drawing, architecture and eventually some sciences. He was busy with these studies for at least twenty years and it is scarcely surprising that he never married. It is a measure of his devotion to engineering that he never had a settled residence before he reached the age of sixty-four.

In 1780, at the age of twenty-three, he left his native district for the first time to work as a journeyman in Edinburgh – Rennie was then starting his attendance at the university there. Early in 1782 Telford rode to London on a borrowed horse and worked for two years on the building of Somerset House. He met Sir William Chambers and Robert Adam, but his first opportunity for architectural design was some alterations to Wester-hall, a house in his native valley. These he designed in London for William Pulteney, a brother of the owner. Pulteney had become the richest commoner in England by his marriage to the heiress of the Earl of Bath, and he was Member of Parliament for Shrewsbury. Telford, according to his letters home, roused some dissension by his methods or his views in each of his appointments but nobody criticised his work and by 1787 he was installed at Shrewsbury by Pulteney's powerful patronage with his great career about to begin. It was a late start, for he was already thirty, but he had acquired a deep personal knowledge of the craft of masons and carpenters which he afterwards considered to be the foundation of his success.

Within months of his arrival in Shrewsbury he began to act as surveyor to the county Justices, but he was also overseeing improvements to a small bridge for the town council[8] and making designs for rebuilding one side of the High Street. His practice as an architect included alterations to Shrewsbury Castle for Pulteney, changes in the plans for the new county gaol, and designs for two churches in the years 1792–6[9] and a third[10] a few years later. In the spirit of prison reform he proposed to the Justices in 1788 a scheme for the use of convict labour and when his first major bridge was built at Montford on the Severn in 1790–2 much of the work was done by convicts, apparently housed in a temporary prison at the site.[11] There were two contracts, one for supply of stone from Nesscliffe Hill quarry and another for the erection of the bridge. The erection was by well-known local builders, John Carline and John Tilley, of whom the former signed the contract but the latter could only make his mark. The cofferdams and piling were excluded from the contract and all the other work was contracted for at measured rates, not by lump sum. Telford was paid a pre-arranged fee of £200 for design and supervision and he imported an old friend from Eskdale, Matthew Davidson, to be his overseer. It was a sandstone bridge of three elliptical arches, a form he seldom if ever used again, and is still in use, though now widened by cantilevered concrete footways.

During the next five years he took responsibility for county bridge repairs and rebuildings, sometimes making the designs himself but on other occasions only writing the 'particulars', or specification of the materials and methods, after a design had been submitted by an 'undertaker'.[12] The enormous floods in the Severn and its tributaries in February 1795 suddenly changed the tempo of this work. Buildwas Bridge over the Severn, which he had just improved for the county, was destroyed along with at least eleven smaller county bridges. He was also asked immediately to design a repair for the borough bridge at Bridgnorth and a replacement for the ancient borough bridge of Bewdley in Worcestershire.[13] There had also been a partial collapse of an unfinished aqueduct over the Tern at Longdon, about seven miles from Shrewsbury on the

Shrewsbury Canal, and Telford was appointed engineer to the canal just after the floods in consequence of the death of its first engineer.[14] The chairman of the canal committee, a banker named Thomas Eyton, suggested that the replacement might be an iron structure[15] and by 18 March Telford had made a design. By 14 April he had designed a single-arch iron bridge for Buildwas and the Coalbrookdale Company had offered to build it. In July a contract was signed for a somewhat altered design.[16] In July also the Bewdley Bridge trustees invited the Company to tender for a single-arch bridge of iron but the Company declined. In August the trustees received a design and estimate for an iron bridge from John Nash but the collapse of his bridge at Stanford in September (see chapter 11) must have put an end to that proposal. In the following year they were pressed to try out the 'tie arch bridge' of iron (or partly of timber) patented by James Jordan of Shepton Mallet[17] and intended for long spans but not used until many years later, when the principle became known as the 'bow-string girder'.

Greater by far than any of these proposals was Telford's own design to carry the Ellesmere Canal across the valley of the Dee near Llangollen in an iron aqueduct of seven spans, later extended to nineteen, at a height of 120 ft. It is still uncertain whether this was designed before or after the other structures mentioned but one of Telford's biographers[18] argues cogently that it preceded all of them and that he probably conceived the project as early as January 1794. If so the three-span aqueduct at Longdon was a convenient trial of some of his ideas. He designed it in collaboration with William Reynolds, the manager of the Ketley ironworks, and Reynolds then made the castings.[19] Reynolds was associated with the Coalbrookdale Company, but Telford also met, on both the Shrewsbury and the Ellesmere Canal committees, the Company's chief rival John Wilkinson, whom he considered to be the 'king of the iron masters'.

The Longdon aqueduct (fig. 146) is a rectangular trough of cast iron plates, the bottom plates cast with ribs to enable them to span 8 ft from side to side, the sides of each span made up of plates cast to the shapes of voussoirs in a horizontal arch, and all the plates, both of sides and bottom, joined by bolts through flanges at their edges. Compressed between the flanges is a caulking material which made the joints perfectly water-tight and is said to have been 'Welsh flannel boiled in sugar'.[20] The supports between the spans are straight cast iron struts of cruciform cross-section, one vertical strut and two sloping ones under each side of the trough. The aqueduct was built in 1795–6 and established the feasibility of a water-holding cast iron trough. Between then and 1805 the astounding Pont Cyssylte (fig. 147) with its nineteen spans was erected over the Dee valley, its trough walls similar to those at Longdon but with four arch ribs underneath each span of 45 ft. The ribs were cast in three sections each and linked by cross frames at the joints, the sections having

145   Thomas Telford, painted by Raeburn.

flanged ends for bolting together. Each section is an open frame with a horizontal top member in contact with the bottom of the trough (fig. 148). The two outside ribs of each span are clad with further iron plates as the photograph shows (fig. 147), but these must have been added only for appearance or to protect the ribs from driving rain. The weight is carried by the ribs and the trough but what proportion is carried by each is quite indeterminate. It is clear that Telford made no complete calculations of the strength of this or the Longdon aqueduct; we can only be sure, from the voussoir shapes he used, that he expected both the arch ribs and the sides of the troughs to act mainly in compression, like stone arches. In March 1795 experiments on the strength of iron bars were made at the Ketley ironworks and in the following month 'ribs' were tested to failure at the Coalbrookdale works, two of the ribs being horizontal and two arched, rising 3 ft in a span of 29 ft 6 in.[21] These were almost certainly trials for some of Telford's works, but whether for one of the aqueducts or for Buildwas Bridge is not clear. The iron for Pont Cyssylte was eventually supplied and erected by another ironmaster, William Hazledine (fig. 161), from his foundry at Plas Kynaston very near the north end of the structure.

The huge stone piers of Pont Cyssylte were built solid up to a height of 70 ft but above that were hollow, thus reducing the load on the foundations and placing the inside of the walls open to inspection. This became a regular practice in Telford's high bridges and also those of others who followed him.

To construct water-tight aqueducts of iron was a tremendous achievement but Buildwas Bridge was a more direct forerunner of Telford's later iron arches of long span. His design (fig. 149) was markedly different from that of the first iron bridge at Coalbrookdale and also from that of Sunderland Bridge (chapter 11). The Coalbrookdale Company, although willing to build the bridge at Buildwas, objected to Telford's low-rise design[22] and the Justices submitted the difference of opinion to Reynolds and Wilkinson as arbiters. Both men favoured Telford's design and the Company then accepted it. Its low profile was similar to Sunderland Bridge but each of its ribs was cast in only three pieces, connected together by bolting through transverse plates like the Pont Cyssylte arch ribs. For the form of the ribs Telford borrowed consciously from timber framing, thinking mainly of Schaffhausen Bridge.[23] His bridge, being only 18 ft wide, had three 'bearing ribs' of 130 ft span and 17 ft rise. On the outside faces two 'suspender ribs' sprang from about 12 ft lower and, having a rise of 34 ft, they crossed the bearing ribs (as some of the struts in Schaffhausen Bridge crossed the line of the road, see fig. 36). The outer bearing ribs were connected to the

146 Longdon-on-Tern Aqueduct.

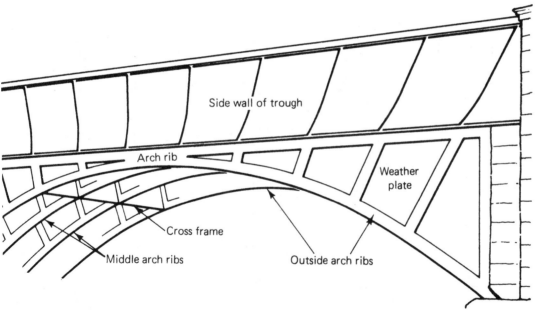

147   Pont Cyssylte, north end span.

suspender ribs by struts, ties and bracing, and the road was supported on flanged cast iron plates bolted to the bearing ribs. As the main arch strength was in the outer ribs the plates were really spanning 18 ft and it was therefore, like Schaffhausen, a system which could not be used for a much wider bridge. The bearing ribs were only 15 in. × 2½ in. in section and the suspender ribs 18 in. × 2⅝ in., the whole weight of iron being 173 tons. It

was all cast iron except for bolts, fixings and perhaps the guard-rails. It was Telford's intention that the large thrust of the low-rise arch should resist the pressure he feared might be exerted on the high east abutment by unstable ground, thus correcting a fault in the Coalbrookdale bridge (see chapter 11). He was aware, at least by 1812, of the criticism that 'by connecting ribs of different lengths and curvature, they are exposed to

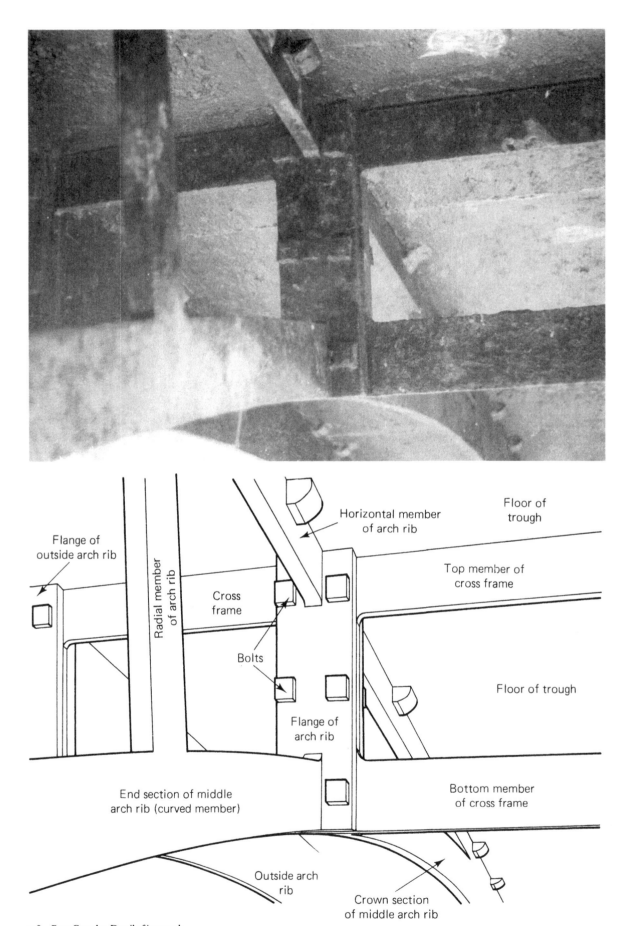

Flange of
outside arch rib

Radial member
of arch rib

Cross
frame

Horizontal member
of arch rib

Floor of
trough

Top member of
cross frame

Bolts

Floor of trough

Flange of
arch rib

End section of middle
arch rib (curved member)

Bottom member
of cross frame

Outside arch
rib

Crown section
of middle arch rib

148   Pont Cyssylte. Detail of ironwork.

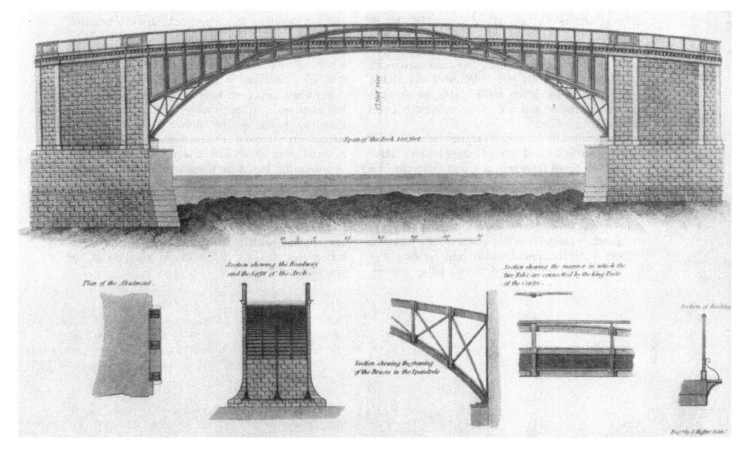

149   Buildwas Bridge (from *Edinburgh encyclopaedia*).

150   John Simpson. Memorial by
Chantrey in St Chad's church,
Shrewsbury.

151   Bewdley Bridge.

different degrees of expansion and contraction', but he had observed no trouble from this. Nevertheless it was a unique design which he never chose to repeat. The abutments, although never quite free from movement, have survived to this day, but the ironwork was cracked and deformed by heavy traffic loads in the late nineteenth century and had to be replaced by steel girders in 1905–6.

The Bewdley Bridge trustees argued for almost two years and finally decided to build a stone bridge of three arches.[24] Telford designed it, John Simpson, his favourite contractor (fig. 150), built it, and it stands unaltered today as one of the most elegant bridges in England. It has segmental arches, classical balustrades, elaborate but delicate cornices, shallow set-back spandrel panels, a careful choice of stone colour – green for the arches and the approach walls, pink for the rest of the bridge – and the road and parapet follow a gentle curve in elevation. It is a perfect bridge in a beautiful setting (fig. 151). The pier foundations were laid on rock 7 ft below the river-bed and 12 ft below low-water within double-walled cofferdams.[25] This, as with Rennie's foundations at Kelso a year or two later, was a depth which could not have been reached in a gravel bed some years earlier – though Telford was happy to found his bridge on the surface of the rock, while Rennie dug 7 ft into it. The readiness of bridge trustees to spend more money had certainly contributed to the improvement but it is possible also that pumping machinery had become more efficient. Within a few more years the general adoption of steam power in deep cofferdams was to improve the pumping even more; and as a consequence the founding of piers by floating caissons was virtually abandoned. It is worth remembering that caissons had been the cheapest of the common methods.

13

IRON AND TIMBER

something like a spider's web . . .

    Local inhabitant's description of Bonar Bridge (1819)

The 1790s saw huge investments in internal trade by the building of canals all over England; by comparison, the sums spent on river bridges were tiny. In London an equally large investment in external trade, by the building of docks, began in 1799; but the river bridges which were needed could not be correspondingly cheap. In the case of London Bridge this was because the rebuilding of the bridge was first proposed as a means of improving the port by admitting sea-going vessels of 200 ton burden to the reach between London and Blackfriars Bridges. The man who first pressed this idea was an engineer with strong leanings towards speculative business who had recently arrived in London from the north-east of England. His name was Ralph Dodd and in his pamphlet[1] published late in 1798 he proposed that a high viaduct should be built mostly on top of the old London Bridge, but with a number of the old arches in the middle of the river removed and replaced by a navigation arch of iron, 300 ft in span and rising 80 to 90 ft above the water. In April 1799 his scheme was taken up by a group of 'merchants, traders and wholesale dealers' of London and Southwark. They petitioned Parliament for a bill to improve London Bridge,[2] independently of the other major port improvements over which committees had argued without reaching any conclusions for three whole sessions. In 1799 a *select* committee was appointed – that is, a committee whose members were free of 'interest' in the matters of debate and could therefore take a judicial view and report it to the House. They were appointed to consider the port improvements, but the merchants' proposal for improving the bridge was also referred to them.

Ralph Dodd[3] had lived in or near Sunderland and watched the construction of Rowland Burdon's iron bridge in 1793–6 (see chapter 11); and Burdon was now a member of the committee considering Dodd's design. They did not take up the question of London Bridge in earnest until the following session (1799–1800), when they advertised for designs which would give 65 ft

clearance above high-water. This stipulation raised the necessary level of the road at each end, after allowing for a tolerable descent from the middle of the bridge, to the level of the roofs of three- or four-storey buildings on the banks, and thus required the approaches to be high, long and enormously expensive, both in the purchase of the ground and the construction of the ramps. The approaches would have cost much more than the bridge itself.

One matter which had ceased to be a serious obstacle to improvement was the waterworks driven by the fall under the old bridge. The company now used a steam engine for added power in pumping water up to the city and the select committee expected to satisfy their demand for compensation by providing the cost of a second engine.[4]

When designs had been submitted the committee gave preference, for the bridge itself, to a three-arch design in iron by Burdon's former assistant Thomas Wilson. He was referred to as the 'architect' of the Sunderland bridge, presumably because Burdon, as a member of the committee, could have no use for the title he had always assumed before. Wilson gave in a very low estimate of the cost of the ironwork of his bridge, £52,561, but no estimate at all for the piers, abutments or approaches. His largest arch was 240 ft span and the arches and spandrels very similar to those of Sunderland Bridge, but iron plates were to be used for the floor.

One designer, namely George Dance, the City's architect, proposed to provide the headroom stipulated by means of drawbridges and thus kept his bridge low, avoiding the high expensive approaches; but in order to have a road open at all times while allowing ships to pass intermittently he proposed to build two whole bridges in parallel with a large area of water between them where the ships could wait, having passed through one bridge, for the second drawbridge to open. High designs were submitted by Ralph Dodd (an ornate five-arch bridge of stone), Samuel Wyatt (a three-arch

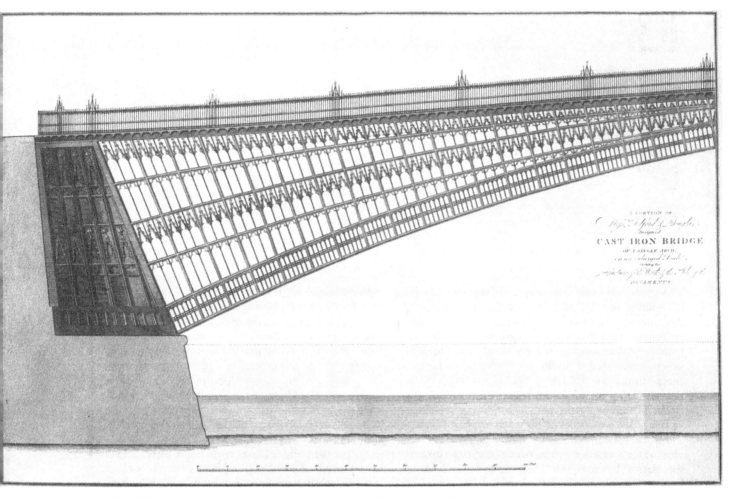

152   Design by Telford and Douglass for a new London Bridge (from *Commons reports*, 1800).

bridge of iron), and Thomas Telford and James Douglass (three designs of three or five iron arches). But a short report by William Jessop to Dance on the prospects for deepening the river above London Bridge also claimed the committee's attention. The channel between there and Blackfriars was typically 6 to 10 ft deep at low-water and Jessop suggested that this should be dredged to 13 ft in midstream. He then advised that, to keep the velocity of flow as high as it had been before and thus avoid silting, the river should be narrowed to 600 ft by embankments or wharves. In the sale of leases of the wharves and the warehouses which could be erected on them the committee saw a means of raising a very large sum (estimated at more than £900,000) towards the cost of the bridge, and they adopted Jessop's proposal with some enthusiasm. Then Telford or Douglass, or Telford and Douglass, had a late thought, that a river of only 600 ft could be spanned by a single arch of iron with 65 ft rise, and they submitted a famous design with those dimensions (fig. 152), springing from just above high-water, tapering from a width of 90 ft at the springings to 45 ft at midspan, and with

the approaches formed of ramps which were also carried on iron arches and extended along the banks over the wharves suggested by Jessop. They had estimated the costs of such approaches for their earlier designs, so when the committee was re-convened in the next session of Parliament they had only one thing to debate, namely the safety and practicality of the 600-ft arch.

They decided to take the opinions of a large number of well-known men and twenty-one questions[5] were sent to each of them with explanatory drawings of the framing of the ironwork (fig. 153). Telford provided these drawings and drafted the questions which he first intended to be in two separate groups, one for 'scientific men' and the other for 'practical men'.[6] In the event all the questions were sent to all the men consulted. Telford corresponded with several of them while they were considering their answers[7] and he was available in London to explain to them his model of the arch.

The committee grouped the replies into those of scientists, including three university professors, Charles Hutton of Woolwich, Neville Maskelyne the

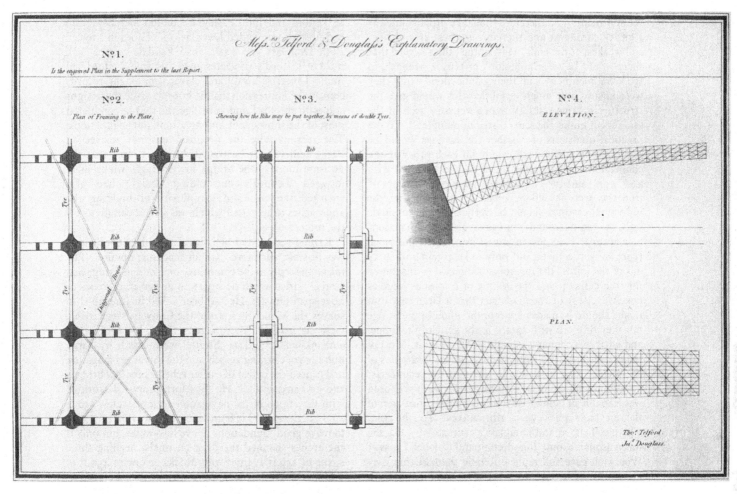

153 Details of Telford and Douglass design (from *Commons reports*, 1801).

Astronomer Royal, and George Atwood who was later to write a well-known treatise on arch theory; those of 'eminent engineers', including William Jessop, John Rennie and James Watt; and those of experts on iron including William Reynolds, John Wilkinson and Charles Bage of Shrewsbury. Messrs Walker of Rotherham were invited to give their views but they declined. Not a single architect was consulted.

The starting point for the theoretical questions was the equilibration theory. It is clear that all the scientists and engineers understood the principles of the theory and the general findings for common shapes of arch which have been given in chapter 5 and appendix 3. Telford was also *au fait* with it, despite a comment by his friend Brewster after his death that he 'had a singular distaste for mathematical studies, and never even made himself acquainted with the elements of geometry'.[8] The questions which he drafted asked how the loads over the arch should be distributed, what the thrust on the abutments would be, and whether the shape of the arch could be improved. Some of the scientists, notably Atwood, responded with tables of

equilibrium loads, but this was too facile and most of the scientists, as well as all the engineers, recognised that the fundamental question was whether the equilibration theory could be held to represent the behaviour of an iron arch at all. The assembly of iron castings, connected together by screws, dovetails, wedges, ties and braces, as shown in Telford's detail drawings (fig. 153), was very different from the arch of smooth stone voussoirs, each loaded vertically by the portion of spandrel over it, which was postulated in the theory (appendix 3). The first of Telford's twenty-one questions put it thus:

What parts of the arch are to be considered as wedges which act on each other by gravity and pressure [i.e. as voussoirs], and what part merely as weight . . . similar to the walls and other loadings commonly erected on the arches of stone bridges; or does the whole act as one frame of iron, which cannot be destroyed but by crushing of its parts?

An extreme answer to this came from Maskelyne, who considered the connections so strong that it would be 'one frame of iron', would exert no horizontal thrust on the abutments, and therefore should be proportioned as a girder, strongest and therefore thickest at midspan, and weakest at the ends. Most of the others agreed that there would be *some* 'cohesion', or tensile strength, in the iron members and their connections – but they were unsure how much – and that this would give the structure greater stability than a voussoir arch of the same form under the same distribution of load. To the practical questions of whether the castings would be accurate enough to bear evenly on each other at the connections, whether the joints should be filled with lead, iron cement or liquid iron, what type or types of iron the arch should be made of, and whether the individual castings should be vertical frames or single bars, the engineers and ironmasters answered in reasonable accord. They favoured cast iron, of the gunmetal type; an iron cement, but only to keep rain and spray out of the joints, the members being cast or machined for true contact; and the casting of frames as large as possible. Most of them advised that a large cast iron model should be made of part or the whole of the bridge but they differed with regard to the scale of the model and what 'experiments' should be made on it. Most of them also suggested strength tests on small samples of the metal or on model-scale ribs. To the second last question, whether an arch of the general form proposed was 'capable of being rendered a durable edifice', ten of the eleven replies were affirmatives. To the last, whether Telford and Douglass's estimate of £262,289 was reasonable, only John Rennie and Colonel Twiss of Woolwich gave full replies. Rennie thought that they had underestimated the cost of labour and materials in London and he also thought all the iron members proposed were too small in cross-section to give adequate strength. William Reynolds thought that the estimated price for supply and erection of the ironwork, at £20 per ton, was adequate, but John Wilkinson thought it much too low. Both these men, though reporting to the committee very briefly, made special efforts to help Telford. Reynolds made strength tests on small cubes of cast iron and Wilkinson offered to make and test a model at a scale of 1 to 16, and also members at full size, for the cost of the labour only, taking back the metal for re-use in his foundry.[9]

The select committee's report for the session 1800–1 contained nothing but a recital of the twenty-one questions and the replies of the sixteen respondents. It offered no new advice to the House about action or finance and although it was presented and printed well before the end of the session it was ordered to 'lie upon the table'[10] and was never mentioned again in subsequent sessions. A large view of the design was published in the summer of 1801 and attracted plenty of attention. In April 1802 Telford wrote to his Langholm correspondent that he had received notes, presumably of congratulation, from the King and three other members of the Royal Family.[11] In June 1802 James Douglass (alias Douglas alias Dalgleish) went abroad without a word to his associate, causing great annoyance

to Telford and John Pasley, a London merchant who had first recommended Douglass to Telford in 1797.[12] Both Douglass and Pasley had Eskdale origins, like most of Telford's associates up to this time. Telford had found Douglass useful and clever, but all the records convey the impression that he himself made the designs and estimates for London Bridge and that he conducted most of the subsequent investigations and negotiations. The reasons why the proposal was not revived in Parliament cannot be traced, but it is clear that the decision to make the bridge so very high, which itself inspired Telford's outstanding design, had also enlarged the cost and complexity of building the approaches to an extent which was all but certain to kill the project.

Rennie's expressed belief that an arch of 600 ft span was feasible was more than an academic opinion. The ink on his reply to the committee on London Bridge had scarcely dried when he began to design iron arches of huge span himself. He had been asked in June 1801 to survey the Menai Straits and the Conway River with a view to making roads across them by bridges or embankments.[13] The Straits were used by large numbers of coasting vessels and the threat to navigation had caused the defeat of earlier schemes for low bridges and embankments.[14] He therefore proposed a bridge with the crown of the navigation arch or arches 150 ft above high-water. He found it easy at two different sites to make good foundations above low-water, but only if the arches spanned the deep channels, needing three spans of 350 ft at the Swilly Rocks or one of 450 ft at Ynys-y-Moch. Because the centring must be such as not to obstruct the navigation, he preferred the Swilly Rocks site where it could be supported on rocks in the navigation channel. The centring at Ynys-y-Moch, where the channel was uniformly deep, would have been much more difficult. Of the practicality of a 450-ft arch, however, he had no doubt at all and this view was supported by William Jessop and Charles Hutton, to whom he sent his designs for comment. Both men suggested that he should reduce the thickness of the piers, designed 100 ft thick in rubble masonry, Hutton having calculated the lateral thrust and found that a hollow pier of 60 ft was more than enough on theoretical grounds. Rennie agreed to make them 70 ft. Hutton also thought the spandrels of the segmental arches too deep and suggested making the arch elliptical, but Rennie rejected this, arguing theoretically himself but without accurate figures and probably without calculation. The iron framing was intended to be similar to what Rennie had first proposed for a small bridge in 1791 and 1794 (see chapter 11) and not dissimilar to what Thomas Wilson built at Staines a year or two later (see below), but with the supporting members from the arch to the road straight and either radial (as in Telford's 600-ft arch) or vertical. There was no suggestion that these spandrel members would assist the arch in spanning between the piers; Rennie thus took an opposite view of the structural behaviour of an iron arch from that of

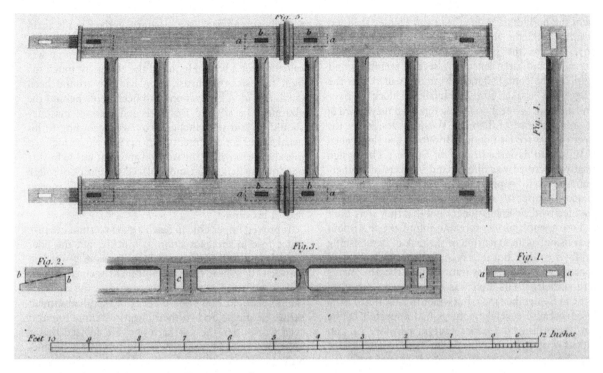

154 Arch blocks and cross frames in T. Wilson's patent, 1802.

Neville Maskelyne. His estimate of the cost of the ironwork, including erection but not centring, was £22 per ton. The scheme, however, went no further, as the Bill met opposition in Parliament and was abandoned.

Two large iron arches were designed by Thomas Wilson after he registered a patent for an improved method of connecting cast iron 'key-stones' of the kind used in Sunderland Bridge (see chapter 11 and fig. 138), but which he now called 'blocks'. Fig. 154 shows his new method as drawn in the patent specification and used in Staines Bridge.[15] Each block was a casting formed of two bars, 7 in. × 4½ in. in section, in the direction of the curve of the arch (though they are drawn straight in the figure) with four connecting radial bars of 3 in. × 4½ in. in section. A block was 4 ft 8 in. long on the extrados and each of the curved bars had a hollow in each end. A dowel piece (a) passed from each hollow to the matching hollow in the next block through the cross frame at c, and folding wedges (marked b) were driven into slots in the bars and dowel pieces to draw the blocks into a tight end-to-end bearing. This was certainly a firmer joint than those between the Sunderland 'key-stones', and the cross frames a great improvement on the tubular ties (fig. 138). There were also cast iron road plates to stiffen the bridge laterally but these were still supported from the arch ribs by rows of decreasing circular rings. The span at Staines was 180 ft and the rise 16 ft, and there were six ribs spaced about 5 ft 2 in. apart, giving a total width of 27 ft. There were thirty-nine blocks in each rib, which was better than the

125 at Sunderland, but still a very large number. The arch ribs built at Yarm on the Tees were of exactly the same span and components, but the spandrel members are shown as vertical bars in a landscape view. Wilson contracted for the ironwork at Yarm which was cast by Messrs Walker, he erected it in 1805, and it crashed into the river on 13 January 1806 before it had been opened to traffic. The failure was attributed to weakness of the abutments, which were only 10 ft thick in the direction of the thrust, and the arch was never rebuilt.[16] The old stone bridge was widened instead and has remained in use to this day.

The story of Staines Bridge takes longer to tell.[17] As early as 1791 an Act had been obtained to replace the old wooden bridge at Staines and after advertising for designs the commissioners adopted one for a three-arch bridge by Thomas Sandby, the professor of architecture at the Royal Academy. Sandby[18] was an accomplished draughtsman but his involvement in construction had been haphazard. He was nicknamed 'Tommy Sandbank' for the failure of the dam by which he had impounded Virginia Water in Windsor Great Park. Like others we have mentioned he had made drawings of a 'bridge of magnificence' to cross the Thames at Somerset House and used them to illustrate one of his annual lectures at the Academy. He was more than seventy years of age when the commissioners for Staines Bridge appointed him their architect. They submitted his design to Robert Mylne and one of the Wyatts, who approved of it; Stephen Townesend, the son of the builder of Maidenhead and Henley Bridges, became contractor and the bridge was completed in 1797. One of the piers began to settle very soon afterwards and after consulting several architects and engineers the commissioners were persuaded that the bridge must be taken down lest it fall and block the river to navigation. While it was being removed they heard of the initial choice of Thomas Wilson's design by the select committee on London Bridge and so they called on Wilson to design a bridge for Staines. The design already described was offered and Walkers of Rotherham agreed to supply and erect the ironwork for £4,900. It appears that the abutments were to be provided by the commissioners and that they were built by Townesend's former foreman John Cooper without supervision, as an extension of the task of demolishing Sandby's bridge; but it is recorded that Wilson 'gave repeated and positive assurances' that the abutments were sufficient. His arch was erected and opened to traffic in September 1803, but a month later it had to be closed because several fractures had appeared in the iron. According to Charles Hutton's history of iron bridges, written in 1812, the fractures were in several of the cross frames and many of the radial members of the blocks at the haunches.[19]

It was quickly discovered that the thrust of the arch had shifted the abutment at the north end horizontally by 3 in. It had moved bodily without cracking, the vertical piles under it bending in the 'peat, moor log and quicksand' which surrounded them for some feet below the masonry. The piles were forty in number, but only four deep in the direction of the thrust (see fig. 155), and when the gravel was removed to reduce the thrust they

actually sprang back, the masonry on their heads again moving bodily to recover some of its 3-in. displacement. The arch was then taken down and John Rennie was called on for advice. He found the abutment much too light to resist the thrust, even had the ground been good, and after excavations had been made behind the abutment he advised that a second mass of masonry should be laid with inclined courses, according to his usual advice for abutments (fig. 141), based on inclined piles between which the soft soil and peat was to be dug out down to firm strata and replaced with brick or rough stone laid in 'water cement', that is, a hydraulic mortar of some sort. Fig. 155 is his sketch of the method in the text of his report.

It proved impossible to reach a good stratum because water rose in the excavation so quickly, but the piles were driven well and a good bed of masonry was laid between their heads, so the abutment seemed secure. Wilson prepared to re-erect the arch. But the commissioners thought they should check the south abutment before he did so and to their chagrin Rennie found it even worse than the northern one. He advised that it must be removed and a new one built in a double-walled cofferdam, which was likely to cost at least £2,000. The commissioners had still not paid Messrs Walker for the ironwork and although they tried to borrow more money and advertised a contract for the cofferdam, nothing was done for nine months. Then on 1 April 1806 they met with the knowledge that the bridge at Yarm had fallen into the river, another iron bridge was reported to have collapsed in Bristol, and Sunderland Bridge was said also to have 'failed'; and they decided formally 'to relinquish altogether the project of a bridge totally of iron'. Their old timber bridge, closed to traffic in 1797 and again in 1803, had both times been re-opened but was now 'in a great state of decay', so they contracted with two local carpenters for £5,215 to erect a new but similar bridge of seven spans with timber beams set to the profile of the road of the iron bridge, and on the beams lay Wilson's iron road plates to carry the road. His guard-rails were also re-erected (fig. 156). It was the ultimate indignity for an iron bridge and what became of the arch and spandrel castings is not recorded.

William Jessop was engineer of the bridge which had collapsed at Bristol, one of two which were built about 1805 with spans of 100 ft and rise of either 15 ft[20] or 12 ft 6 in.[21] (fig. 157). The arches were of similar framework to those at Staines, each rib composed of two curved bars linked by radial bars spaced 1 ft 3 in. apart. The curved bars were 8 in. × 2 in. in cross-section and the radials 3 in. × 2 in., all cast together as a perforated plate, the overall depth of the rib being only 2 ft 4 in. However, each rib was formed of only two large castings, each one more than 50 ft long. These were the longest castings used since the first iron bridge and almost certainly the heaviest yet used in any bridge. They were cast at Coalbrookdale. The cast iron road plates were supported on bearers carried on vertical

155   Sketch of proposed repair to north abutment of Staines Bridge by J. Rennie, 1804.

156 Staines Bridge. Engraving by Cooke, 1814.

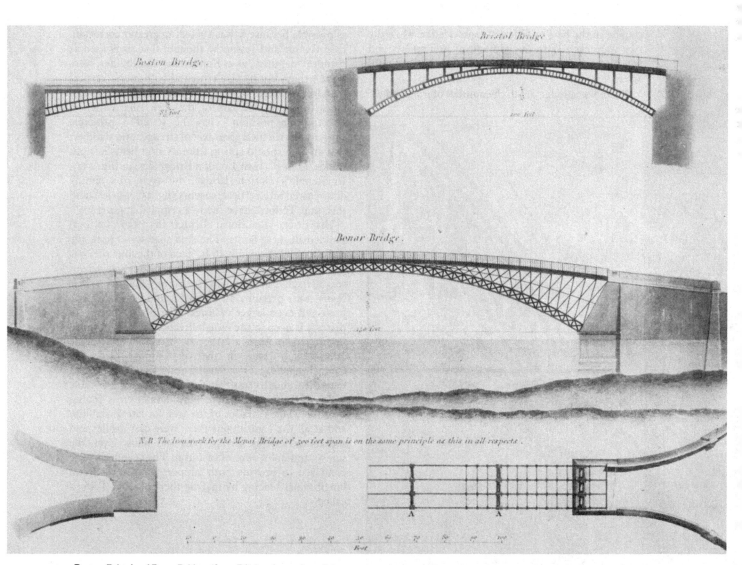

157 Boston, Bristol and Bonar Bridges (from *Edinburgh encyclopaedia*).

posts rising from the arch ribs. Both the posts and the bearers were T-shaped in cross-section, which was a new form.

Rennie's first large iron bridge to be built was of similar appearance but of smaller castings. He had made four designs, two of iron and two of stone, for a bridge over the Witham in the town of Boston in 1800.[22] To avoid the need for high approaches from the existing streets, he kept the arch very low in all the designs. The iron arch eventually built[23] rose only 5 ft 6 in. in a span of 85 ft, and this kept the whole structure above high-water. It was 36 ft wide, of eight ribs 3 ft deep and each made up of eleven castings, each casting about 8 ft long. The spandrel members were vertical like those at Bristol. At the joints of the ribs there were cross frames or 'gratings', probably as deep as the ribs, like those of Pont Cyssylte (fig. 148), and therefore unlike the narrow cross frames at Staines (fig. 154). A number of fractures occurred in the ironwork as soon as it was erected and loaded, most if not all of them in the radial bars of the rib castings. These bars were 4 in. × 3 in. in section, while the curved bars were 7 in. × 4½ in., almost exactly the same as those of the rib castings at Staines, and both Telford in 1812 and Robert Stephenson many years later thought the fractures were mainly due to residual stresses caused by unequal rates of cooling in the bars of different cross-section when the ribs were cast. With the addition of iron splices and fish-plates over the fractures, the bridge survived until 1912.

There was a reduction in the number of iron bridges

built for a few years after these troubles[24] and the attitude of the Staines Bridge commissioners in 1806, as quoted above, is probably typical of what bridge owners were thinking. Of the leading designers Telford had the cleanest record of success and it was he who designed the next large iron arches. He had been appointed engineer to the Commissioners for Roads and Bridges in the Highlands of Scotland in 1803 and designed many stone bridges for them in the next five years. The expected road traffic was never large and river traffic non-existent, so there were few sites which demanded spans of unusual length. But at the narrowest part of the Dornoch Firth he was bridging the flow from a large tidal basin and to give it free passage he decided on one or two iron arches of 150 ft span.[25] Only one iron arch was actually built, but its design embodied both the form and the methods of construction which he used in large iron arches to the end of his life. As it happens, he wrote his article on bridges for the *Edinburgh encyclo-paedia* just two or three years later and his description of the design can be read there together with his comments on most of the other bridges we have described, particularly those at Buildwas, Sunderland, Bristol and Boston.[26] He was critical of arch ribs with many joints and of ribs with bars of different thickness in a single casting. He believed in excluding wrought iron as much as possible because it was subject to greater corrosion. Like Rennie and Jessop he thought that rings used as spandrel members, as at Sunderland and Staines, were quite wrong, but he also criticised both men's vertical spandrel members because they were not made to line up with the radial bars of the arch ribs, either in direction or position. Here he stated a principle concerned with the *appearance* of strength and stability; that an arch should spring from its abutments at right angles, as in his own London Bridge design (fig. 152). In Rennie's Boston Bridge the arch ribs simply disappeared into vertical masonry (fig. 157, top left) and this, said Telford, gave them 'a crippled appearance'.

His design for Bonar Bridge (fig. 157, bottom) avoided all these faults. The arch span was 150 ft, the rise 20 ft and the width 16 ft. Each of the four ribs was cast in five pieces 'for the conveniency of distant sea-carriage' from William Hazledine's foundry at Plas Kynaston – presumably Telford would have otherwise preferred even larger castings. The rib castings were plates 3 ft deep in the radial direction and 2½ in. thick, perforated to form a lattice of bars all of the same cross-section (seen in the very similar bridge at Craigellachie, fig. 158), flanged at their ends and bolted together through cross plates which extended the whole width of the bridge. There were thinner cross frames over the whole extrados of the arch for lateral stability and at all the connections there were cast joggles and sockets as well as screws, pins and bolts. Even the screws and bolts were of cast iron. The spandrels were contrived to provide both support for the road and longitudinal bracing by making them of what Telford called

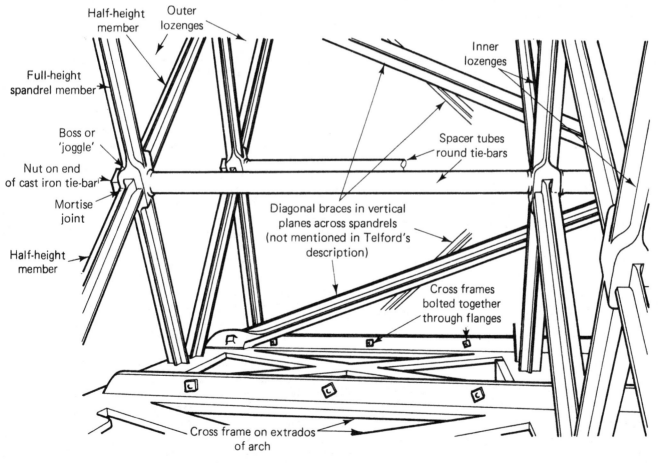

Half-height member

Outer lozenges

Inner lozenges

Full-height spandrel member

Spacer tubes round tie-bars

Boss or 'joggle'

Nut on end of cast iron tie-bar

Mortise joint

Diagonal braces in vertical planes across spandrels (not mentioned in Telford's description)

Half-height member

Cross frames bolted together through flanges

Cross frame on extrados of arch

159 Original spandrel construction of Craigellachie Bridge. Photo 1963.

lozenge, or rather triangular forms . . . each cast in one frame, with a joggle at its upper and lower extremities, which pass into the sockets formed on the tops of the ribs, and in the bearers of the roadway. Where the lozenges meet in the middle of their height, each has a square notch to receive a cast-iron tie, which passes from each side, and meets in the middle of the breadth of the arch, where they are secured by forelocks . . . By means of these lozenge or triangular forms, the points of pressure are preserved in the direction of the radius.

The individual bars were of cruciform cross-section (see fig. 159). The road plates were composed of a reticulated system of ribs with holes between them to obtain strength with minimum weight. Because of the form of the spandrel members there had to be a bearer under the road plates along the line of each arch rib and spandrel. Telford expressed the opinion that 'the disposition of the ironwork, especially in the spandrels, also greatly improves the general appearance'. The total weight of ironwork was 180 tons. The spandrel members in particular were very light, but they seem to have carried the direct loads from the road perfectly well. Examination of the sister arch at Craigellachie in 1963–4 (fig. 158)[27] showed that each vertically elongated X was composed of one member the full height of the spandrel and two members of half that height which were mortised (or 'joggled') into the rectangular boss (presumably what Telford meant by the 'square notch') half way up the long member. There was also a hole through the boss for the horizontal tie-bar. By 1963 the mortise joints, both at the crossings of the X's and between the spandrel bars and the road bearers, had worked loose and the whole structure above the arch was removed and rebuilt in welded steelwork. Fig. 160 shows the old road plates after the gravel had been removed.

160   Roadplates of Craigellachie Bridge with gravel removed, 1963.

161   William Hazledine. Bust by Chantrey in St Chad's church, Shrewsbury.

Telford persuaded John Simpson and his partner Cargill to contract for the masonry of Bonar Bridge and Hazledine (fig. 161) to supply and erect the iron, although he found the former were 'miserable about undertaking Bonar Bridge at a closed sum' because of the remoteness and difficulty of the site. But although unexpected ground conditions caused changes in the design after the bridge was started,[28] and even necessitated the use of caissons for some of the foundations – a rare and possibly unique event in Telford's practice – the arch was completed in 1812 and soon proved the strength of its framing. Early in 1814 a great mass of ice came downstream carrying many logs of felled timber. 'Those logs which were in an upright position struck the iron arch with such violence that the crash of the timber was heard at considerable distance; but the bridge stood firm without suffering either crack or flaw.'[29]

The arch at Craigellachie on the Spey, of identical span, rise and total weight though with members of different sizes, was already under construction and it still stands, though now relieved of vehicular traffic. So do two arches of 170 ft and 150 ft span built over the Severn with very similar frameworks and one over the canal at Galton in Birmingham of 150 ft. The 170-ft arch, built near Tewkesbury in 1823–6, and the Galton bridge (1829) have been changed very little and now carry only restricted traffic; the 150-ft arch at Holt Fleet, Ombersley, built in 1827, was greatly strengthened with steel and concrete in 1928.[30] Another bridge with iron arches of similar framing, two of 105 ft span and one of 150 ft, was designed by Telford and built over the Esk near Carlisle in 1820; it was demolished in 1916,[31] but the site is still known as 'Metalbridge'. Waterloo Bridge at Bettws-y-coed on the Holyhead road was built in 1815; it was of the same structural form but with marvellous decorative castings on the façades to celebrate the victory over Napoleon in that year (fig. 162). The decoration is unchanged but the inner ribs were encased in reinforced concrete and a concrete road slab added in 1923.[32] Most if not all of these arches were supplied by William Hazledine and erected by his foreman William Stuttle.[33] Some wrought iron bolts and other fixings were used in the later bridges but all the large parts were of cast iron. In at least two of the later designs, namely Tewkesbury and Galton, the spandrels were made to meet vertical walls over the abutments by a satisfactory alignment of the lozenge members, but the arch in each case sprang from small skewbacks radial to the arch curve.[34]

Of all this work only the design of Bonar Bridge preceded the writing of the article for the *Edinburgh encyclopaedia* in 1812, but Telford also reported in that article two designs made for much larger spans. He had been appointed in 1810 to survey the road to Holyhead and propose improvements and had already reconsidered the two designs made by Rennie for iron

bridges across the Menai Straits. He made new designs for both sites[35] using framing of exactly the type and components proposed for Bonar Bridge. At the Swilly Rocks site he proposed to place a shorter arch of stone between each two of the three iron arches. The stone arch, Telford wrote, was 'to add weight and stability to the piers'. It was to be 100 ft in span and, most surprisingly, to have iron spandrels and road plates over it. For the narrower crossing at Ynys-y-Moch he proposed a single iron arch of 500 ft span and a newly invented method of erecting centring for it without any supports in the channel. He would use a centre in the form of an arch of braced timber frames as shown in fig. 163 and it would be erected from the abutments outwards. The first frame at each end, of four ribs 50 ft long and 25 ft deep, would be built *in situ*, supported on the masonry offsets at the base of the abutment and tied by iron bars into the masonry at the top. A floor would be laid on top of this frame and the next 50-ft frame assembled upside down on it. This would be fixed to the first frame by iron hinges at the outer end and swung over as shown in the lower right-hand diagram to take its position. This would be repeated until a complete arch of timber frames spanned the

162   Half arch of Waterloo Bridge, Bettws-y-coed.

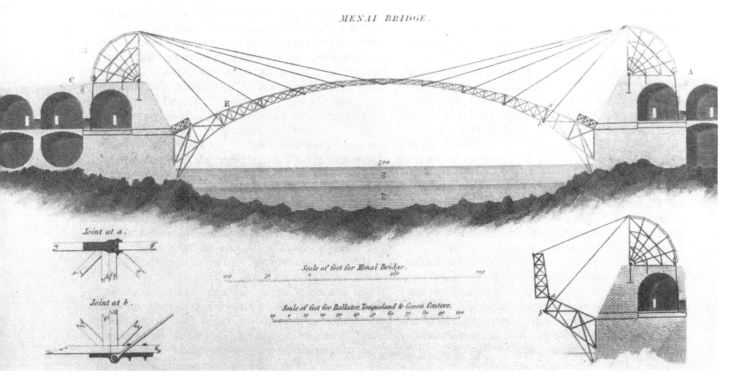

MENAI BRIDGE.

Joint at *a*.

Joint at *b*.

Scale of feet for Menai Bridge.

Scale of feet for Ballater, Tongueland & Conon Centers.

163   Method of erecting centring for an arch over the Menai Straits (from *Edinburgh encyclopaedia*).

Straits, but as the frames were constructed each was to be also tied by an iron 'chain-bar', as shown in the lower left-hand diagram, to the top of a quadrant framework 50 ft high erected on top of the abutment. The chain-bars were to pass over 'strong blocks or rollers, and to be acted on by windlasses or other powers'. The iron arch, similar to that of Bonar Bridge but with nine ribs each of twenty-three castings, would be erected on this centring and the centring released by slacking the chain-bars and also slacking wedges under the feet of the first timber frames at the abutments. The weight of the arch and bracing, without the spandrels, was calculated to be less than that of the middle arch of Blackfriars Bridge and the weight of the centring was only one-twelfth of the known tensile strength of the chain-bars, this rough check being enough to satisfy Telford that the scheme was feasible.

But this was his last proposal for a record-breaking iron arch. The project was shelved for another seven years and by then he was ready to offer a design for a suspension bridge.

Rennie designed the largest cast iron arch to be built in Britain, namely the middle arch of his Southwark Bridge in London. It was one of three toll bridges built by joint-stock companies over the Thames within the short period of ten years and the distance of three miles. None of them brought any profit to the investors and the Southwark Bridge Company failed to pay the builders or the engineer in full. Two of the three

companies, those of Vauxhall and Waterloo Bridges, were inspired by the speculative engineers Ralph and George Dodd, but both appointed Rennie their engineer when the time for construction came.[36] His design for Vauxhall Bridge, of seven stone arches, proved too expensive and work was stopped in 1811. It was later re-started, to a design of nine iron arches of 78 ft span by James Walker, and completed in 1816. Waterloo Bridge, being a stone bridge, will be described in chapter 14. Southwark Bridge was the third to begin and Rennie was its engineer from the outset.[37] The site was downstream from Blackfriars, where there was a prospect of navigation by sea-going vessels if London Bridge should be rebuilt; it was also the narrowest and deepest stretch of the river within the city, 30 to 36 ft deep at high-water, making foundations in the channel very difficult. Rennie therefore designed long arches, two of 210 ft and one of 240 ft span, rising 19 ft and 24 ft respectively and springing from above high-water. The two piers were founded in cofferdams similar to those already used by Rennie at Waterloo Bridge (see chapter 14) and because they were isolated monoliths without the lateral support of stone arches he built them of very large granite blocks, some of them set as vertical bond-stones between the courses. There were similar bond-stones in the abutments. The weight of the stone blocks was required to be 15 to 20 tons each, a size almost if not entirely unprecedented in London masonry. Rennie's son John was sent to Aberdeen where he found such blocks unobtainable, but he went on to Peterhead and directed personally the extraction of a block of 25 ton weight from a quarry, transport of fourteen miles to the harbour and loading on a brig by improvised equipment including piles specially driven on the seaward side of the vessel and a platform laid right across its deck. It was clearly expensive in equipment, horses and labour, the workmen being encouraged with 'good wages and whisky', but the block arrived safely in London and further supplies were then forthcoming. It was said to be 5 ft square and 10 ft long, though a block of the stated weight of 25 tons would be considerably larger. For pumping from the cofferdams Rennie erected a single steam engine on one bank of the river, as he had done for the Lune Aqueduct (chapter 10). The transmission from these engines to the pumps in the cofferdams out in the rivers must have been by long reciprocating spars of timber such as were used much later for pumping from the Kilsby tunnel works on the London and Birmingham Railway (fig. 164; see also fig. 179).

For the ironwork Rennie made two designs,[38] one with the arch ribs perforated as in Telford's Bonar Bridge (fig. 157) and vertical spandrel members, the other with solid ribs and 'lozenge' spandrels. The second was adopted with the arch ribs 6 ft deep at the crown and 8 ft at the springings, $2\frac{1}{2}$ in. thick in general

but slightly thickened at the top and bottom edges and with flanges on the ends of each casting. There were thirteen castings, about 20 ft long, in each rib and cross frames at the joints between them. The joints were made 'by dove-tailed sockets and long cast-iron wedges, so that bolts for holding the several pieces together are unnecessary, although they were used during the construction of the bridge to keep the pieces in their places until the wedges had been driven'.[39] The arches were assembled on centring from the crown outwards to the springings and finally eased off the centres by driving iron wedges, 9 ft long and very carefully machined, between each rib and a springing plate set in the face of the abutment or pier. These details show an intention on the part of the designer that the arch castings should act like stone voussoirs, free to bear on each other at any part of the contact surfaces in accordance with the position of the line of thrust. This flexibility might avoid damage of the kind experienced in earlier bridges (especially Boston) due to stresses caused by restraint; and Rennie was aware that the ordinary seasonal changes of temperature would cause the line of thrust to move up and down to a considerable extent in low arch ribs which were fixed between rigid abutments. Rennie, or perhaps his son George (fig. 165), made some calculations for the stability of the arches and the effects of temperature, and these were

164   Drive rod from steam engine (in distance) to reciprocating pumps in shaft of Kilsby Tunnel on London and Birmingham Railway (from J. C. Bourne, 1839).

165   George Rennie, painted by J. Linnell, 1824.

checked by Dr Thomas Young[40] who was probably the ablest physical scientist of the day. The calculations which Young published showed that the line of thrust would not deviate more than 2 in. from the centre-line of any part of the ribs when the structure was carrying its own weight and that of the road. These loads would cause a uniform stress on the cross-section of the arch ribs of 2 tons per square inch; and the maximum increase and decrease of stress (at bottom and top of the ribs at the crown) caused by a change of temperature of 32° Fahrenheit would be 3 tons per square inch, giving maximum total stresses of 5 tons per square inch compression and 1 ton per square inch tension. Young revealed that Rennie had proposed to shape the end surfaces of the castings next to the abutments to a very slight curve so as to allow the arches to change their angle slightly without developing temperature stresses there, but this was not carried out. Temperature stresses would still have occurred at the crown, but whether Rennie and Young calculated them is not clear. Young concluded that the arch ribs were much deeper than was necessary; and the iron framework was certainly much heavier and stronger than in earlier bridges. The weight of iron in the arch ribs and spandrels of the middle arch, which was 44 ft wide, was almost 1,300 tons, compared with 260 tons in Sunderland Bridge, of similar span and 32 ft wide.

Measurements of the rise and fall of the arch crowns were made as the bridge was built, as a check on the predictions, and satisfactory agreement was obtained.

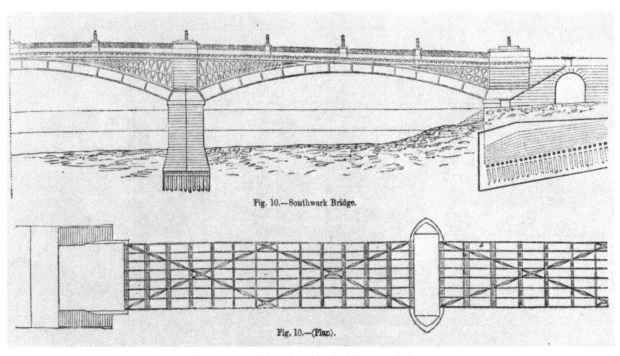

166   Plan and elevation of Southwark Bridge ironwork (from *Encyclopaedia Britannica*, 1856).

At one stage the spandrel members were found to be causing spalling of stones in the piers when the arch rose, but this was corrected by slacking the wedges between the spandrels and piers to allow some relative movement. The rise or fall of the crown of an arch was about $\frac{1}{4}$ in. for 10°F change of temperature and the total movement probably never exceeded 1 in. After he built a masonry bridge of three low-rise arches at Staines some years later, George Rennie extended his studies of temperature movement to stone arches and again obtained satisfactory agreement between predictions and measured movements.[41]

There appear to have been some plates or frames laid on the extrados of the Southwark Bridge arches as in Telford's arches (see fig. 159) but details are lacking. There were also diagonal braces extending in four lines across each arch (fig. 166), each brace being of cruciform cross-section and forming the diagonal of the rectangle enclosed by two rib castings and two cross frames. Similar braces are still to be seen in High Bridge, an arch of 140 ft span and 14 ft rise over the Trent at Handsacre in Staffordshire designed by Joseph Potter, the county surveyor, and built in 1830[42] (figs 167 and 168). The three-arch bridge at Chetwynd in the same county is of similar construction. Rennie was well known to Potter and concerned for many years with bridges in the county so a coincidence of details between Southwark and High Bridge is not surprising. In Potter's arch there were seven castings, each about 20 ft long, to each rib, and the spandrel work (fig. 168) was also very similar to Southwark Bridge. The 'lozenge' shapes in both were obtained by forming triangular holes in cast plates of considerable size, and so were much heavier and more rigid than Telford's lozenges. The plates were fitted to the tops of the arch ribs by sockets and joggles at Southwark, but by flanges and bolts at High Bridge (fig. 168). In High Bridge the plates are stayed at half their height by tubular ties, but at Southwark they were braced by cross frames in the spandrels. On top of the spandrels in both bridges were iron road plates.

Southwark Bridge stood until 1913. The arches were still acting like voussoir arches, the piers having tilted slightly towards the banks so that the crown of the middle arch settled and the joints between castings opened by about $\frac{1}{4}$ in. at the intrados.[43] But the extra strength and safety in Rennie's heavy design gave little comfort to the Southwark Bridge Company or to Walkers of Rotherham who supplied and erected the ironwork. The cost to the company was £666,000 instead of the £287,000 first estimated, and Walkers had been induced to take many shares in the enterprise, shares which were worthless by the time the bridge was opened. Their foundry had been strained to the utmost to produce the large castings required, once having to charge and tap four furnaces all at once to cast a single plate weighing 19 tons.[44] Their loss on the contract led almost directly to bankruptcy and the closure of their foundry in 1829.[45]

167   Underside of High Bridge at Handsacre.

The drawings and models of long, low timber arches, which have been described in chapter 3 and provided some of the ideas for the early iron bridge designs, also led to the construction of timber bridges of long span at the end of the eighteenth century and later. Three of the earliest were designed by James Burn, an 'architect' who acted also as a contractor. His home town was Haddington in East Lothian but most of his known work was within reach of Aberdeen where he practised from about 1797 to 1809. The first of his timber bridges[46] was of three arches, one 58 ft span and two 48 ft, with rise about 10 ft. It was built in the grounds of Brechin Castle in 1797. The second was a single arch of 109 ft span and 13 ft rise crossing the Don at Dyce near Aberdeen and built in 1803. The third was a two-arch

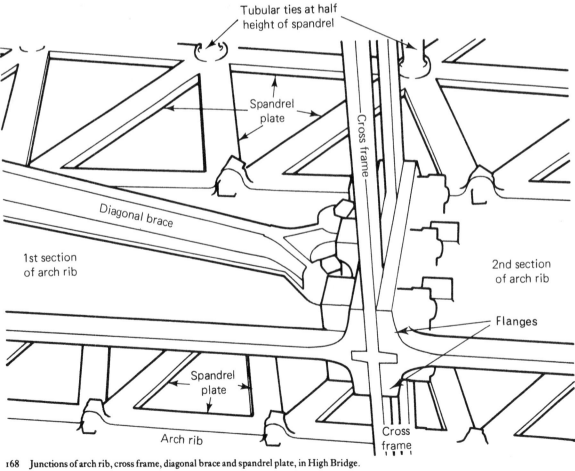

168   Junctions of arch rib, cross frame, diagonal brace and spandrel plate, in High Bridge.

169   Piers of Brechin Castle bridge.

bridge with spans of 71 ft and 10 ft rise, built for access to the flax mills at Grandholm, also over the Don, at about the same date. The Brechin Castle bridge was a grand affair on stone piers with life-size sculptures (fig. 169), the sides of the spandrels boarded and painted in imitation of stone. The sockets in the masonry for the butts of the arch ribs can still be seen but the arches have been replaced by steel girders. The Dyce bridge gave good service but became rickety by 1851 and was

replaced.[47] The Grandholm bridge, being mainly used for foot traffic, lasted until 1911 or later.[48]

Burn's method of building was to prefabricate rectangular frames from short lengths of timber, the thickest members of the frames being in the direction of the arch with thinner members lying across it and diagonally across the frames (as in the later design shown in fig. 170). The frames were laid flat to the profile of the arch, presumably supported on scaffold-

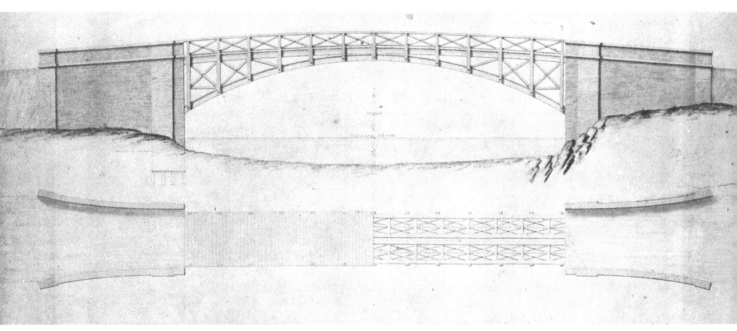

170   Design for Laggan Kirk Bridge by J. Mitchell, 1827.

ing or centring, and connected together. A manuscript design for the Brechin Castle bridge[49] shows three sets of frames in the width of the bridge, apparently bolted together through the rib members at every 5 ft of their length; each of the six rib members is drawn 1 ft 6 in. deep and 6 in. wide. On the arches were erected vertical posts with diagonal bracing, and bearers and planking to carry the road. There is no record of how the ends of the frames were connected together, but it is likely that the end members were bevelled to a plane radial to the arch curve and then bolted together. The Dyce bridge at least was built entirely of native timber from Braemar, which was quite unusual.

Telford summed up the advantages of Burn's designs in his 1812 article. 'Short pieces of timber may be employed. The principal pieces abutting endwise, little change can take place from shrinkage. The principle will admit of carrying an arch to a very great extent; and by judicious arrangement, the parts may be taken out and renewed separately.' Several long timber spans were built in the later years of Telford's Scottish road work. In 1827 Joseph Mitchell made a design at the office of the Commissioners for Roads and Bridges in Inverness[50] for a single-arch bridge (fig. 170) across the Spey at Laggan Kirk in which the framing was an exact copy of that of Burn's bridge at Dyce. There were four arch rib members each 1 ft 6 in. × 11 in., the span was 100 ft and the rise 13 ft 6 in. This design was altered to a truss-like form,[51] presumably at Telford's wish, and a similar design was used for the five-span bridge at Ballater. But in 1833 a timber bridge with curved arch ribs was built across the Findhorn at Freeburn, under Mitchell's supervision and probably to his design.[52]

There were two spans of 95 ft and the framing appears to have differed from that used by James Burn, although full details are not shown in the published engraving.

A quite different wooden arch had been erected to span the Mersey at Warrington in 1813-14.[53] It was designed by Thomas Harrison who, as noted in chapter 9, was the son of a carpenter. It had a span of 140 ft and only 6 ft 6 in. rise. There were only two main ribs, each made up with five square balks to be 5 ft deep at midspan but with extra tapered pieces inserted to increase the depth to 7 ft at the springings; the balks 'fixed together with long spikes or screws' and 'bound at

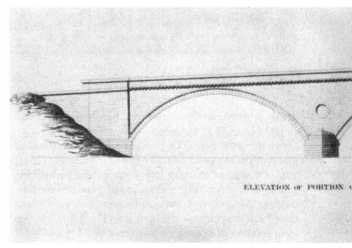

171   Elevation of Fochabers Bridge design, 1832.

every six foot six inches distance by the best beaten iron screw bolts one inch and a half diameter, of sufficient length to go through the rib'. The balks were 30 to 50 ft long, with staggered joints, and before they were assembled each balk was

sawn across at every six foot six inches distance to the depth of one half of the balk or more; into these cuts [were] inserted thin wedges of cast iron or hard wood, by the force of a proper screw or machine for the purpose, so as to give the balks the curve of the arch.

The road, 19 ft wide, was laid on beams spanning between the ribs, and the footways, each 5 ft wide, were carried by oak corbels fixed on the outsides of the ribs. The ribs were of 'best American red pine' or 'best Riga or Dantzic timber', and the rest of the bridge was of English oak.

Although the contract was with Harrison's regular associate, William Cole of Chester, soon after it was finished the bridge aroused so much alarm that a pier was built under it at midspan with a truss of some kind to stiffen or support the arch.[54] In that crippled condition it was used until 1837 and then demolished. It was apparently the first attempt in Britain to build a large bridge with laminated arch ribs and it may have been inspired by a book published in 1810 by Carl F. von Wiebeking,[55] describing his construction of nine bridges in Bavaria in the years 1807–9, all with laminated arches of span more than 100 ft and rise 7 to 20 ft. Wiebeking had used balks more than 12 in. thick and powerful levers to bend them to the gentle curves he needed, and then he clamped three or more bent balks together by bolts driven through them and also drove oak keys or joggles into notches cut in the contiguous sides of every two balks. There were from two to five ribs in each arch, with various systems of bracing between the ribs. Harrison's arch at Warrington was similar in its overall dimensions to the arches in two of Wiebeking's bridges, but his method of curving the balks was original and apparently unsuccessful.

There is no further record of the design or construction of laminated arches until John Green, architect, of Newcastle-upon-Tyne, began designing a bridge to cross the Tyne at Scotswood in 1827–8.[56] He gave serious consideration to timber arches but chose instead a suspension bridge design. In 1831 Robert Stevenson, engineer, of Edinburgh, proposed an arch of 120 ft span and about 9 ft rise, to be made with 'thin planks bent to a circular form'.[57] In 1834 Green made and tested a one-twelfth scale model of a laminated arch of 120 ft span and he offered a bridge of five such spans in a competition for a railway crossing of the Tyne, also at Scotswood. He won the competition but his bridge was not built. In 1837–41, however, he did build five railway viaducts with laminated arches. He made up the arches from planks only 3 in. thick and was therefore able to use less clumsy methods for bending the timbers than Wiebeking or Harrison. He bent them to sharper curves, most of his arches rising 30 to 33 ft in spans of 110 to 120 ft. His were the first of many railway viaducts built with thin-laminated arches in the next fifteen years.

After Green's first thoughts but well before his first viaduct was built a very big arch was constructed at Fochabers near the mouth of the Spey. A bridge of four large stone arches had been built there in 1802–6, designed by George Burn, brother of James, and built by both brothers in partnership; but the western half of it was destroyed by the flood of August 1829. The responsibility for rebuilding it was contested between private interests, the county authority and the Commissioners for Highland Roads and Bridges; and a suspension bridge to replace the collapsed arches was proposed by Joseph Mitchell,[58] another of longer span to replace the whole bridge by Telford, and an iron arch to span the breach by a 'Mr Hughes'.[59] The private parties eventually repaired the bridge in 1832 at much lower cost by erecting a single timber arch of 184 ft span and about 19 ft rise. The designer was Archibald Simpson,

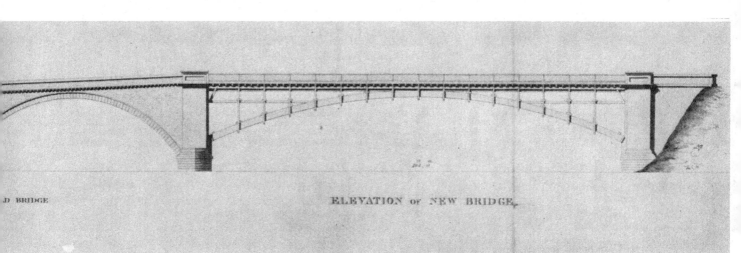

D BRIDGE   ELEVATION OF NEW BRIDGE.

the leading architect in Aberdeen. It was similar to Wiebeking's arches in its proportions and curvature, composed of three ribs which tapered from 38 in. deep at the springings to 34 in. at the crown and were 16 in. wide throughout (fig. 171). Each rib was made up of six balks about 12 in. deep × 8 in. wide and the method of bending or shaping the balks was not stated in the surviving specification. They were only required to be 'scarfed, overlapped and bolted together . . . oak keys to be also inserted at the abutting parts'. Over the ribs were posts and braces of considerable size and beams and joists to carry the road. All the timber was to be best 'Memel or Riga red fir', all bolts and other wrought ironwork 'best Swedish iron'; both iron and timber to be given 'six coats of good oil paint; . . . all the joining and abutting parts of the woodwork . . . well soaked in hot oil, and covered with white lead, and . . . thoroughly dry when joined'; and 'a covering of lead of 6 lb per foot laid over the back of the ribs' for protection against rain. These apparently thorough precautions were unsuccessful, for the ribs were suffering from dry rot by 1853 and the timber was then replaced by an iron arch.[60]

172 Union Bridge, Aberdeen, during construction.

14

CLIMAX IN MASONRY

Grandeur of effect, power of resistance, and eternity of endurance, are to be sought in masonry.

William Hosking, in Weale, *Architecture of bridges* (1843)

There were three developments in the design of masonry bridges in the years 1800–35: a lengthening of the spans of the largest arches, changes of form and architecture derived partly from French practice, and improvements in foundation work consequent on the introduction of steam power.

Although there were practical reasons for long spans in many bridges, a competitive impulse to build a record span can be suspected in other instances. It was quite explicit at several stages of planning for Union Bridge in Aberdeen,[1] where a competition for design of several improvements to the streets of the city was won by David Hamilton, architect in Glasgow, in February 1801, with James Savage, architect in London, in second place. As the first phase of the improvements Hamilton's design of Union Bridge, of three arches, was contracted for and Thomas Fletcher was appointed 'superintendent' or overseer for the city. He had been superintendent of the Aberdeenshire Canal under John Rennie and had also been employed on the Lancaster Canal. The site of the bridge was not difficult, having only a small stream at the bottom of a steep-sided valley, and by December 1801 the abutments and piers were finished. But errors in the estimate caused the contract to be cancelled and Fletcher found mistakes in the levels fixed by Hamilton. Rennie was asked for a new design and submitted three, one for a cast iron arch of 120 ft span, one for an elliptical stone arch of 116 ft and a third, the cheapest, for three stone arches of 42 ft.[2] However, Thomas Telford had passed through the city on his first survey of the Highlands in October 1801 and, by his own account, was

desired by the magistrates to examine their intended bridge. On considering the excellent granite stone which was used he prevailed with them to abandon the scheme of having three arches. At their desire he gave a plan of one arch of 150 ft span, being larger than any stone arch in Britain, and otherwise containing many singular features calculated to prove what could be performed with Aberdeen granite.[3]

This being rejected as too expensive, he 'afterwards made a simpler design' which was presumably the 'rough sketch of a masonry arch bridge of 150 ft span', dated 5 February 1802, which was still at Aberdeen in 1908.[4] But in November 1802 the trustees were considering only Rennie's three-arch design and a single-arch design by Thomas Fletcher of 130 ft span. Telford was asked to comment on Fletcher's design, which he believed to be a scaled-down version of his own, and he recommended several minor changes, adding again that 'I am only sorry the span is not 150 feet instead of 130 feet, as there is a rough whinstone arch in South Wales (Pont-y-pridd) 140 feet span.' A new contractor was found and the bridge finished in 1805 (fig. 172); but just before the contract was signed the contractor offered to increase the span to 142 ft for an extra payment of only £280, doubtless hoping that his name would supplant that of William Edwards as the builder of the longest arch in Britain. The trustees showed interest in the proposal but their chief need was for an early completion and Fletcher's design was adhered to. The bridge is still in perfect condition, carrying the main street in Aberdeen. It was widened on both sides by steel arches in 1906–7.

The building of roads and bridges in France was in the hands of the Corp des Ingénieurs, all of whom were trained at a single school founded in 1747. Jean-Rodolphe Perronet was the first head of the school and later Premier Ingénieur of the Corps, and hundreds of bridges were built all over France in the style which he developed. The style can be characterised by three regular features and one occasional one. Normally, there were pointed cutwaters, rounded at the shoulders and often from point to shoulder also. They extended up the spandrels to well above the springings of the arches but not as far as the parapet, and were sloped at the top as shown in fig. 173. The faces of the spandrels were plane and vertical and their interior was normally filled solid with masonry. Secondly, the arch curves were either

173   Bridge at St Loup in France.

circular segments or ellipses and frequently of very low profile; in Perronet's bridge at St Maxence the ratio of span to rise was more than 12, which was quite as flat as most of the iron arches in Britain. When the arches were low Perronet designed only the abutments to resist horizontal thrusts; at piers the thrusts of adjacent arches were assumed to balance each other and the piers, designed only to support the vertical load, were made unusually narrow. In Perronet's masterpiece, the bridge of Neuilly at Paris (fig. 174), the ratio of arch span to width of pier was 9 (cf. table 3). When building by this principle it was necessary to erect centring for all

the arches at once and no centring could be eased until all the arches were built. Thirdly, the lines of the roadway and parapets were generally horizontal or very nearly so. The fourth, and occasional, element gained special fame from its use in the bridge of Neuilly. This was the sloping or chamfering of the edges of arches, which appeared as tapering crescents on the elevations and were therefore known as *cornes de vâche* or cow's horns (the name 'eyes' applied to cylindrical voids in the spandrels had originated in French as *oeils de boeuf*, which must have been related to the *cornes de vâche*). *Cornes de vâche*, which were found on several older

174   Half of bridge of Neuilly (from Perronet, 1783).

French bridges, gave a hydraulic benefit in that more water would pass through an arch with this somewhat stream-lined shape than through a sharp-edged opening. The French engineers' scientific education made this clear to them, though it has often been denied by so-called experts since.[5] At Neuilly, flat segmental arch profiles were seen on the façades (fig. 174), rising only $14\frac{1}{2}$ ft in 128 ft span, while the bodies of the arches, usually in shadow, were of elliptical form and sprang from $17\frac{1}{2}$ ft lower; the whole segmental line on the

benefit from *cornes de vâche* was presumably uppermost in the mind of its designer.

A year or two earlier James Black, civil engineer, had submitted a design of the Neuilly type for the rebuilding of London Bridge.[8] It had three main arches of 220 and 230 ft span and was to be built of Aberdeen granite and mortar of Aberthaw lime. The arches were ellipses with *cornes de vâche* of similar proportions to those of Neuilly. It seems to be the first such design to have received publicity in Britain but the parliamentary

175   Ringsend Bridge, Dublin.

façade was above high-water level and so the unwanted appearance of partly submerged arches was avoided.

When Perronet published his *Description des projets* in 1783 with sixty-seven plates showing all the details of his bridges to a large scale, his methods and style became available to British designers and elements of both began to appear in British practice. An early instance was not British but Irish, namely St Patrick's Bridge at Cork,[6] which was designed by Michael Shanahan, the Bishop of Derry's erstwhile draughtsman (see chapter 3), and completed in 1790. The cutwaters and spandrel masonry were perfectly French, though the arches were higher and the road and parapets curved in elevation; there were no *cornes de vâche*. But thirteen years later slim *cornes de vâche* were built in Ringsend Bridge at the mouth of the Dodder in Dublin, a single arch of elliptical shape, the curve continuing into an inverted arch, or paved river-bed (fig. 175). It was built after the destruction of the former bridge by a flood in December 1802[7] and the hydraulic

committee made no comment other than ordering it to be printed with their third report in July 1800.

In John Rennie's masonry bridges there are three elements which commonly conform to the French style, namely arches of low profile, cutwaters with curved shoulders and a horizontal or near-horizontal road and parapet line. But in other respects they differed sharply from French bridges. His Wolseley Bridge over the Trent near Rugeley, built about 1800,[9] has cutwaters of the curved and pointed plan but surmounted by thick pilasters with apsidal niches. The three arches are low segments. His two masonry designs for Boston Bridge, also made in 1800, were low single arches.[10] The Esk Bridge at Musselburgh (fig. 176), designed in 1803, has pilasters and niches like Wolseley and five low segmental arches springing from high-water, the end arches having spans of 37 ft with only 3 ft rise, a ratio of 12 like that of the Pont St Maxence. The cutwaters, however, are semicircular in plan. The profile of the road is a gentle curve, thus descending to the low river-banks at a smaller cost than if it had remained horizontal to the ends of the arches. The largest span is 45 ft and the width of piers 7 ft, giving a ratio of 6½, which is more conservative than several of Perronet's designs. The spandrels are hollowed by longitudinal voids, which was also unlike French methods. Of Rennie's later designs, Cree Bridge at Newton Stewart, built about 1812–14, has similar proportions to the Esk Bridge, and Darlaston Bridge (fig. 141, designed in 1805), Stoneleigh Bridge (built 1819) and Wellington Bridge in Leeds (designed 1817) are all single arches with span-to-rise ratio of about 7. All have substantial pilasters on the faces of the spandrels and abutments and general classical detailing; apart from the shape of the arches and parapet lines there is nothing at all French in their appearance. Moreover, Rennie made several designs in which there is even less of the French style. In Kelso Bridge (see chapter 12) there is only the horizontal road and in Waterloo Bridge only that and the plan of the cutwaters.

His first engagement in connection with Waterloo Bridge was to assess a thoroughly French design. The designer was George Dodd, son of Ralph, who had gathered sponsors for a bridge across the Thames from the Strand at Somerset House much as his father had done for his London Bridge scheme in 1798–9 and for Vauxhall Bridge in 1806–9 (see chapter 13). The sponsors, known as the Strand Bridge Company, used this design to obtain an Act of Parliament in June 1809,[11] but six months earlier they had sent Dodd's design to Rennie and William Jessop for their opinions of it. It was a close copy of Neuilly Bridge with nine arches of 130 ft span (Neuilly had five of 128 ft) and 30 ft

rise (Neuilly rose 32 ft), all with *cornes de vâche*.[12] Rennie and Jessop observed in a joint report[13] that since such a bridge stood in Paris it could obviously stand in London if properly built. They approved of the large arches but wrote of the *cornes de vâche*:

We do not wholly approve of M. Perronet's construction. It is complicated in its form and we think wanting in effect. The equilibrio of the arches has not been sufficiently attended to, for when the centres of the bridge at Neuilly were struck their top sunk in a degree far beyond anything that has come to our knowledge, while the haunches retired or rose up [in fact the centring had been depressed 13 in. at the crown by the weight of masonry before de-centring and the crown of one arch sagged a further 10½ in. when the centre was removed] . . . Our opinion therefore is, that the bridge over the Thames should either be a plain ellipsis without the slanting off in the haunches so as to deceive the eye . . . or it should be the flat segment of a circle . . . But whether the ellipsis or circular segment is adopted, the arches should be those of equilibration, by lengthening the voussoirs towards the haunches, and by such other contrivances as will effect this.

This shows that Rennie thought of an equilibrated arch as one which would support its own weight when its centre was removed without undergoing a large change of shape. The condition for such behaviour, in the theoretical terms explained in chapter 5, is that the line of thrust must lie at the middle of the thickness of the arch throughout its length; this would be achieved, as the report observes, by a correct lengthening of the voussoirs towards the haunches of a low segmental arch. The two engineers also criticised the large flat expanses of spandrel wall and advised that 'some form of building should . . . be placed on the saliant projections of the piers [i.e. the cutwaters] for the treble purposes of

176   Esk Bridge, Musselburgh.

giving an equality of weight as far as may be on the foundation, strength to the spandril walls, and ornament to the bridge'. There was scant respect here for French practice. Criticising Dodd's choice of stone and the incompleteness of his drawings, Rennie and Jessop also checked his estimate and found it necessary to add considerable sums. These included the cost of a small steam engine for pumping.

They did not disagree with Dodd's intention to lay the pier foundations in caissons and after Rennie himself became chief engineer to the company in June 1810 he confirmed this intention for six of the piers in the deepest part of the river;[14] but he proposed cofferdams for the abutments and the first pier at each end, which would have made his foundation techniques an exact repetition of those used by Robert Mylne at Blackfriars Bridge (see chapter 7). Work on Waterloo Bridge began at the south abutment and first pier where the river was no more than 3 or 4 ft deep at low-water. The dams were double walls of timber piles, 'caulked with hemp or oakum' and with the spaces between them filled with 'gray stock bricks laid on puddle'. As the work moved northwards into deeper water the shape of the dams in plan was changed from a rectangle to a figure with rounded ends and finally to a true oval, and frames of horizontal struts had to be placed at three different levels to support the walls against the pressure of the water outside. Success with the early cofferdams encouraged Rennie to try one for the nearest pier to the north end where the river was 17 ft deep at low-water. Draining this dam proved very difficult and he threw earth round the outside and then drove a third line of sheet piles to retain it. They were shorter piles not reaching to high-water. For the remaining piers he drove three lines of piles at the outset and filled both the spaces with puddle and bricks; and it was just this type of cofferdam that he used successfully about two years later for the even deeper piers of Southwark Bridge (see chapter 13) and that his sons used to found the new London Bridge in 1824–30.[15] Fig. 177 shows George Rennie's notebook sketches of the cofferdam and fig. 178 one of the dams used for London Bridge. There was so much more certainty about driving piles and laying foundations on the dry river-bed within these dams that the use of floated caissons for founding bridge piers became obsolete. James Walker made the last extensive use of caissons on the Thames for his Vauxhall Bridge which was completed in 1816, a year before Waterloo Bridge. Since the timber platforms at the base of these piers were not connected to the piles under them, they floated off whole when the bridge was demolished eighty years later, carrying several courses of masonry with them;[16] this could not happen when the platforms were laid in cofferdams because of the close packing of masonry between the pile heads and in the interstices of the grating. At London Bridge there were also jagged iron spikes 18 in. long driven through the platforms into the pile heads.[17]

Steam power was important to the success of

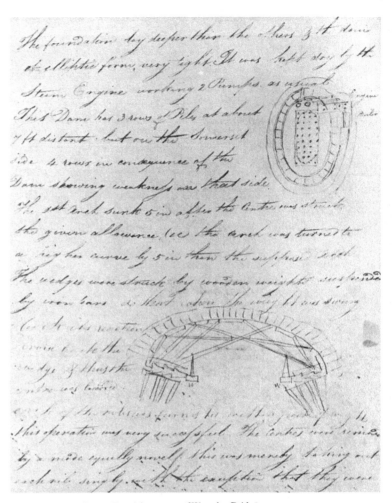

177 Page from George Rennie's account of Waterloo Bridge, describing the cofferdams and steam engines.

Rennie's cofferdams at Waterloo Bridge. Two steam engines were bought at the start of the work, one of six horse-power and the other of ten. The first of these could be used either to operate two pumps or to drive piles, but details of the machinery used for pile-driving have not survived. After it was decided to found all the piers in cofferdams both engines were generally occupied with pumping and many if not all of the piles were driven by manual labour. Rams of 7 to 11 cwt were used, falling 28 to 30 ft and raised by four men, obviously through a system of gears and probably a capstan. The steam engines were moved from dam to dam as they were needed and stood sometimes on the wall of the dam as shown in Rennie's sketch (fig. 177), and sometimes on a special piled platform, linked by a chain drive and gears to a shaft and crank which moved the pump rods. Alternative machinery was used later but the description of it is less clear. No contemporary illustration of such pumping gear has been found, but fig. 179, which was published with a French description of American engineering practice in 1843, shows the necessary parts of a system of steam-driven pumps (see also fig. 164).

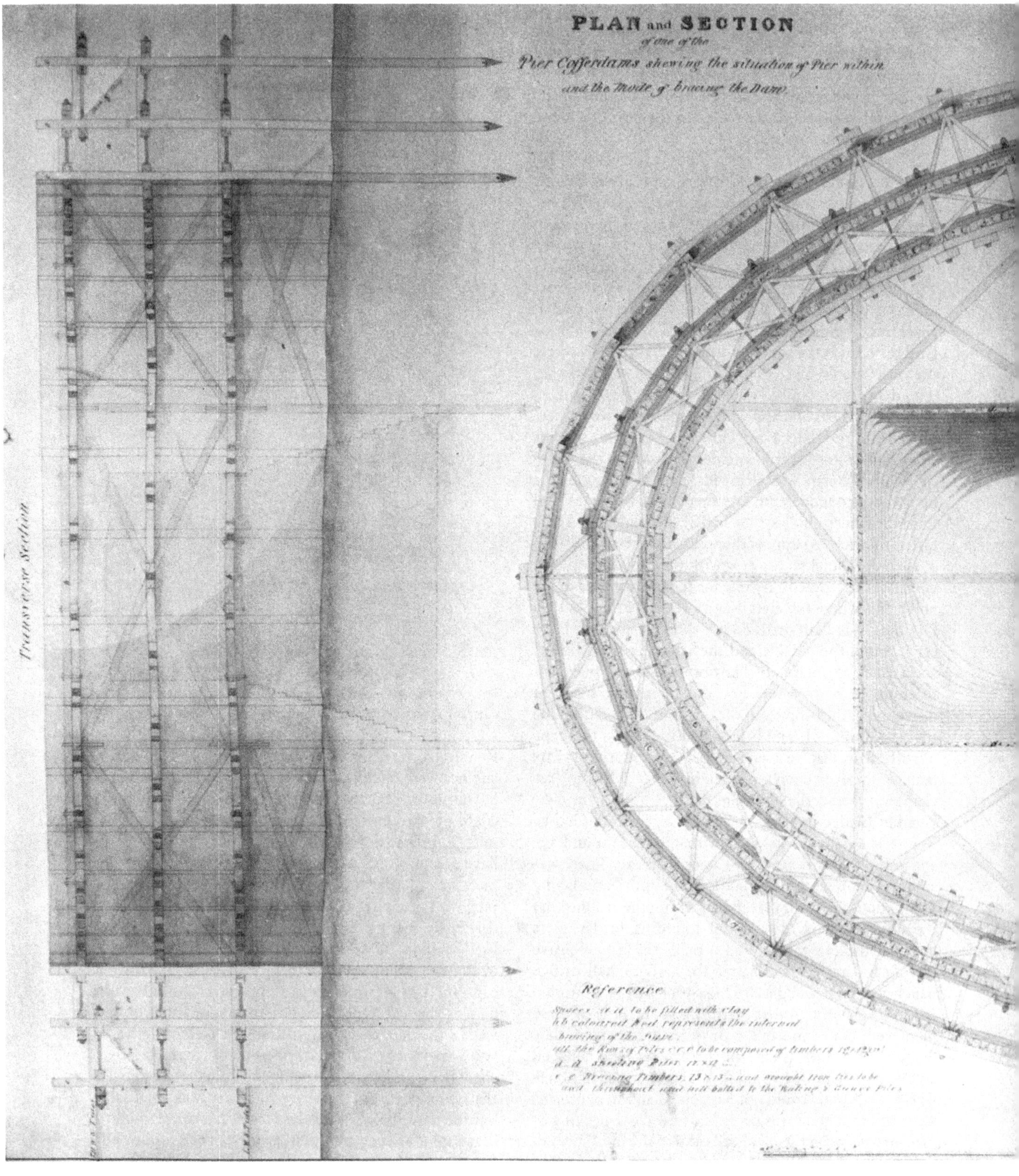

PLAN and SECTION
of one of the
Pier Cofferdams shewing the situation of Pier within
and the mode of bracing the Dam.

Transverse Section

Reference

Spaces a a to be filled with clay
b b coloured Red represents the internal
bracing of the Dam
all the Running Piles &c to be composed of timbers 12×12"
d d sheeting Piles 12×12"
e e Bracing Timbers 13×13 and wrought Iron ties to be
used throughout and well bolted to the Walings & Gauge Piles

178   Contract drawing of cofferdam for a pier of London Bridge, 1824.

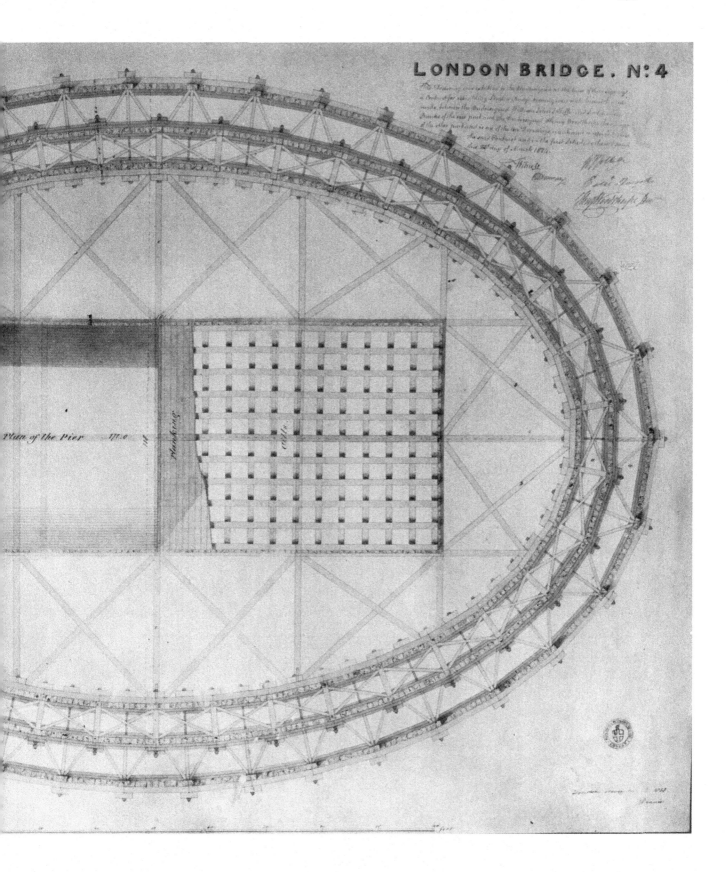

LONDON BRIDGE. N° 4

Plan of the Pier

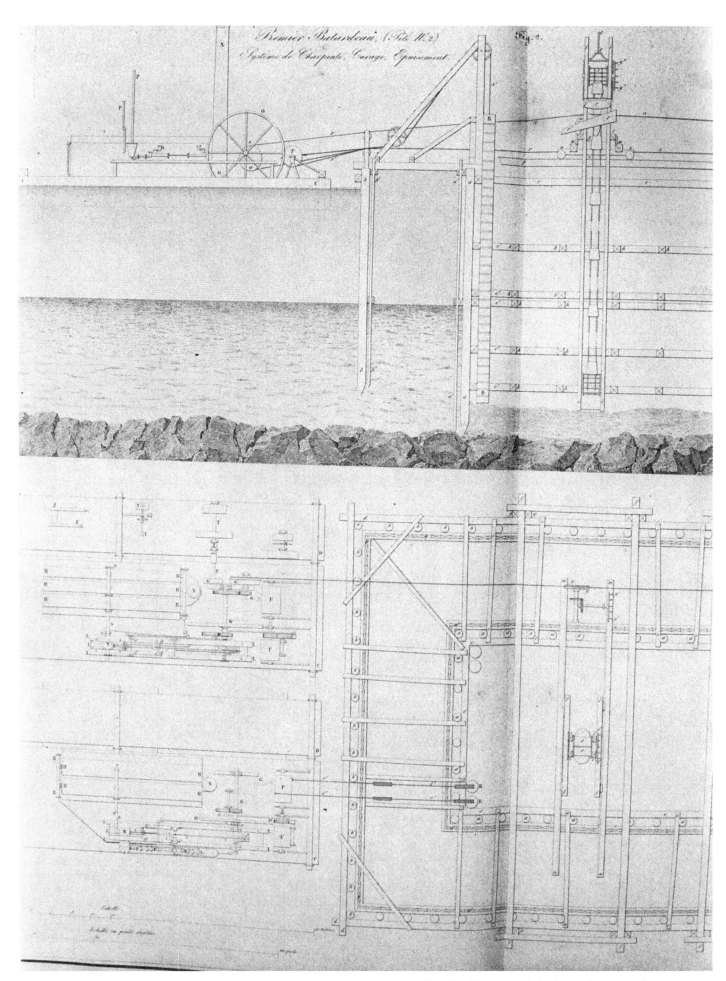

179 Pumps in a cofferdam driven by steam engines on floating platforms (from Chevalier, *Histoire et description des voies de communication aux États-Unis, 1843*)

For erection of the nine arches, each 120 ft span, Rennie designed trussed centres (fig. 180) not unlike those used by Robert Mylne for Blackfriars Bridge (chapter 7) but with more members, which must have given them greater stiffness. Where main compression members met end to end he had their ends inserted in a shaped cast iron box to ensure good bearing and prevent any splitting of the ends. He made a daring decision to assemble each rib of the centre on shore and then employ cranes to place it upright on a floating stage composed of three barges lashed together. Each rib weighed 40 tons, stood 120 ft long and 30 ft high and was only 2 or 3 ft wide. By the rise of the tide and the use of screw jacks it was seated on the sliding wedges shown in fig. 180. A whole centre consisted of eight ribs erected one by one, braced together and covered by 'lagging', or cross timbers, on which the arch stones could be laid. The removal of ribs when an arch had been completed and their re-erection for another arch took one week per rib at first but eventually a rate of one rib per day was achieved.

Most of the exterior masonry was of Cornish granite from Penryn but a little Craigleith sandstone from Edinburgh and 'Yorkshire stone' (sandstone or gritstone) were used externally and much more internally. Almost all the structural details above water level were similar to those used in the Lune Aqueduct twenty years earlier (chapter 10). Four flat bars of iron were set in selected joints of each arch – and were found to have corroded very little when the bridge was taken down after 120 years – the spandrels were hollowed by longitudinal brick walls, there was a 2 ft 6 in. layer of clay under the road metal and the road and parapet line were perfectly horizontal. Only the iron ties laid from side to side at the top of the aqueduct were omitted, as there was no outward pressure at the top of Waterloo Bridge. Fig. 181 shows the similar spandrel construction of London Bridge, photographed during its demolition in 1970. The iron bars in the arch joints, as well as all the other details described, can be seen in fig. 180.

Waterloo Bridge was by far the largest bridge contract yet undertaken in Britain, involving a huge labour force as well as large quantities of machinery and river craft for transport and lifting of the materials and for the driving and extraction of piles. It was undertaken by the first great British contracting *firm*, which had been formed in 1807 by the partnership of a provincial canal contractor, Edward Banks, with a former clergyman, William Jolliffe.[18] They had already built Croydon court house, Dartmoor prison, a lighthouse at Heligoland and much of Howth harbour in Ireland. Waterloo Bridge established them as major bridge-builders and they went on to build the masonry of Southwark Bridge and the whole of London and Staines Bridges, all of which were designed by John

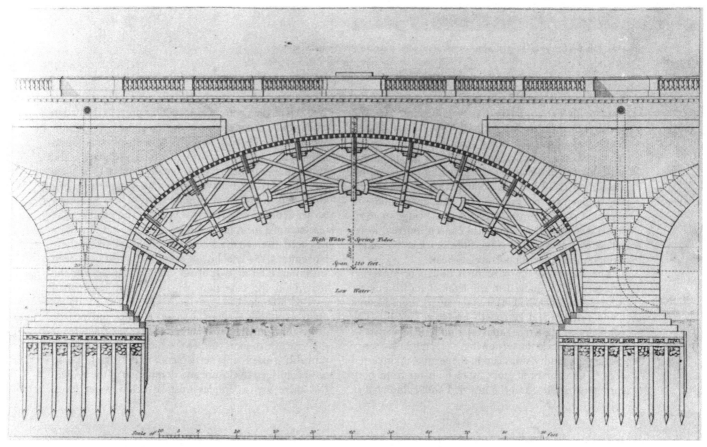

180 Rib of centring of Waterloo Bridge, with section of piers, arch and spandrels. Engraving by Taylor, 1827.

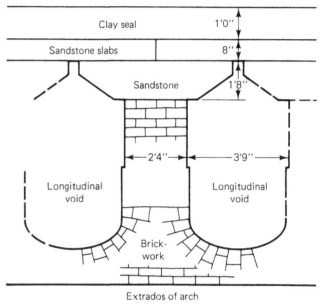

181   London Bridge (1824–31): cross-section of spandrel and road photographed in 1970.

Rennie and his sons. Banks (fig. 182) was knighted in 1822 for his 'extraordinary exertions, industry, skill and perseverence in the execution of the Waterloo and Southwark bridges', being the first engineer so honoured since Sir Hugh Myddleton in 1622 and probably the first man ever knighted for bridge-building. John Rennie had refused knighthood when Waterloo Bridge was opened in 1817 but his second son accepted it at the opening of London Bridge in 1831. Banks's memorial in Chipstead church, Surrey, bears reliefs of arches of Waterloo, Southwark and London Bridges, the latter supporting his bust.

The success of the structural design of Waterloo Bridge must be judged in the light of the comments Rennie and Jessop had made at the outset about George Dodd's design and Perronet's bridge at Neuilly. The voussoirs of the arches of Waterloo Bridge varied in depth as shown in the section (fig. 180) and the masonry inside the spandrels was laid in the form of an inverted arch. The courses of stone were built up from both springings at once and the inverted arch completed as

soon as the arches at each side of it were up to that level. The top of the centre was loaded to prevent distortion while the first courses near the springings were built. When the arch was keyed the centring was eased immediately and the settlement which took place at the crowns of the arches varied from only $\frac{1}{4}$ in. to $1\frac{5}{8}$ in. This was the sort of behaviour Rennie and Jessop had attributed to arches of equilibration. There had been earlier depression of the crowns under the weight of the masonry of 4 in. at most and, as the curve of the centres had been given an extra height of 5 in. to allow for these movements, the final line of every arch must have been within an inch or two of its intended curve. For the weight on the cutwaters which he and Jessop had advised Rennie had provided only a pair of three-quarter columns against the spandrel walls, and the lack of weight there may have contributed to the fractures at the base of two piers (fig. 183) which accompanied and probably caused large settlements in the 1920s. These settlements led to closure and demolition of the bridge. But there was a simpler mistake below water level, namely that the steps by which the piers were widened at the foundations (fig. 180) were too wide and shallow. If the widening of each pier from its minimum area of 1,300 square feet to its maximum of 2,720 square feet at the base had been accomplished gradually in a height of 20 ft, as was done at London Bridge ten years later, instead of in only 6 ft, the cracking might never have occurred and the bridge might still be standing. However, the weight of the bridge was also too great for the areas of foundation provided and this was an error of judgement which Rennie and his sons repeated at London Bridge.[19] It was known at the time that the pressures exerted on the river-bed were close to the limit of safety, but it could not be said with confidence that they had overstepped the limits of good design until the second half of the twentieth century, when modern methods of site investigation and the science of soil mechanics made possible an accurate assessment of the bearing capacity of the piles and footings.

When the threat of demolition arose Waterloo Bridge was acclaimed as the finest of London's bridges, and perhaps rightly. It was certainly the nearest thing to a 'bridge of magnificence' ever built in Britain. But it was Rennie's last use of the twin-column decoration and his sons never chose to use it; his greatest contemporary, Thomas Telford, thought it bad architecture.

Much of our knowledge of the construction of Waterloo Bridge comes from the manuscript account of the site work written month by month by Rennie's

182   Sir Edward Banks, painted by George Patton.

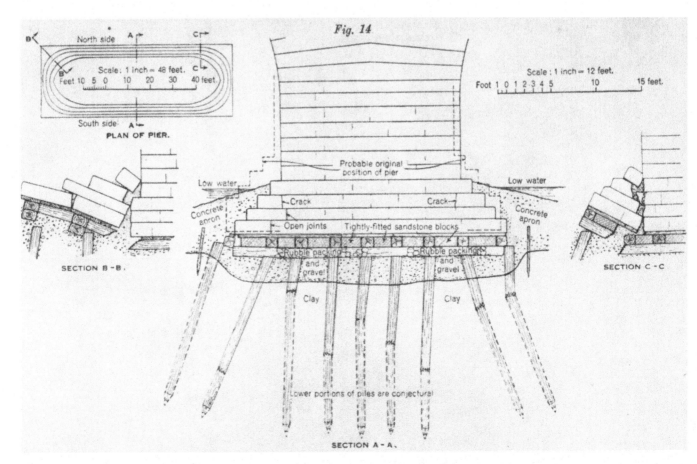

183   Fractures at base of a pier of Waterloo Bridge (from Buckton and Fereday, 1936).

eldest son, George. George had been placed by his father at Edinburgh University in 1807 and for two of his four years there he lived in the house of John Playfair, Robison's successor as professor of natural philosophy.[20] George studied mathematics, physical sciences and classics before returning to London in 1811 to learn practical engineering in his father's office and workshop. Rennie's second son John was not sent to university and so it was George who applied theoretical methods to the brothers' designs in later years. By his own report,[21] George made the design for the new London Bridge for his father in 1820. The five arches were elliptical, the largest having a span of 150 ft and rise of $29\frac{1}{2}$ ft and being supported by piers 24 ft thick. Because he held a government appointment as inspector of machinery at the Royal Mint he could not be engineer for the bridge, and so that honour went to John and earned him a knighthood. But it was George again who visited Chester in 1825[22] to report on the practicality of the architect Thomas Harrison's design for a masonry arch of 200 ft span. Several documents imply that George designed Staines Bridge,[23] whose

form is as near to those of Perronet's bridges as any of his father's, with low segmental arches and slim piers, but the ratio of span to pier thickness is still no greater than 7. Hyde Park Bridge, of similar proportions, was built to the Rennies' design in 1824–7.

It seems certain that George employed theoretical calculations in designing arches. He was not alone amongst engineers and architects in this but he was perhaps the first to apply theory to large structures which were actually built. For Harrison's Grosvenor Bridge design at Chester he proposed changes in 'the dimensions of the voussoirs and form and dimensions of the abutments' to make them conform to 'the correct principles of equilibrium'[24] and the changes in the abutment masonry from fig. 184[25] to fig. 185[26] show that he made it a typical Rennie structure (compare fig. 141). The arch proper was made 4 ft thick at the crown and 6 ft at the springings but there is a large thickness of radial masonry over almost the whole of the extrados, as well as the radiated courses in the abutments (fig. 185). The body of the arch is of Cheshire sandstone but the springing courses are granite.

184   Section of Grosvenor Bridge as designed by T. Harrison.

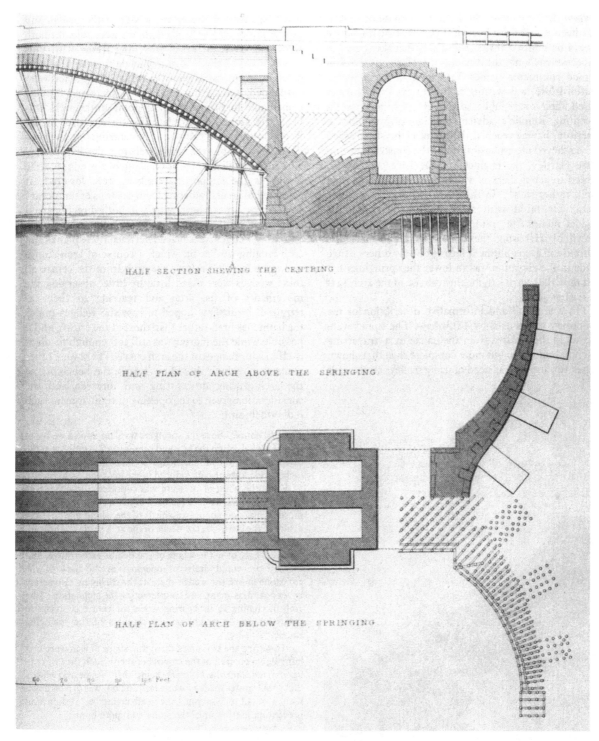

HALF SECTION SHEWING THE CENTRING

HALF PLAN OF ARCH ABOVE THE SPRINGING

HALF PLAN OF ARCH BELOW THE SPRINGING

185   Sections of Grosvenor Bridge as built (from *Trans.InstCE*, 1836).

In the extended negotiations which preceded the start of the construction of Grosvenor Bridge the practicality of an arch of 200 ft span was never questioned; the matters of argument were the cost and the nature of the ground. One of the Brunels heard of the proposal and claimed that such an arch could be built of rubble masonry for only £10,000; and the trustees had to investigate the claim before going ahead with Harrison's design. George Rennie strongly advised that both abutments should be founded on rock although it would require at least one very deep excavation. Telford, after thorough investigation, advised the Exchequer Bill Loan Commissioners, who had been asked for a loan of £35,000, that no bridge should be built on that site and also that the span was unnecessarily long; the loan was therefore refused. But a good contractor, James Trubshaw of Haywood in Staffordshire, was waiting to build it and the trustees raised their own public loan and set him to work, accepting Rennie's advice and ignoring Telford's. Harrison, having reached the age of eighty-two, asked to be relieved of responsibility for the construction and Jesse Hartley, the engineer of the Liverpool docks, agreed to act as clerk of works. The bridge was then built, rather slowly, in the years 1827–33. The abutments are faced with massive pilasters containing apsidal niches (fig. 186), one of the two treatments drawn by Harrison; the other was a pair of Doric columns on each abutment face. The only other change made in the elevation was to lower the springings 1 ft and raise the crown 1 ft, making the rise of the arch 42 ft instead of 40 ft.

There is a full and informative description of the temporary works used by Trubshaw. The stones were carried to their places from the banks by a 'traversing machine' similar to but more complete than that shown in figs 195 and 196. Instead of using trolleys on the rails which spanned the river at the side of the bridge and lifting stones off the trolleys, there were rails on both sides and on them ran a travelling frame with rails which spanned the bridge from side to side. On these cross rails ran carriages from which the stones were suspended, and by moving both carriage and frame a stone could be placed over any point on the arch. It is likely that similar devices had already been used on the Rennies' bridges in London. The centring (fig. 185) was devised to enable the huge arch to 'take its bearing' without large distortion or cracking of the edges of the stones as the supporting timbers were lowered. It consisted of struts radiating from the tops of temporary stone piers, braced horizontally in both directions and supporting planks bent to the curve of the arch. Six sets of folding wedges were placed between these planks and each lagging timber on which a course of arch stones was to be laid. To release the arch from the centre all these wedges were eased little by little, observing the movements of the arch and reacting to them as required. Trubshaw hoped to ease the wedges under the haunches first, rather than those at the crown, and if possible while the mortar was still soft enough to adapt itself to any changes of the arch curve. There were other precautions which show how clearly the behaviour of the arch during de-centring was foreseen and the consideration given to the opening or tightening of each individual joint:

The first course above the springers was laid upon a wedge of lead 1½ inch thick on the face and running out to nothing at the extremity of the bed, and strips of sheet lead eight or nine inches wide were also introduced in the joints on each side, up to where the point of pressure was considered to change its position from the front to the back of the archstones, or in the present case over about two-thirds of the whole soffit . . . the easing of the centre let the whole of the arch settle on the lead, which from its yielding nature then caused the pressure to be spread evenly over the whole of the bed of each course, and thereby prevented drafts or openings at the back of the archstone joints; the wedge-piece at the springing also acting by way of adjustment, and counteracting the inclination of the arch in coming to its bearing when the centre is struck to throw an undue weight on the intrados of the springing course.

. . . In setting the keystones three thin strips of lead were first hung down on each of the stones between which they were to be inserted, and the keystone being then besmeared with a thin greasy putty made of white lead and oil, was driven down with a small pile-engine, the lead acting as a slide and preventing grating until the stone was quite home.

The effect of all these precautions was that the crown dropped only 2½ in. during de-centring, 'the joints remaining perfectly close and no derangement of form being perceptible'. The good condition of the bridge today, after 140 years of main road traffic, reflects great credit on the skill of Hartley and Trubshaw and perhaps also on George Rennie's science. Grosvenor Bridge was the largest stone arch in the world for about thirty years, being overtaken by Cabin John Bridge at Washington, DC, in 1864.

187   Tongueland Bridge, Kirkcudbrightshire.

While Rennie's largest bridges were built in London, he also built about fifteen of reasonable size in the provinces, and a number of aqueducts. Telford was responsible for more than forty large bridges scattered throughout England, Scotland and Wales, as well as hundreds of small ones. History is quite unfair to the men who directed these works as overseers and contractors, many of them on remote sites where they were visited only once or twice a year by the engineer; and it is a fact that the influence of the engineer was always strongest at the time of the initial design, when decisions were made about the form and style of the bridge. It is therefore logical to confine our treatment of Telford's bridges largely to their form and style. The construction methods used for his bridges were generally similar to those already described with reference to Rennie's work and will only need occasional comments.

That Telford's views on bridge design were markedly different from Rennie's is clear from his article on the 'Practice of bridge-building'[27] written in 1812. On the choice of sites, he considered that a bridge should be placed on a straight stretch of river and at right angles to the current, but he also noted that 'where practicable, from the approaches being in a curve, the general outlines of the bridge are seen to most advantage'. He thought that there should be an odd number of arches, all springing from the same level and increasing in height and span towards midstream, so as to give a gradient on the road of about 1 in 24. He noted a wide range of practice in the ratio of arch span to pier thickness, with the highest values at Neuilly Bridge and his own Conon Bridge, 9 and 8 respectively. Of Rennie's work he quoted Kelso Bridge with a ratio of 6. On the ornament of bridges he wrote, 'The decorations should be varied, according to the situation and accompanyments. In the country, the utmost simplicity, consistent with distinguishing the essential parts . . . and even in the most splendid cities . . . the decoration should be kept perfectly subservient to . . . the essential parts.' He criticised columns and entablatures as improper because they were derived from Greek temples where there were no arches, and also because they deprived the spandrels of the heavy buttresses which could be placed on top of the cutwaters. The 'affectation' of making the entablature and road horizontal he condemned as 'a false taste . . . introduced by some of the French engineers', making an imperfect road because the rain could not run off it and also giving 'an appearance of feebleness'. He showed interest in the *cornes de vâche* at Neuilly, noting that the form, 'besides affording facility for the passage of flood waters, gives a great appearance of lightness to the whole fabric'.

These quotations suggest that Telford still thought himself something of an architect and this is confirmed

188   Potarch Bridge, Aberdeenshire.

by the fact that he undertook the writing of the article on 'Civil architecture'[28] for the encyclopaedia very soon after he wrote 'Bridge'. It contains a long survey of architectural styles with no expression of general preference for any one style but a distinct display of interest in the re-birth of castellated and Gothic design for houses, for instance in Wyatt's Ashridge Hall in Hertfordshire. There is a separate section on the 'principles of architecture', derived almost entirely from Alison's *Essays on the principles of taste*,[29] where the sensation of beauty is held to depend on recognition of the fitness of an object for its purpose. The dominant purpose in Telford's view of architecture seems to be the support of weight. In a discussion of the beauty of line and ornament he holds, with Alison, that straight lines and heavy lines suggest hardness or strength, curved lines and faint lines weakness or delicacy.

At the start of his practice in Shropshire most of his designs were conventional. The larger bridges, such as Montford (1790–2, see chapter 12) and Chirk (1793),[30] were built of hewn stone throughout, V-jointed on the faces and soffits of the arches and with projecting string courses and copings. However, Chirk Aqueduct,[31] built between 1796 and 1801, is all of rubble except for quoins, string courses and keystone courses. Bewdley Bridge, not being a county bridge, allowed him to add some ornament and he chose with excellent judgement from the classical vocabulary (fig. 151). He seems to have felt the need for an architectural style in his great London Bridge design and so added Gothic tracery (of

iron) to the framing of the spandrels (fig. 152). Moving in the next few years into Scottish landscapes that were considered romantic and wild he turned to the architecture of castles for most of his embellishments, as Robert Adam and Robert Mylne (fig. 81) had sometimes done. At Tongueland in Kirkcudbrightshire he bridged the Dee just where it enters its estuary with a large arch framed by semicircular turrets at the abutments and with castellated battlements for parapets (fig. 187). To suit the site the road is horizontal and he hollowed the high approaches by series of Gothic arches on tall thin piers. All the walls are of rough-faced grey granite. Even the corbels under the battlements are large and rounded and he rightly thought it had 'a bold effect'[32] when seen spanning choppy water at high tide. In 1812 he considered Dunkeld Bridge, which he had designed about 1805 and was built by 1809, 'the finest bridge in Scotland'.[33] It has five main arches and two small ones, all of them framed by solid semicircular turrets. The keystone courses protrude right across the soffits, which was generally an architect's conceit but is also found on some of Telford's early bridges, such as Chirk and Montford. He built turrets on the spandrels between the masonry arches of Bonar Bridge (fig. 157), and at Craigellachie where there was only one iron arch and no need for such turrets he erected four free-standing castellated towers at the corners of the bridge[34] (fig. 158). They have excited many comments, some critics finding them clumsy or irrelevant while others have called them romantic and 'an effective contrast'. A late instance of Gothic detail is the small pointed arches on high, thin piers in the approaches to the iron bridge at Tewkesbury (see chapter 13) which recall those of Tongueland Bridge.

In the same years he built another series of quite different bridges at Scottish sites.[35] They included bridges over the Dee at Ballater and Potarch, the Don at Alford and the Beauly and the Conon in Ross-shire. All but the first and last are still standing. They were all of three or five segmental arches, the road and parapet curved in elevation, and trapezoidal buttresses or pilasters over the cutwaters providing recesses at the level of the road. They are practical structures with no added ornament (fig. 188). Even less pretentious are the rubble stone bridges built in hundreds for the Highland Roads Commissioners of the same materials as the military bridges described in chapter 2, but with several improvements in form and methods of construction. The chief of these were that piers and abutments could now be founded below low-water level and larger waterways could be provided by building arches of greater number and longer span. The roadways were also wider. Telford's way of 'designing' the many small single-span bridges was to write a standard specification for the bridges on any one road, describing the masonry of the arch, abutments, spandrels and parapets and giving the dimensions to be used in bridges of various different spans.[36] The materials, at least for work above water level, seem to have been very similar to the older bridges; but there is one structural improvement which seems never to have been omitted even in the smallest bridges and which is an almost foolproof mark of distinction between pre-1800 and post-1800 bridges of this type, namely the battered form of the wingwalls in the latter. The form of the battered walls varies on the Commissioners' roads, but after some time the arrangement shown in fig. 189, with vertical spandrels and battered wingwalls, became the commonest, and it continued to be the standard construction for bridges in the Highlands, both large and small, until at least the end of the nineteenth century. Good examples of Telford's design which are still standing are the single-arch Calder Bridge near Newtonmore, the three-arch asymmetric Contin Bridge, and Altnaharra Bridge, also of three arches, on the road from Lairg to Tongue.

At all the ordinary sites the road and parapets rose in a smooth curve with a gradient of at least 1 in 24, the arches were segments but higher segments than Rennie favoured, and there were no inverted arches over the piers or radiated courses in the abutments. There were always solid pilasters of considerable bulk over the cutwaters, and never columns.

During the course of the work in the Highlands there was more change in the contractual procedures[37] than in the style of the designs. In the early contracts the contractor and his securities were required to maintain the bridge for five years but this was soon reduced to three years, partly because the risk was recognised to be greater than in other places, but also because the best guarantors would be debarred from undertaking further contracts if their liability under earlier ones lasted for five years, and the whole programme would therefore be delayed. Moreover, Telford was so often confounded by the magnitude of the spates in the Highland streams that he confessed with remarkable frankness in 1811 that he could not ensure safety of the bridges in the maximum floods he had learned to think possible; if he did, it would cost more than occasional repair and rebuilding. He reported in 1812 that 'where . . . the stream has not before been noticed with any correctness it is only by degrees that the proper dimensions for a bridge can be ascertained . . . The bridges now built will serve for a scale of measurement each for its own river.'[38] Some later contracts removed the responsibility for damage by 'abnormal floods' from the contractors, but the definition of 'abnormal' was very difficult.

190   Glasgow Bridge widened by iron arches (from Rickman, 1838).

Each of the types of bridge we have already described was established in Telford's practice before he wrote his 'Bridge' article in 1812, but there is another series of designs which were only about to begin and which must be considered his most original work. He referred to most of them in his autobiography[39] but by then he was more reticent about his architectural intentions and so we can only speculate about his motives. But his interest in the arch form of Neuilly Bridge and his approving references to an appearance of lightness give a strong clue to the central reason for the forms of these bridges.

The old bridge at Tenbury[40] on the borders of Shropshire and Worcestershire needed to be widened in 1812. Telford submitted a scheme for taking down part of the big medieval cutwaters and spanning iron arch ribs between their points, then laying a floor from the ribs to the old spandrel walls to provide an extra 5 ft of carriageway. The new ribs were of low rise and therefore sprang from points much higher than the masonry arches. There were also new iron guard-rails. The scheme was carried out by John Simpson in 1814, the year before his death. About five years later Telford responded to a similar problem with a similar design and because the bridge was better known, namely the Old Bridge of Glasgow, it attracted more attention. In his autobiography he remarked that 'thus improved, the bridge was rendered not only sufficiently commodious, but even ornamental; the external appearance having an air of originality and lightness by the projection of the ironwork, and the shadow thereby thrown on the masonry below it' (fig. 190).[41] It was not the first time bridges had been widened by arches set higher than the original arches, the widening of London Bridge in 1756–63 (see fig. 54) and of Newcastle Bridge in 1801 (see fig. 197) being prominent examples. But Telford appears to have been so pleased with the effect that he went on to apply it to new bridges. First, however, he

191   Over Bridge, Gloucester.

the piers and abutments by an unbroken curved line. The segments which appear on the façade are of 40 and 50 ft span and 6 ft and 8 ft rise.

There is nothing in Telford's writings which links his designs for Tenbury, Glasgow, Over and Morpeth with those he made for Dean Bridge in Edinburgh, Pathhead Bridge in Midlothian and Stonebyres Bridge in Lanarkshire.[44] But in these three designs, all made in the years 1827–30, he deliberately created shadows on the spandrels and an appearance of lightness by placing slim projections on the faces of the tall piers and springing subsidiary arches from the projections at points well above the springings of the main arches. Telford gave no aesthetic reason for this form as from his own pen, but he included in his autobiography an account by the resident engineer for Dean Bridge, Charles Atherton, who wrote that the form was devised to overcome the criticism of high viaducts and aqueducts that 'the pillars have a heavy and clumsy appearance, or that the great mass of masonry being uppermost, the superstructure appears too massive'. It certainly enhances the visual interest of Dean Bridge (fig. 192), the forward projection of the higher arches being 6 ft and the footpath and gardens under the bridge being much used. In Pathhead Bridge the projection is only 2 ft, the arches of smaller span, and the effect is rather fussy; but it is seldom seen because there is no public route under the bridge. The Stonebyres Bridge was never built.

To appreciate the fertility of Telford's architectural design it should be recalled that in these late years he also designed the two iron arches over the Severn (chapter 13); and in 1826 the Don Bridge at Aberdeen,[45] of five arches but similar in elevation to Potarch and other Highland bridges, while in the new Broomielaw Bridge at Glasgow[46] he reverted to the elegant

designed a bridge of a single arch 150 ft in span to cross the Severn at Over near Gloucester.[42] Because the magistrates insisted on a stone bridge he designed the arch of the Neuilly form, an ellipse of 35 ft rise in the body of the arch with a segment of only 13 ft rise on the façade. The reason he gave in his autobiography for choice of this form was the need to ease the passage of the Severn's great floods, but when the bridge was finished a local newspaper commented that the form gave 'a character of airiness and lightness to the arch' and Telford echoed this sentiment in his autobiography. The shape of the chamfers on the arch differs from those of Neuilly and they look rather clumsy to a near view (fig. 191). When he next designed arches of this form it is easier to believe he did it solely for hydraulic benefit, for the bridge over the Wansbeck at Morpeth in Northumberland had to conform to low street levels and the width between the existing river walls.[43] The chamfers there are normal *cornes de vâche* running into

192   Dean Bridge, Edinburgh.

193   Interior of spandrel of Dean Bridge.

classical details he had last used in Bewdley Bridge in 1798. But the highest tribute to his skill in the design of masonry bridges is that in Dean Bridge (fig. 192), the farthest development of his architectural experiments, he also achieved an excellence of construction and abiding strength which can seldom have been equalled. The whole interior, both of piers and spandrels, was built hollow and has been recently inspected,[47] when it was evident that there has scarcely been any movement of the structure and only the slightest of decay in the 145 years since the bridge was opened to traffic. A few of the tie-stones which connect the outer spandrel walls to the inner parallel walls (fig. 193) have failed in tension, otherwise everything is perfect. It is a measure of the solidity of the work that while city buses and lorries roar along the roadway absolute silence reigns within the spandrel voids. The road construction itself was novel, for in addition to the clay seal used by earlier bridge-builders and by Rennie at Waterloo and London Bridges, Telford used hydraulic concrete as a base for the gravel or broken-stone pavement. The cross-section found in the recent inspection (fig. 194)

does not correspond exactly to the original design but the layer of concrete on top of the clay is as designed. Its purpose was often to provide a waterproof layer but in Dean Bridge the intention was more likely to prevent the coarse whinstone from penetrating the clay.[48]

Construction of the arches required some daring.[49] The subsidiary arches extend inwards over the main arches a distance of 3 ft 2 in. To allow both sets of arches to settle freely when their centres were eased, they were built independently of each other and the centres eased before any spandrel walls could be built. The subsidiary arches, which are only 2 ft 6 in. thick, had therefore to be built and very gently de-centred, all four spans together and each arch spanning 96 ft between supports only 5 ft thick. This meant that the ratio of span to arch thickness was 38 and the ratio of span to pier thickness 19, which is more than double Perronet's maximum value of 9. The arches subsided by about $4\frac{1}{2}$ in. at each crown in the course of a month and the spandrel walls were then built up between them and the main arches, providing much increased stability.

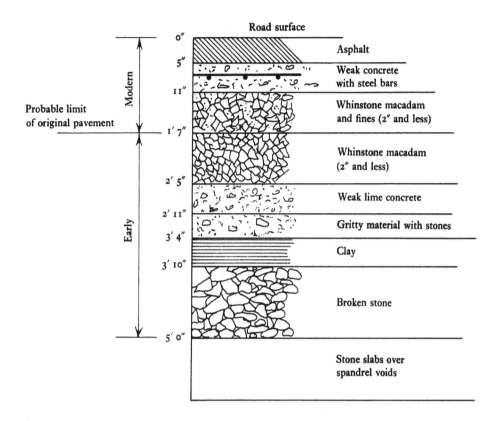

194   Cross-section of roadway of Dean Bridge (after Paxton, 1976).

A climax of design in masonry comparable with that of Dean Bridge, but quite different from it, had been achieved a little earlier by one of Telford's protégés, Alexander Nimmo; but before describing it we must look briefly at a few of the contemporaries of the two great engineers, Rennie and Telford. Thomas Fletcher, who designed Union Bridge at Aberdeen, was also supervisor for a time at Over Bridge; like many others he learned his engineering on the canal projects of the 1790s. Hugh Baird, the engineer of the three large aqueducts on the Union Canal, built in 1816–22 with iron troughs on piers and spandrels of masonry, had begun in the same way. James Walker, who designed Vauxhall Bridge, started his career in dock-building on the Thames just after 1800. Towards the end of Telford's career Walker reached the top rank of civil engineers and on Telford's death he succeeded to the presidency of the Institution of Civil Engineers. He directed extensive repairs to Blackfriars Bridge in the years 1834–40,[50] replacing much of the external masonry and using cofferdams similar to those of the Rennies to expose the bases of the piers and surround

them with extra piling. Similar work round the piers of Westminster Bridge[51] had been started by Telford in 1823, to forestall the scour which was expected from the increased velocity of the tides when Old London Bridge was demolished, and Walker continued the attempt to stabilise the foundations and modernise Westminster Bridge. Work went on from 1829 to 1846, but ended in failure due to no fault of Walker but to the precarious foundations laid by Labelye a hundred years earlier.

Robert Stevenson had begun his practice as engineer to the Commissioners of Northern Lights.[52] He built four large bridges, as well as some small ones, by methods similar to those of Rennie and Telford. In some, including his largest bridges at Stirling and Glasgow, he used segmental arches of low rise, but not so low as some of Rennie's. Full records survive of the building of Hutcheson Bridge at Glasgow (completed in 1832) and the engravings of work in progress show the equipment used for delivery and hoisting of stones better than any other contemporary source. Rails ran across the river on one side of the bridge and trolleys carried the stones to the arch under construction (fig.

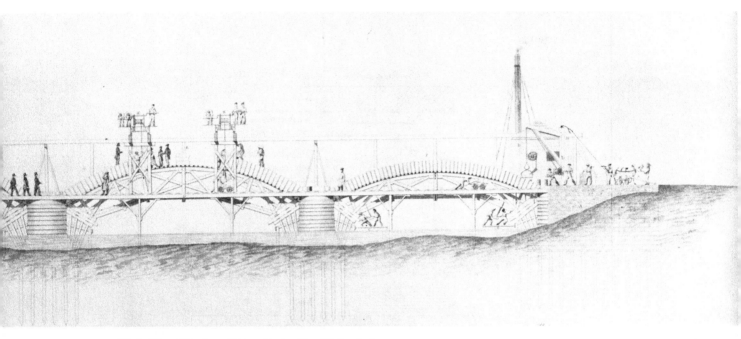

195   The building of Hutcheson Bridge: side view (from Weale, 1843).

195). The stones were then lifted one by one (fig. 196) by cranes which were themselves able to travel on rails across the width of the arch and place the stones in their required positions.

Stevenson's prophetic design of a laminated-timber arch in 1831 has been mentioned in chapter 13. He was also prophetic when in 1826 he proposed to the Corporation of Newcastle-upon-Tyne a design for iron arches to support a second road over their overcrowded bridge (fig. 197 and see chapter 8). The double-deck principle was adopted by his near-namesake Robert Stephenson many years later for the High-Level Bridge at Newcastle, with the road underneath and a railway on top.

An architect who had some success in bridge design was James Savage (1779–1852).[53] He was a student at the Royal Academy from 1798 and after taking the second premium for the Aberdeen improvements competition in 1801, as mentioned above, he won the first premium in the competition of 1805 for rebuilding Ormonde Bridge in Dublin. The project was delayed but in 1808 another design for a different site was adopted and called Richmond Bridge (but now re-named O'Donovan Rossa Bridge). Construction took place in the years 1813–16. It is a symmetrical three-arch bridge of granite, with a classical balustrade in which the balusters are of cast iron, moulded arch rings with sculptured heads on the keystones, and other classical details (fig. 198). It is well composed in general but the arches appear to be truncated ellipses or else irregular curves and a newspaper report at the date of opening said, 'The mistake, or whatever it is to be called, in respect to the height of the arches . . . has

deprived the edifice of all pretences to beauty.'[54] The contractor, George Knowles, himself designed another bridge only a few hundred yards away,[55] of three more correct elliptical arches but also of granite and with similar details. It was called Whitworth Bridge (now Father Mathew Bridge) and completes an elegant trio with Richmond and Queen's Bridges (see chapter 9). None of the three has been widened or otherwise altered.

Savage answered another advertisement for bridge designs in 1815 and again obtained the commission. It was Tempsford Bridge over the Ouse in Bedfordshire.[56] His estimate of £7,300 was much less than any contractor's offer to build it and the first contract made was for £9,550. There was trouble and dissatisfaction with Savage as architect, with the contractor and with the foreman, ending with the latter going to jail. The bridge was finished in 1820 and by 1823 was in need of repairs. Somewhat surprisingly, it has lasted well and still carries the northbound carriageway of the Great North Road. It has an interesting form. The arches are segments, but with slight chamfers at the haunches forming something like *cornes de vâche*; and the piers seem to be also shaped for ease of flow, being smoothly curved from the river-bed up to the spring of the arches instead of having the usual offsets near the bottom. The points of the cutwaters have a similar sweep.

Alexander Nimmo (fig. 199) became a bridge designer without undergoing any training as architect or engineer. He is little remembered today but in 1843 Hosking advised young engineers to study the 'various works executed by Mr. Nimmo, in Ireland, for some very fine examples of skill, taste, and engineering

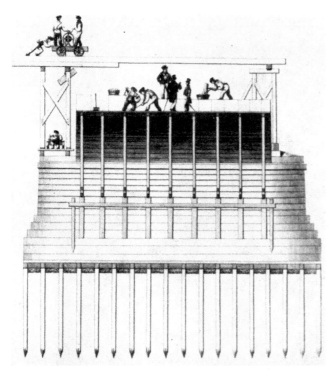

196   The building of Hutcheson Bridge: cross-section (from Weale, 1843).

judgement'.[57] Nimmo[58] was born in Kirkcaldy, Fife, in 1783. He was brilliant at school and attended St Andrews and Edinburgh Universities for three years. He then went north to be second master of Fortrose Academy where it is recorded that he taught 'arithmetic, bookkeeping and drawing; the elements of Euclid, navigation, land surveying and other mensurations; architecture, fortification and gunnery; also the elements of chemistry and of natural philosophy'. He is also said to have read seven languages. At the age of twenty-two he moved to Inverness to be rector of the academy, and there Telford met him and employed him in school vacations to make surveys of some of the county boundaries and some new roads. Four years later he resigned his post – according to Joseph Mitchell because the governors had censured him for failing to attend church with the boys. Telford recommended him for a government appointment as an engineer in Ireland and thus began a short career of what must have been frantic activity. Nimmo supplied the section on theory for Telford's encyclopaedia article 'Bridge'. It contains the theory of equilibration of

197   Design for a two-level bridge at Newcastle-upon-Tyne by R. Stevenson (from D. Stevenson, 1878).

198   O'Donovan Rossa Bridge, Dublin (formerly Richmond Bridge).

arches, of stability of piers and of the flow past piers of various shapes and size, with long discussions of the validity of theories in practice. He later contributed the article 'Carpentry' and probably provided the extensive material on Irish canals for Telford's article on 'Inland navigation'. In Ireland he made surveys of bogs and reported on reclamation schemes, built harbours and piers all round the coast and directed the spending of large sums on public works to relieve poverty and famine.

The bridges Nimmo is known to have designed show that he was as conscious of style as Telford, but they suggest that he wished to be master of all styles. His road construction in the West of Ireland was similar to Telford's work in Scotland, but there is evidence of much more attention to the style of individual bridges, even when they were small. On remote roads, he built Toombeola Bridge, three elliptical arches of almost black limestone over a sea inlet; Glenocally Bridge, which is a pointed arch dated 1826 on the keystone; Kilhale Bridge by Lough Beltra which is a segmental arch dated 1829, of hewn sandstone and with the wingwalls skilfully curved to form a bend in the road; and the sculptured abutments of Maam Bridge, the arch of which has been replaced by a steel girder. Beside

199   Alexander Nimmo, bust by J. E. Jones.

Maam Bridge, as a personal investment, he built an inn[59] of quite original architecture. But the flamboyance of his bridge design is best illustrated by three structures which still carry main road traffic. At Poulaphouca the turnpike route from Dublin to Baltinglass crosses a very deep, tree-filled gorge by a bridge which Nimmo designed in the late 1820s. His response to the romantic setting was to build 'a picturesque bridge of one pointed arch springing from rock to rock . . . in an antique style'[60] (fig. 200). The width of the road is generous at 30 ft, the span of the arch is 65 ft and the height from the bed of the river is 150 ft. A quarter of a mile further south is an even less likely structure (fig. 201), never attributed to Nimmo in print but obviously built to match the Poulaphouca bridge, being of the same width, the same stone, and with a Gothic arch. It is only a farmer's accommodation through the embankment which carries the road across a dry valley, but designed like the gateway of a medieval fortress, with a large vaulted room in one of the abutments. These structures say much about Nimmo's personality but are not evidence of great engineering skill. That evidence is in the Wellesley Bridge[61] at Limerick, now known as Sarsfield Bridge (fig. 202).

It is the last bridge over the Shannon before it enters its estuary. A government grant of £60,000 was made for the building of the bridge and enclosure of a portion of the river to form a wet dock. The whole scheme was designed by Nimmo and begun in 1824, but it was not completed until 1835, three years after his death, and the total cost was £89,000. The bridge itself is of five

201 Arch under road near Poulaphouca.

202 Sarsfield Bridge, Limerick (formerly Wellesley Bridge).

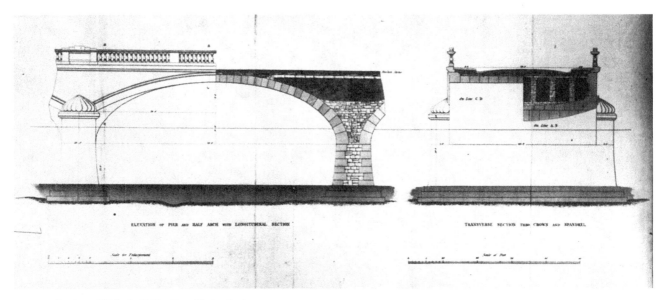

ELEVATION OF PIER AND HALF ARCH WITH LONGITUDINAL SECTION

TRANSVERSE SECTION THRO CROWN AND SPANDREL

203   Sections of Wellesley Bridge (from Weale, 1843).

equal arches with horizontal roadway 42 ft wide between the parapets. It is built of pale grey limestone, has a continuous open balustrade on one side and balustrade alternating with solid parapet on the other, perfectly plane faces of arches and spandrels, and the caps of the cutwaters carved like sea-shells. The line of each arch on the façade is a high, flat segmental curve, conveying the lightness that Telford often sought; it is 70 ft in span and of 8 ft 6 in. rise. But the curve changes to an ellipse of 17 ft 6 in. rise at the middle of the width. Nimmo was not content with ordinary *cornes de vâche*; he curved the whole soffit to a double bell-mouth shape as can be seen in the cross-section (fig. 203) and is just noticeable in the photograph (fig. 202). The masonry of the piers is also dressed to smoothly curved surfaces past which the water flows with the least possible turbulence (fig. 203). Moreover these special forms are not just a proud display of ideal scientific shapes, for the

tide rises high on the piers and every tide that is higher than ordinary springs goes part of the way up the streamlined soffits. The design was therefore economical as well as scientific and elegant, and the condition of the bridge today shows that the construction was excellent; there is only one sign of structural movement, a slight distortion of the lines of the end arch nearest to the city.

In everything except size Nimmo's bridge is the equal of any of Perronet's and in its basic form at least a little nearer perfection. Like Telford's Dean Bridge, George Rennie's Staines Bridge, Harrison's Grosvenor Bridge and Potter's High Bridge it carries today's enormous traffic as readily as the carts and coaches for which it was designed. Together these bridges represent the best of the art and skill of bridge-builders at the opening of the railway age.

APPENDIX 1

CONTINENTAL 'TEXTBOOKS', 1714–1752

A. A. Palladio

I quattro libri dell' architettura (**Venice, 1570**). Extracts are from the second English translation, published by Isaac Ware (London, 1738) and in a facsimile edition by Dover Publications (New York, 1965). The first English translation was published by G. Leoni in London in 1715–16.

The place . . . for building bridges ought to be . . . where the river has a direct course, and its bed equal, perpetual, and shallow.

Book 3, chapter 4

The pilasters [i.e. piers] . . . ought to be in number even; as well because we see that nature has produced all those things of this number, which being more than one, are to support any weight, as the legs of men, and all other animals can justify: as also because this same compartment is more agreeable to be looked at, and renders the work more firm; because the course of the river in the middle, (in which place it is naturally more rapid, as being farther from the banks) is free, and doth not damage the pilasters by continually shaking them. . .

The pilasters ought not to be thinner than the sixth part of the breadth of the arch; nor ordinarily thicker than the fourth. They must be made with large stones, which are to be joined together with cramps, and with iron or metal nails, that by such concatenations they may come to be all of one piece. The fronts of the pilasters are commonly made angular, that is, that they have in their extremity a rectangle [i.e. a pointed end of 90° angle]; and some are also made sometimes semicircular, that they may cut the water, and that those things which are carried down by the impetuosity of the river, may, by striking against them, be thrown off from the pilasters, and pass through the middle of the arch.

The arches ought to be made firm and strong, and with large stones, which must be well joined together. . . Those arches are very firm that are made semicircular, because they bear upon the pilasters, and do not shock [thrust?] one another. But if . . . the semicircle should offend by reason of the too great height, making the ascent of the bridge difficult, the diminished must be made use of, by making arches, that have but the third part of their diameter in height [Designs in the book show that a ratio of height to *span* of one-third is meant]; and, in such case, the foundations in the banks must be made very strong.

Book 3, chapter 10

B. H. Gautier

Traité des ponts (**Paris, 1714**). From extensive experience as a government engineer Gautier describes methods of diverting rivers and founding piers. If possible piers should be founded on rock, which should first be levelled and excavated several inches to receive the bottom of the masonry. For other ground there are six methods:

(a) joists across the width of the pier and planking,
(b) the same with piles below the joists,
(c) 'simple grillage',
(d) grillage of 10–15 in. square timbers with openings 2–2½ ft square and two piles driven in each opening,
(e) piles with sheet-pile surround, to compact the ground and prevent scour,
(f) 'encaissements' and materials dumped through the water to form mounds on which piers may be founded.

For draining cofferdams and excavations machines are subject to breakdown and Gautier favours baling by men with buckets.

Downstream cutwaters should be acute-angled, upstream cutwaters of any of the common shapes. For structural dimensions, the experience of local materials must be allowed to moderate the rules based on books, existing structures or theory.

Dissertation sur les culées, voussoirs, piles et poussées des ponts (**Paris, 1717**). For the proportion of pier thickness to arch span, he quotes the advice of Alberti ($\frac{1}{4}$), Palladio ($\frac{1}{4}$ to $\frac{1}{6}$), Serlio ($\frac{1}{2}$, in the ancient bridges at Rome) and Blondel ($\frac{3}{8}$, in his one bridge design), and the dimensions of the Pont du Gard ($\frac{1}{4}$), the Pont Royal in Paris ($\frac{1}{3\frac{1}{3}}$) and the Pont Neuf at Toulouse ($\frac{1}{4}$); for the proportion of arch thickness to span, the rule of Alberti (not less than $\frac{1}{15}$), Palladio's drawings of three ancient bridges ($\frac{1}{10}$ to $\frac{1}{12}$), one of Palladio's designs ($\frac{1}{14}$ to $\frac{1}{17}$), and Serlio's figures for the bridges at Rome ($\frac{1}{10}$ to $\frac{1}{12}$).

He then writes his own table of recommended dimensions based on the following rules:

$$\frac{\text{thickness of abutment}}{\text{span of arch}} \quad \frac{4}{15} \text{ for spans 50 ft and upwards}$$

thickness of pier / span of arch $\frac{1}{5}$ for spans 20 ft and upwards

thickness of arch / span of arch $\frac{1}{15}$ for spans 40 ft and upwards (with strong stone)

$\frac{1}{15}+1$ ft for spans 40 ft and upwards (with weak stone)

For shorter spans the ratios should be increased, because the load of traffic is proportionately greater relative to the weight of the structure.

C.   B. F. de Bélidor

Architecture hydraulique, vol. IV (Paris, 1752), book 4, chapter 11. The length of a bridge (including piers) should be at least equal to the existing width of the river. The number of arches should be uneven, the gradient of the road not steeper than 1 in 24, and where the banks are high the line of road should be horizontal and the arches all equal. The width between the parapets should be 27 ft, but in a city bridge 48 ft, as at the Pont Royal.

The springings of arches should be normally at and never below low-water, except in pointed arches. The crown of the middle arch should be 3 to 4 ft, and other arches 2 ft, above high-water. Arches should be semicircles or ellipses with rise equal to $\frac{1}{3}$ span (as in

Pont Royal). The plan of cutwaters should be two arcs of 60° with radius equal to the pier thickness, tangent to the long sides of the piers. Radial courses in the outside of the spandrels are commended (as in the bridge of Compiègne and Pont Royal). The interior should be filled with masonry.

Rules for thickness of piers:

Semicircular arches $\frac{\text{span}}{6}+2$ ft, for spans up to 48 ft,

reducing to $\frac{\text{span}}{6}$ at span of 96 ft and greater.

Elliptical arches (rise $=\frac{1}{3}$ span) $\frac{\text{span}}{5}+2$ ft, for spans up to 48 ft,

reducing to $\frac{\text{span}}{5}$ at span of 96 ft and greater.

Thickness of abutment to be $\frac{7}{6}$ times thickness of first adjacent pier.

Rules for thickness of arches at keystone:

Semicircular arches $\frac{1}{24}$ of diameter or span $\Big\}$ irrespective of

Elliptical arches $\frac{1}{12}$ of radius at crown $+1$ ft strength of stone

For Bélidor's advice on machines and construction see above, chapter 4.

APPENDIX 2

CALCULATION OF THE FALL UNDER A BRIDGE BY WILLIAM JONES'S METHOD

This is a précis of Robertson, 'Concerning the fall of water under bridges' (1758)

Principles

1. Since the quantity of water passing through the arches in one second is the same as that flowing through a cross-section of the unobstructed channel (for instance, a cross-section a short distance upstream), the velocities in the unobstructed channel and in the waterway through the arches are in inverse proportion to the cross-sectional areas.

2. 'Water forced out of a larger chanel thro' one or more smaller passages, will have the streams thro' those passages contracted in the ratio of 25 to 21.' This contraction was measured by Newton for streams emerging from sharp-edged orifices (*Principia*, book 2, prop. 36).

3. The bed of the river is assumed of a constant level both upstream and under the arches.

Derivation of the fall

Let b = breadth of unobstructed channel,
  c = breadth of waterway through the arches,
so that $b - c$ = breadth of obstructions (i.e. piers).
By Newton's observation, the *effective* breadth of the obstructed channel = $\frac{21}{25}c$

If v = velocity of flow in the unobstructed channel,
$v\dfrac{b}{\frac{21}{25}c}$ = velocity through the arches.

Height of fall from rest to cause velocity v

$$= \frac{v^2}{2g}, \text{ say } h_1.$$

Height of fall from rest to cause velocity $v\dfrac{b}{\frac{21}{25}c}$

$$= \frac{1}{2g}\left(v\frac{b}{\frac{21}{25}c}\right)^2, \text{ say } h_2.$$

(g = gravitational constant)

Then fall of water surface from level upstream to level under the arches = $-(h_1 - h_2)$

$$= \frac{v^2}{2g}\left[\left(\frac{25}{21} \cdot \frac{b}{c}\right)^2 - 1\right]$$

Examples

1. London Bridge in 1746, as measured by Labelye.

  $b = 926$ ft
  $c = 236$ ft at time of greatest fall
  $v = 3\frac{1}{6}$ ft/sec. (measured)
  $2g = 64.4$ ft/sec.^{2}

Reduce c by one-sixth to allow for the obstruction caused by 'drip-shot' piles (see chapter 6). Revised $c = 196$ ft.

$$\text{Fall} = \frac{(3\frac{1}{6})^2}{64.4}\left[\left(\frac{25}{21} \cdot \frac{926}{196}\right)^2 - 1\right] = 4.74 \text{ ft.}$$

The greatest fall measured about the year 1736 was 4 ft 9 in.

2. Westminster Bridge as built.

  $b = 994$ ft
  $c = 820$ ft
  $v = 2\frac{1}{4}$ ft/sec. (measured)

$$\text{Fall} = \frac{(2\frac{1}{4})^2}{64.4}\left[\left(\frac{25}{21} \cdot \frac{994}{820}\right)^2 - 1\right]$$
$$= 0.084 \text{ ft}$$
$$= 1 \text{ in.}$$

The greatest fall measured by Labelye was about $\frac{1}{2}$ in.

APPENDIX 3

THEORY OF THE EQUILIBRATED ARCH

After de la Hire, Emerson and others

Given the curve of the soffit of an arch (fig. 1a), the joints between the voussoirs being perpendicular to the curve and the surfaces of the joints frictionless: To find the weight required in each voussoir for equilibrium of forces.

Since the contact of voussoirs is frictionless, the action of one voussoir on another can only be a force perpendicular to the surface of contact. Hence the three forces on the keystone of the arch are those shown in the figure. If a value is assumed for the weight W_k, the magnitude of the other two forces can be found from the triangle of forces outlined in heavy lines in fig. 1b, the sides of the triangle being parallel to the directions of the forces and the lengths of the sides proportional to the magnitudes of the forces. If a triangle of forces is drawn to the same scale for each voussoir in turn, working outwards from the keystone, the triangles fit

together to make up the force diagram in fig. 1b. As the joints between voussoirs approach horizontal, the required weight of voussoir increases; for instance, the weight (W_x) of voussoir X, shown in heavy lines, must be very much greater than that of the keystone (W_k). In the limit, if the joint at the springing is horizontal, the required weight of the last voussoir is infinite.

To provide the required weights – or an approximation to the required weights – with stone of uniform density mathematicians suggested either

(i) voussoirs extended radially, as in fig. 1c, or

(ii) voussoirs with horizontal tops carrying columns of masonry of varying heights, as in fig. 1d.

When calculus was used, the curve of equilibrium load was found to be a smooth curve corresponding in height to the columns in fig. 1d. Such a curve is shown in the first diagram of fig. 47 (chapter 5), taken from Emerson's *Mechanics*, 2nd edn (1758).

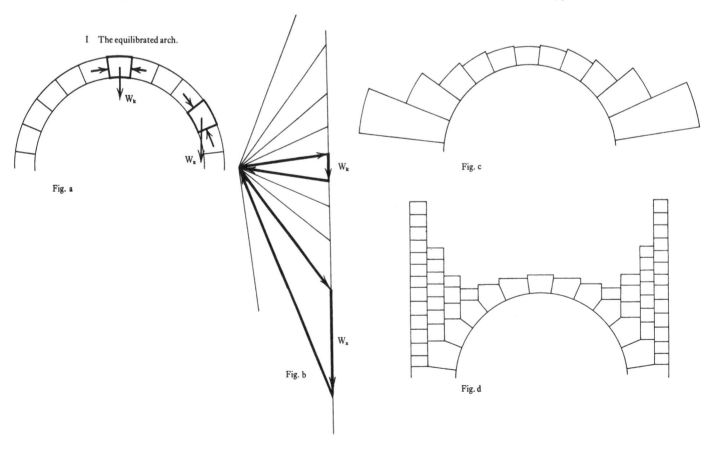

I The equilibrated arch.

Fig. a

Fig. b

Fig. c

Fig. d

APPENDIX 4

SURVEYS OF COUNTY BRIDGES, WEST RIDING (1752) AND NORTH RIDING (c. 1805)

The sources are 'Book of the bridges belonging in whole or part to the West Riding of the County of York . . . drawn by order of the General Quarter Sessions . . . 1752. By Robert Carr and John Watson' and 'Plans and elevations of the North Riding Bridges . . . [by] John Carr' (c. 1805).

Much of the following data is derived from notes on some of the drawings. Since other drawings lack such notes the data are incomplete and record the number of times a feature was noted, not necessarily the number of times it occurred. The data for the histograms of bridge widths, maximum arch spans and angles of arc of largest arches, are complete (fig. 11), being measured from the drawings.

WEST RIDING SURVEY 1752	NORTH RIDING SURVEY c. 1805
	Materials
1 bridge of timber	1 bridge of timber
1 bridge with brick arch	1 bridge with brick arches
114 bridges of stone	110 bridges of stone
	1 bridge with iron arch
	Type of masonry
53 bridges of hewn stone	13 bridges of hewn stone
10 bridges partly of hewn stone	5 bridges partly of hewn stone
	16 other bridges 'of good stone' (4 of these 'ill executed')
	Form of arches
7 bridges with all main arches ribbed	No ribbed arches mentioned
4 bridges with some main arches ribbed	
(only one of the 11 had arches ribbed *and* pointed)	
9 bridges with all main arches pointed	4 bridges with all main arches pointed
7 bridges with some main arches pointed	5 bridges with some main arches pointed
(two of the 16 rebuilt and one widened before 1768)	
No bridges with elliptical arches	2 single-arch bridges with elliptical arches
No bridges with skew arches	2 bridges with skew arches
	Number of main arches
42 bridges of 1 arch	63 bridges of 1 arch
30 bridges of 2 arches (12 asymmetrical)	14 bridges of 2 arches (4 asymmetrical)
22 bridges of 3 arches (12 asymmetrical)	21 bridges of 3 arches (4 asymmetrical)
10 bridges of 4 arches (6 asymmetrical)	6 bridges of 4 arches (1 asymmetrical)
4 bridges of 5 arches (2 asymmetrical)	2 bridges of 5 arches (1 asymmetrical)
2 bridges of 6 arches (both asymmetrical)	2 bridges of 6 arches (1 asymmetrical)
1 bridge of 7 arches (asymmetrical)	1 bridge of 7 arches (asymmetrical)
1 bridge of 8 arches (asymmetrical)	1 bridge of 8 arches (asymmetrical)
3 bridges of 9 arches (2 asymmetrical)	1 bridge of 11 arches

(Asymmetry is evidence that the bridge was built or rebuilt at two or more different times, and therefore usually an old bridge.)

Approaches

16 bridges with narrow or angled approaches at both ends	13 bridges with narrow or angled approaches at both ends
21 bridges with narrow or angled approach at one end	21 bridges with narrow or angled approach at one end

Foundations

15 bridges founded on rock	9 bridges founded on rock (one on 'shivery rock')
2 bridges with the foundations 'on the river-bed'	23 bridges on shallow or suspect foundations
4 bridges on 'deep foundations' (one of these also 'very bad')	4 bridges on 'bad' or 'ill executed' foundations
	1 bridge with foundations on 'strong clay'
1 bridge on a timber grate on the river-bed	1 bridge with foundations on 'gravel'
	19 bridges founded on piles

Protection of river-bed

14 with whole river-bed framed and sett	3 with whole river-bed framed and sett
1 with whole river-bed 'sett and cramped' (without timber frame)	
18 others framed and sett round all piers and abutments	1 other framed and sett round all piers and abutments
6 others framed and sett round some piers	
2 with dams downstream	

Administration

5 bridges shared with Westmorland	1 bridge shared with Westmoreland
1 bridge shared with Derbyshire	6 bridges shared with Durham
2 bridges shared with North Riding	2 bridges shared with West Riding
1 bridge shared with York City	5 bridges shared with East Riding
1 bridge shared with Tadcaster Town	1 bridge shared with Richmond Corporation
105 bridges maintained entirely by the Riding	100 bridges maintained entirely by the Riding

(The figures for administration include several minor approach bridges, etc., which are omitted from the other statistics.)

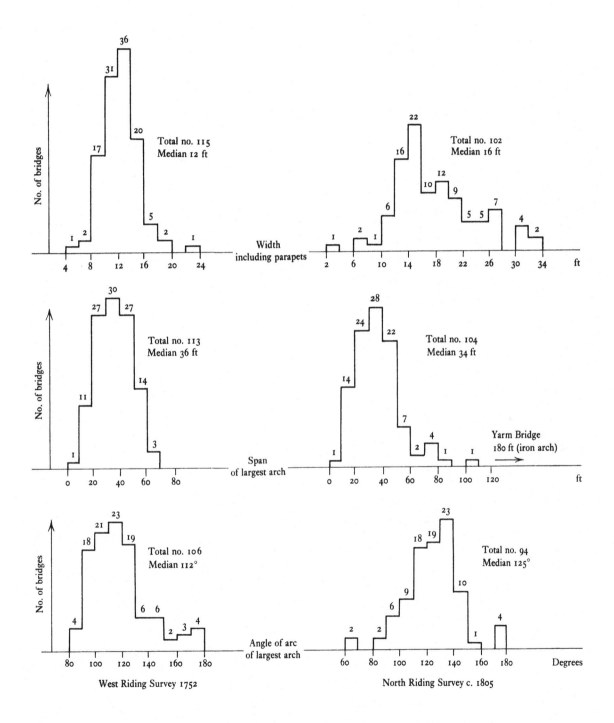

II Histograms and median values of bridge dimensions.

GLOSSARY

To keep this glossary to a reasonable length it has been written as a complement to the *Illustrated glossary of architecture 850–1830*, by J. Harris and J. Lever (Faber and Faber, London, 1966 and 1969), which should be consulted for all terms not listed here and also for illustrations of a number of the terms which are included here.

All the meanings given are valid in this book, but they are not in all cases the most common meanings. Some of the terms have different meanings when used at other periods or in other contexts.

Cross-references are indicated by the use of italics.

abutment: the resistance offered to the horizontal thrust of an arch or, more generally, the body (usually of masonry) which provides the resistance; and hence the end support of a bridge at the river banks. See fig. III.

arch: a curved structural member. Typically it supports vertical loads across an opening. See fig. III.

arch ring: the vertical face or edge of an arch showing on the elevation of a bridge. See fig. III.

Archimedean screw: see *endless screw*.

archivolt: a projecting moulding which follows the curve of an arch on top of the *extrados*. See fig. III. Several writers have used the word more loosely to mean the *arch ring* on the façade, or the shape of the arch curve.

balk or *baulk*: a piece of timber of a large, square cross-section. It was generally 6–12 in. square and was often a *whole timber*.

ballast: gravel obtained by dredging from a river-bed. It was probably a mixed material containing some rather large stones, and also some sand.

barge: a boat used for cargo on inland waters. Generally wide and flat-bottomed, it might be driven by sail or towed by a horse.

battardeau: French term for *cofferdam*.

batter: the slope of the surface of a wall (expressed, for instance, as 1 horizontal to 10 vertical); or the sloping surface itself. See fig. 189.

beam: a straight, solid, structural member which supports loads and transfers them to its supports by its resistance to bending. It is contrasted with an *arch*, which is not straight, and a *truss*, which is not solid.

bearing piles: piles which transfer loads (such as the weight of a bridge) to the ground at some depth below the surface. They are contrasted with *sheet piles*.

bond-stone: a stone, generally long, which extends from one course or other division of masonry into another, thus bonding the two together.

boring: a hole of small diameter made in the ground with an auger or drill to identify the materials underground and estimate their strength.

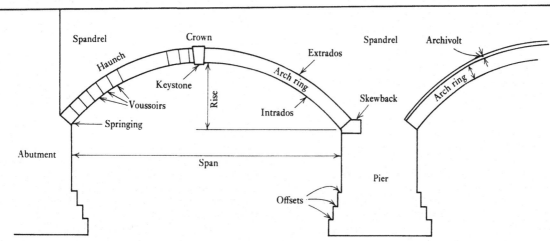

III   Parts of an arch bridge.

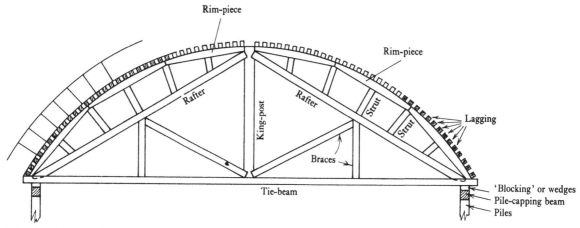

IV One rib of a trussed centre.

borough bridge: a bridge for which the responsibility of maintenance lay with the council of the town or borough in which it stood.

brace: a secondary structural member which serves to maintain the form or the stability of the primary structure.

bracing: the system of braces in a structure, e.g. the diagonal members of a truss or centre. See fig. IV.

butment: see *abutment*.

caisson: water-tight, open-topped box for founding *piers*. See p. 8 and figs 5, 67, 68 and 86.

carriage: in the eighteenth century the word was generally used to describe a horse-drawn cart used to transport coal and other materials in bulk.

cast iron: iron which is poured molten into moulds (usually of sand) to form castings of the shape required. The metal has a high carbon content and is brittle, but is less subject to corrosion than *wrought iron*.

casting: the unit of cast iron made in a single mould.

catenary or *catenaria*: the name of the type of curve formed by a chain hanging freely from two points of support. See p. 46.

caulk: to seal the joints of a ship or water trough, usually by packing them with a fibrous material such as hemp.

cautioner: one who guaranteed the performance of a contract. For most bridge contracts the builder had to find two cautioners who joined him in a bond for completion of the contract before an agreed date.

centre, centring: the temporary structure used to support an arch during its erection. See fig. IV.

chain-bar: a long bar of *wrought iron* used as a tie between the parts of a structure, e.g. spandrel walls of a bridge, or ribs of a centre. See fig. 94.

chain pump: see pp. 38–9 and fig. 39.

cofferdam: a temporary dam forming an enclosure from which the water can be removed in order to construct foundations on the dry river-bottom.

cohesion: the ability to resist tensile stress. (The term is seldom used nowadays with this meaning.)

column: a vertical structural member carrying downward load. For the meaning in classical architecture see Harris and Lever's *Glossary*.

cornes de vâche: French term for tapered chamfering of the edges of arches. See p. 176 and fig. 174.

counter arch: sometimes used to mean an *inverted or reversed arch* in a spandrel (fig. 62), but sometimes a secondary arch laid with radial joints over the primary arch (see p. 13 and fig. 11).

county bridge: a bridge for which the responsibility of maintenance lay with the Justices of the Peace in a county, meeting in Quarter Sessions.

cramp: a metal bar by which large blocks of stone are held together, usually by insertion of turned-down ends of the bar into holes in both blocks. May be seen in the copings of bridge parapets. The cramp was generally of *wrought iron* and its ends were secured in the holes by running in molten lead around them.

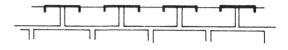

crosswall or *cross wall*: a wall (for instance in a spandrel) running across the bridge from façade to façade.

crown (of arch): the highest part of an arch, about the keystone. See fig. III.

cutwater: the end of a bridge pier protruding beyond the face of the spandrel and shaped to divide the stream.

day labour: work paid for according to the time and number of men employed. It was favoured in circumstances which made the estimation of costs difficult, e.g. foundation work.

direct labour: a system of construction in which all the workmen are employed and paid wages by the promoter or owner of the bridge, and work under his supervisor or surveyor. Contrasted with construction by contract in which the workmen are paid wages by the contractor, but he is paid agreed sums by the owner on the completion of agreed parts of the bridge.

dovetail: a piece of timber, stone or iron formed to a splayed shape like the tail of a dove and fitting into a socket or hole of similar shape to make a tight connection.

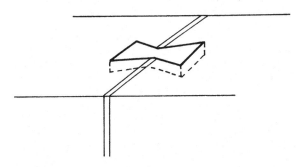

dowel: a pin of wood or metal which connects two stones or structural members by being inserted into matching holes in the members.

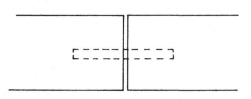

drag: a small platform attached by a rope to a carriage and dragged like a sledge up one side of a steep bridge to act as a brake on the descent of the carriage on the other side. See pp. 50–1.

drawbridge: a span which could be raised on hinges at one end to allow the passage of tall vessels through a bridge. Always built of timber until after 1800.

dredging bag: a bag fixed on the end of a long pole for dredging mud and sand from river bottoms.

ease (a centre): to reduce the supporting forces under a centre by a small movement of the wedges (see fig. IV), so that the arch is made to support its own weight for a time before the centre is removed.

endless screw or *Archimedean screw*: a helical tube or helical screw in a straight tube which, when placed on an incline with its lower end in water and rotated, raises water constantly to the top of the tube. See fig. 40.

equilibration: the mathematical determination of the distribution of loads which will ensure equilibrium. See appendix 3.

equilibrium, equilibrio: a state of balance between the forces on a body. Refers in this book to the equilibrium of the *voussoirs* or sections of an arch. See appendix 3.

extrados: the convex surface of an arch. See fig. III.

fall (under or through a bridge): the difference of level of the water surface upstream and downstream of a bridge. See appendix 2.

fill: material (such as gravel, earth or rubble from demolition) used to fill the space in *spandrels*, behind retaining walls, etc., and to construct embankments.

fire engine: the common name for atmospheric engines and steam engines before about 1800.

fish-plate: a connecting plate placed across a joint between two members and bolted to each, or similarly across a crack for repair.

flange: a wide flat projection on a structural member, usually at right angles to the main member, e.g. the end flanges on sections of an iron arch, see fig. 148.

flush (with mortar): refers to the pouring of liquid mortar on to completed masonry (e.g. on the extrados of an arch) to ensure that all the joints are filled.

flush (of arch stones): refers to the spalling or cracking of the corners of stones at points of high stress on the extrados or intrados. Prevented by ensuring that sharp corners do not come into contact, i.e. by thick mortar or *V-joints*.

fluxions: Newton's term for the mathematical method now known as the differential calculus.

folding wedges: two wedges driven against each other to adjust the level of the structure they support and retracted to remove the support. Often placed under centring, see fig. IV.

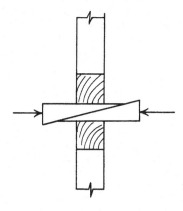

forced earth: compacted fill. See *fill* and p. 65.

framed and sett (of a river-bed): see p. 26 and figs 21 and 24.

girder: in the eighteenth century, a large beam of wood. Gradually changed its meaning to any large beam, but generally a *trussed beam* or *truss* shaped to match the varying bending moment. See p. 158.

gothic or *Gothic*: in bridges, refers to (i) the use of pointed arches or (ii) decorative mouldings derived from Gothic architecture, as in fig. 54.

grating: a frame of large timbers laid on the river-bed or on the heads of piles, as a base for the masonry of *piers* or *abutments*. See fig. 85, but more commonly the grating was of balks equally spaced in both directions.

guard works: piles, booms, etc., surrounding construction works to protect them from collisions by river craft.

half timber: piece of timber approximately half as thick as it is wide, made by sawing a *whole timber* into two along its length.

haunch: the part of an arch midway between its springing and its crown. See fig. III.

hewn stone: stone worked to smooth joint surfaces to give fine mortar joints. The external surfaces are also worked but may be smooth, rusticated, rock-faced, etc.

high-water: short form of 'high water level of ordinary spring tides', a level which is established by averaging many observations of the levels of spring tides.

hollandoise: see p. 39 and fig. 41.

hoop: a wrought iron band fitted round the head of a pile to prevent it splitting under the blows of a *monkey*. Also used for similar bands of iron in other positions.

horizontal arch or *platband*: a horizontal structure which is not curved but bears load as an arch because it is formed of wedge-shaped stones or sections. See fig. 146.

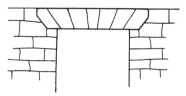

hydraulic architecture: French term covering all construction work in and under water.

hydraulic lime: lime which is manufactured by burning a limestone which contains siliceous impurities and therefore burns to produce calcium silicates which harden by hydration in water. It therefore forms mortar which will set and harden under water. (Pure lime never hardens while it remains wet.)

hydraulics: the science of moving water.

hydrostatics: the study of the pressures and forces of water at rest.

intrados: the concave surface of an arch. See fig. iii.

inverted arch: (i) an arch with its concave surface upwards, as in spandrels (see fig. 180); (ii) masonry laid from pier to pier on a river-bed or on the bottom of a water channel with its surface concave upwards (see fig. 175).

iron cement: a mixture of iron filings and certain chemicals, used to fill cracks or joints in iron pipes, etc.

ironmaster: proprietor of an iron works or manufactory.

joggle: an interlock between stones in *hewn masonry*. See figs 62 and 98.

journeyman: a workman hired by the day.

keel: a *lighter* or flat-bottomed *barge* of the type used in the north-east of England, especially on the Tyne.

key: a block of stone, wood or iron which is inserted into cavities so as to prevent sliding at a joint between two structural members. Generally thicker than a *dowel*.

keystone: the *voussoir* placed last at the crown of an arch, see fig. iii. It was made slightly too wide so that in driving it downwards the haunches were forced outwards and the arch tended to lift off the centring.

king-post: in a roof truss, the vertical timber at midspan connecting the ridge to the tie-beam. The word became used more generally for vertical posts in framed structures, with characteristic notches to receive the rafters as shown in figs 27 and 78. See also fig. iv.

lagging: the horizontal timbers which are laid on top of a *centre*, bridging between the *ribs*, to support the masonry. See fig. iv.

lamination (of timber): the use of a number of pieces of timber laid face to face and held firmly together by bolts, clamps, etc., to make a structural member of large cross-section. See figs 26 and 37.

landbreast: a term used rather loosely to indicate the *abutments* or other construction where a bridge reaches the bank of a river.

lighter: a boat, usually open and flat-bottomed, used in loading and unloading ships.

line of thrust: the line through the points of resultant force between *voussoirs* at all the joints of an arch (see appendix 3 and pp. 47–8).

load: the term used in structural mechanics for forces applied to a structure. For arch bridges the loads are the weights of all parts of the structure, the weight of traffic, and forces caused by the impact of ships, ice and flowing water.

lock: the word used for one of the passages through an arch of old London Bridge and a few other bridges, by analogy with the locks used for navigation on canals.

low-water: short form of 'low water of ordinary spring tides', established as for *high-water*. At bridge sites without tides it means the lowest water level which commonly occurs and is often called the 'summer level'.

lozenge: a diamond-shaped panel. Used of the form of spandrel members of iron bridges, see p. 164 and fig. 158.

malleable iron: used as a synonym for *wrought iron* at the period of this book. 'Malleable iron' has a different meaning today.

maul: a heavy wooden hammer.

millwright: a maker of industrial machinery.

monkey: the falling weight or hammer in a *pile-driver*.

newel: a pillar, often ornamental, which terminates the *wingwall* of a bridge.

offset: the horizontal projection of a lower course outside the plan of a higher course of masonry. See fig. iii

Palladian bridge: the commonest meaning is a bridge in the style of Lord Pembroke's bridge at Wilton House (see p. 11), but in public bridges the term means a bridge with aedicules like that of Rimini (fig. 2).

patternmaker: a maker of wooden patterns for the moulds in a foundry.

penning: the paving of stones in a *framed and sett* river-bed.

perpendicular: almost invariably meant vertical.

philosophy: commonly meant science.

pier: the support between two arches. See fig. iii.

pig-iron: iron as run from the smelting furnace. Not always differentiated from *cast iron* at the period of this book.

pile: a vertical member (always of timber at the date of this book) driven into the ground for vertical support. (See also *sheet piles*.)

pile bridge: a bridge supported on piles whose heads are above water level.

pile-driver or *piling machine*: see pp. 8–9 and figs 6, 7 and 44.

pin: a wooden connector similar to a *dowel*.

piston pump: a pump in which water is raised by the movement of a piston in a tube (called the barrel). Includes the common force pump and lift pump.

platform: the horizontal frame or floor of wood on which the masonry of a pier is built. Usually consists of a *grating* covered by boards. See fig. 104.

pozzolana: volcanic dust found at Pozzuoli in Italy which contains active silica and can be mixed with lime to give it the properties of *hydraulic lime* or *water cement*. It was imported to Britain at a considerable price.

puddle or *puddle clay*: clay, sometimes mixed with sand, and worked with water (i.e. 'puddled') until it is uniformly plastic and impermeable.

puncheon: a short timber post. See, for instance, p. 15 and fig. 12.

quicksand: a sand which is incapable of bearing any weight because its particles are made buoyant by an upward flow of water.

ragstone: a type of coarse-grained sandstone. There are well-known quarries near Maidstone in Kent.

recess or *refuge*: a space surmounting a *cutwater* and buttress, where pedestrians can retire out of the line of moving traffic. See, for instance, figs 10, 46, 102.

reversed arch: see *inverted arch* (i).

rib (of a masonry arch): a band of masonry projecting from the soffit of the arch. See fig. 99.

rib (of a timber arch, iron arch, or centre): one of several plane structures in parallel vertical planes which together form the spanning structure of the arch or centre. See figs 28, 167, and IV.

rim-piece (of centre): timbers shaped on one side to the intrados curve of the arch. See fig. IV.

rise (of arch): the vertical height from the springings to the crown of the intrados. See fig. III.

rise (of tide): the vertical height from *low-water* to *high-water*.

road plates (of iron bridge): iron plates placed on top of the spandrels to carry the gravel, etc., of the roadway.

rubble stone(s): quarried stones of angular shape and random size.

saliant or *salient angle* (of a pier): the *cutwater*.

scarf: an end-on joint between two timbers of the same cross-section formed by bolting through long tapered ends.

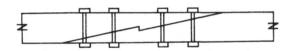

scour: erosion of river-bed or banks by the flow of water.

scrabled or *scappled masonry*: stones with their external surfaces dressed plane but not smooth, generally with the pick of the mason's hammer.

screw: generally meant a screw jack at the date of this book.

security: refers either to a person bound with the main contractor in a contract to guarantee his performance, or to the guarantee, or to the amount of the guarantee. See *cautioner*.

sheet piles or *sheet-piling*: piles (usually rectangular) driven edge to edge so as to form a continuous sheet or wall below the ground. See figs 85, 87 and 178.

shoe (of a pile): an iron point on the foot of a *pile* which helps it to penetrate the soil.

shoulder (of a pier): the vertical line in which the angled or curved face of the *cutwater* meets one of the parallel sides of the *pier*.

sluice: a small door covering an opening in a gate, wall or vessel, by which water can be allowed to pass in a fully controlled manner.

soffit: the under surface of any structural member.

sough: a tunnel in a mine which contains a water-course and is used for transport of the coal by boat.

sounding: measurement of the depth of water by a sounding line.

spall: to splinter or flake off a surface, as in damage to masonry by frost or cracking of cast iron subjected to high compressive stress.

span (of arch or beam): clear distance from face to face of the supports. See fig. III.

spandrel or *spandril*: the space between the extrados of an arch, or two adjacent arches, and the bottom of the road metal or road plates. See fig. III.

spate or *speat*: a violent river flood.

specification: a detailed description, item by item, of the work to be done under a contract.

spirelet: a free-standing decorative feature of conical or similar shape.

springing: precisely, the surface of contact between the first course of voussoirs of an arch and the masonry of the abutment or pier. Often simply means the level at which an arch 'springs' from its supports. See fig. III.

square pump: a *piston pump* whose barrel and piston are square.

stage: a jetty or platform raised above water level (usually on piles).

starling: a protective structure surrounding a pier, rising to at least a little above low-water, see fig. 54. Also written 'sterling' and sometimes used loosely to mean only *cutwaters*.

stress: the effect of *load* on a structure. The measure of stress is the force per unit of resisting area and is not necessarily constant at all points of a cross-section.

strike (of centring): to take down or remove.

strut: a structural member transmitting a compression (or 'push') force.

superstructure: the upper part of a bridge, excluding the foundations and at least the lower parts of the piers.

temperature stress: *stress* caused by a change in temperature. If the temperature of a structure rises and it is unable to expand, compressive stress is caused.

terras: ground basaltic rock which was imported from Holland and mixed with lime to form a *water cement*.

thrust (of an arch): the resultant force acting towards the *abutment* in any cross-section of an arch. At the abutment the thrust of the whole arch acts on the abutment. See appendix 3.

tie: a structural member transmitting a tension (or 'pull') force.

tie-stone: similar to *bond-stone*. See fig. 193.

traversing machine: an arrangement of rails and trolleys for conveying materials from delivery on the river bank to their final positions in the structure. See figs 195 and 196.

triumphal bridge: a bridge in the neo-classical style made popular by Piranesi and others about 1740–50. The reference is to the ancient Ponte Trionfale over the Tiber but it is unlikely that the form of this was known accurately.

truss: a framework (of timber at the period of this book) in which the individual members experience only tension or compression although the truss as a whole supports loads like a beam or girder (see fig. 25).

trussed beam: a *beam* which is stiffened by a pair of sloping *struts* (see fig. 30) or *ties* connected to one or more short vertical posts.

turnpike: a road on which tolls were collected and used to supply the costs of maintenance. The word refers to the type of toll-gate first used.

underpin: to strengthen the foundations of an existing structure, usually by placing new masonry or piles below the former foundations so as to obtain support over a wider area and/or at greater depth.

V-joints: V-shaped incisions at the joints of ashlar, obtained by dressing the edges of both the adjacent stones with equal chamfers.

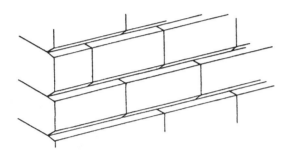

voussoir: an arch stone of tapered or wedge shape. Sometimes more particularly those arch stones which show on the façade of a bridge.

waling: a horizontal timber connected to a wall of *sheet piles* to support them against the pressure of the ground or water, and which is itself supported by struts or frames across the excavation. See fig. 178.

water cement: cement (meaning mortar) for use in or under water. Made with *hydraulic lime* or with lime mixed with *pozzolana*, *terras* or some other additive such as ground slag.

water wheel: a wheel driven by flowing water to provide power for machinery such as pumps.

waterman: a boatman carrying passengers or goods for profit.

waterway: a passage through a bridge through which water flows and, more specifically, the width or sum of widths of such passages.

whole timber: a *balk*, roughly square, made from a whole tree-trunk by trimming off the curved faces only.

wing or *wingwall*: a side wall extending landwards from the abutment of a bridge and retaining the fill under the approach road. See, for instance, figs 122 and 189.

wright: Scottish term for a carpenter at the period of this book.

wrought iron: iron made in bars by beating out the slag and impurities in a semi-molten state on the furnace hearth. It had a low carbon content and so was ductile and non-brittle, with good tensile strength but subject to rapid corrosion. At the period of this book it could not be made in large units. Sometimes referred to as 'bar iron'.

A NOTE ON REFERENCES

Where only a short title, or only the author's name, is given in the notes the full title, place and date of publication will be found in the bibliography at the end of the notes. Every reference to a manuscript source gives the full title and reference number of the document(s) and the name of the archive. In the reorganisation of local government which took place during the period of the author's research many record offices and other departments of local authorities changed their names and in some cases documents were moved. It has not been possible to re-check the locations of all the documents quoted and where no check has been made the name of the record office or department is that which obtained before the reorganisation. With the old reference available the present location of any document will be known to the record office or appropriate department of the new local authority.

General abbreviations

CRO	County Record Office.
MS(S)	Manuscript(s).
QS	Quarter Sessions.
DNB	*Dictionary of national biography*. 22 vols. London, 1908–9.
GM	*Gentleman's magazine*. London, monthly (eighteenth and nineteenth century).
HCJ	*House of Commons journal*.
ICJ	*Irish House of Commons journal*.
LM	*London magazine*. London, monthly (eighteenth century).
OSA	*The statistical account of Scotland*. 21 vols. Edinburgh, 1792–9.
CL	*Country life*. London, weekly.

Archives and libraries referred to by abbreviations

BM	The British Library, Reference Section (including prints and manuscripts), the British Museum, London.
GLCRO	The Greater London Record Office.
InstCE	The Institution of Civil Engineers, London.
LRO	The Corporation of London Records Office.
NLW	The National Library of Wales, Aberystwyth.
NLS	The National Library of Scotland, Edinburgh.
NMRS	The National Monuments Record of Scotland, Edinburgh.
PRO	The Public Record Office, London (Chancery Lane, London).
PRO (Transport)	The Public Record Office, Transport Section (Porchester Road, London).
RIBA	The Royal Institute of British Architects, London.
SRO	The Scottish Record Office, Edinburgh.

Private collections referred to

Sir John Soane's Museum, Lincoln's Inn Fields, London.

The Duchess of Roxburgh's muniments, Floors Castle, Roxburghshire.

Miss J. M. H. Mylne's archive, Great Amwell.

The Earl of Pembroke's muniments, Wilton House, Wiltshire.

The Marquess of Bute's muniments, Dumfries House, Ayrshire.

All the other archives and libraries referred to can be located by their names.

NOTES

1. Westminster Bridge

1. For finding many of the sources for this chapter I am indebted to Carson, 'Provision and administration of bridges over the lower Thames 1701–1801' (1954). See also Carson, 'The building of the first bridge at Westminster, 1736–1750' (1957–8). All the minutes of the Westminster Bridge Commissioners and their committees, with accounts, reports, etc., from 1736 to 1853, are at PRO, reference WORK 6/28–62. Facts from the minutes at obvious dates are not separately noted here.
2. *HCJ*, 22 and 25 March 1736.
3. Hawksmoor, *Account of London-Bridge* (1736).
4. He was one of five men appointed by Parliament to draft the Bill for Westminster Bridge, and was a frequent attender at Commissioners' meetings until his death in 1748.
5. Colvin, *Biographical dictionary* (1954), pp. 272–6, 634–9.
6. Gautier, *Traité des ponts*, 3rd edn (1728). Earlier editions did not contain these specifications, but had a small-scale elevation of the Pont Royal.
7. *HCJ*, 16 February 1736. Other biographical details are from Labelye, *Short account* (1739), preface, and *DNB*, vol. XI, pp. 365–6.
8. Kings Maps III 64.1, at BM.
9. *DNB*, vol. V, pp. 850–1.
10. Hawksmoor, pp. 18–19.
11. Labelye, *Short account*, p. 13.
12. Hawksmoor, p. 9. For the accuracy of this figure see chapter 6, note 6.
13. Hawksmoor, p. 19.
14. Labelye, *Short account*, p. 12, and Robertson, 'The fall under bridges' (1758).
15. *HCJ*, 16 February 1736.
16. Price, *Some considerations* (1735). Dated 25 February 1735.
17. *HCJ*, 20 May 1736.
18. James, *Short review* (1736). Dated 30 July 1736.
19. Langley, *Design for the bridge* (1736). Dated 5 June 1736.
20. A. J. Rowan, 'Batty Langley's Gothic', in *Studies in memory of David Talbot Rice* (Edinburgh, 1975), gives a good biographical sketch of Langley, as does Colvin, but both contain errors about dates of his bridge designs.
21. The full elevation is reproduced in Ruddock, 'Hollow spandrels' (1974), p. 281.
22. B. Langley, *The builder's compleat assistant*, 3rd edn (London, 1750), p. 69.
23. Colvin, pp. 273, 314, 475.
24. Letters concerning Westminster Bridge, June–July 1736,

HER/1–21 at RIBA Drawings Collection, and C. Marquand, *Remarks on the different constructions of bridges* (London, 1749), p. 7.
25. 'Book of particulars relating to Fulham Bridge' (MS), PRO 30/26/13, at PRO.
26. Anon., *A short narrative of proceedings* (1737), pp. 13–16.
27. Labelye, *Short account*, preface, p. iv.
28. These and subsequent details of the caisson are from Labelye, *Short account* and *Description of Westminster Bridge* (1751).
29. Anon., *Gephyralogia: an historical account of bridges including the new bridge at Westminster* (London, 1751), p. 96.
30. Labelye, *Short account*, p. 52.
31. Bélidor, *Architecture hydraulique*, vol. IV (1752), pp. 198–206 and pl. XXVIII. The source of this plate is explained in chapter 4 below.
32. Labelye, *Short account*, pp. 21–3.
33. Bélidor, *Architecture hydraulique*, vol. III (1750), pp. 183–8 and pl. XXV.
34. J. T. Desaguliers, *A course of experimental philosophy*, vol. II (London, 1744), p. 417.
35. Committee of works minutes (see note 1), 8 February 1744.
36. Commissioners' minutes (see note 1), 9 June 1738, 23 April 1740 and 26 February 1751.
37. For a charming architectural and personal biography, see J. Lees-Milne, *Earls of creation* (London, 1962), pp. 60–100.
38. Sir J. Hawkins, *Life of Samuel Johnson* (London, 1787), p. 376.
39. MS papers of Ninth Earl of Pembroke at Wilton House, some dated 1739–43, others n.d.
40. The details given here are from engravings of both designs, from Labelye, *Short account* and *Description of Westminster Bridge* and from Gayfere's notebook at InstCE. The only full elevation in the latter corresponds to the drawing of the middle arch in the Pembroke papers and not to either of Labelye's designs.
41. Commissioners' minutes (see note 1), 10 May 1738.
42. The source is clear from a later quotation, see note 47.
43. Commissioners' minutes, 11 June 1740.
44. *Ibid.* 28 December 1743.
45. *Ibid.* 10 December 1745.
46. Labelye, *Description of Westminster Bridge*, p. 81.
47. The reference is to P. de la Hire, *Traité de mécanique* (Paris, 1695) which contained a new mathematical treatment of the equilibrium of arches. See chapter 5.

48. Commissioners' minutes (see note 1), 5 December 1749.
49. Quoted in Lees-Milne (see note 37), pp. 99–100.
50. B. Langley, *A survey of Westminster Bridge as 'tis now sinking into ruin* (London, 1748).
51. Commissioners' minutes, 26 February 1751.
52. Robertson, p. 499.
53. V. Moschini, *Canaletto* (London and Milan, n.d., twentieth century), reproduces them all.
54. Bélidor, *Architecture hydraulique*, vol. IV, p. 191.
55. *Ibid*, vol. III, pp. 183–4.
56. E. M. Gauthey, *Traité de la construction des ponts*, 2nd edn (2 vols, Paris, 1832), vol. I, pp. 70–1.
57. Smeaton, *Mr. Smeaton's answer* (1760), reports his watching of the work.

2. Bridges in the country

1. The historical facts about Wade and the military bridges are mostly taken from Salmond, *Wade in Scotland*, 2nd edn (1938) and are not referenced individually. A shorter but, in some respects, more thorough account is in two papers by Sir K. Mackenzie in *Transactions of the Inverness scientific society*, v (1895–9), 145–77 and 364–84. For the primary sources see Salmond, pp. 303–7. Description of the bridges is from the present author's inspection of surviving bridges, reinforced and refined by discussions with Mr Ronald Curtis, the present authority on Wade's roads and bridges. For a good short description of Highland roads and bridges before 1800 see Haldane, *New ways through the glens* (1962), pp. 1–14.
2. Salmond, pp. 135–56.
3. Contract between Lt-Gen. George Wade and John Stewart of Canogan, Esq., for a bridge over 'the River of Tumble', 25 July 1730, GD1/53/97, at SRO (see fig. 16 above). Mackenzie, p. 158, quotes E. Burt, *Letters from the North of Scotland* (2 vols, London, 1754), as saying that 'artificers' were employed for building bridges, but I have not found the reference in modern editions of Burt.
4. Report of General Mackay, quoted by Haldane, *New ways through the glens*, p. 9, and Mackenzie, pp. 377–9. Less than fifty were described as 'capital bridges' and the remainder were probably very small.
5. Petition, letter and contract, 1–4 May 1717, GD248/22/5(1–3), in Seafield Papers at SRO.
6. T. D. Lauder, *The great floods of August 1829 in the province of Moray* (1830), p. 109.
7. See note 5.
8. Mr Curtis considers that all the arches built in Wade's time were probably of semicircular shape or near it, and I concur with this. The lower arches are in later military bridges.
9. Hawksmoor, *Account of London-Bridge* (1736), pp. 41–3.
10. Referred to in a report of Robert Reid in 1785 printed in Smeaton, *Reports* (1812), vol. III, pp. 245–6. Further 'pileing and shoeing' of the bridge cost £230 in 1774 (*HCJ*, 1774, p. 559).
11. Gautier, *Traité des ponts*, 3rd edn (1728), pp. 37–9.
12. Fleming, *Robert Adam and his circle* (1962), pp. 33–75, has a good assessment of William Adam's architecture. Many of his designs were engraved in Anon., *Vitruvius Scoticus* (n.d.), including Aberfeldy Bridge.
13. R. Southey, *Journal of a tour in Scotland in 1819* (London, 1929), p. 44.
14. The description of Welsh bridges is almost entirely based on the author's inspections. Documentary study is hampered by lack of indexing in many of the Quarter Session records.
15. For an instance, see Ruddock, 'William Edwards's bridge' (1974).
16. 22 Hen VIII c.5, 12 Geo II c.29, 14 Geo II c.33.
17. *Report on the public bridges in the county of Middlesex* (London, 1826); 'Report of John Green, civil engineer, on state of county bridges' (1809), at Devon CRO; MS 295/94 at Brotherton Library, University of Leeds, and 'Book of bridges belonging in whole or in part to the West Riding' (1752), at the office of the County Engineer and Surveyor of West Yorkshire; abstract of dates of adoption of county bridges, from County Archivist of North Yorkshire.
18. 'Book of bridges' (1752). Many of these are signed by John Carr (see chapter 9). A second set of the drawings, also in manuscript, is in the archive section of Leeds Central Library.

3. Timber bridges

1. Palladio, *Four books* (New York, 1965), book 3, pp. 62–8, pls II–VI.
2. Gautier, *Traité des ponts* (1714), pp. 13–14, 116–23, pls XI–XVII.
3. C. Perrault, 'Receuil de plusieurs machines de nouvelle invention', in P. Vander (ed.), *Oevres diverses de physique et de méchanique de Mrs. C. et P. Perrault* (5 vols, Paris, 1721), vol. V, pp. 712–14 and pls X and XI. Perrault died in 1688.
4. Minutes of the Commissioners and Trustees for Fulham Bridge 1726–9, and 'Book of particulars relating to Fulham Bridge', both MS, PRO 30/26/12 and 13 respectively, at PRO. The references are to 30/26/12, fols 7–12, and 30/26/13, fols 12, 36–9, 51–2. Short histories of the bridge are in A. Chasemore, *History of the old bridge at Fulham and Putney* (London, 1875), and J. F. Wadmore, 'Old Fulham Bridge', *Transactions of London and Middlesex archaeological society*, VI (1890), 401–48.
5. Weale (ed.), *Architecture of bridges* (1843), vol. II, pp. xcvi–xcvii, and vol. III, pl. xx.
6. Committee of Works minutes, 9 August 1737, WORK 6/39, at PRO. This is the only design mentioned there which can be identified as the single-arch design referred to in Anon., *Short narrative of proceedings* (1737), p. 29.
7. B. Langley, *The builder's compleat assistant*, 3rd edn (London, 1750), p. 136 and pl. LXII.
8. Commissioners' minutes, 10 May 1738, WORK 6/28, at PRO. Details of the design are from the published engraving (fig. 28) and Gayfere's notebook at InstCE.
9. 20 Geo II c.22 (1747).
10. These and subsequent details are collected from a description in *GM*, xx (1750), 587–90, a letter by Dicker in *GM*, xxiv (1754), 116–18, Etheridge's published engraving (fig. 29) and Smeaton's 'Report on the state of Walton Bridge in July 1778', in Smeaton, *Reports* (1812), vol. III, pp. 371–5. For a general history see J. A. Stonebanks, *The Thames bridges at Walton*, Walton and Weybridge Local History Society paper no. 4 (Walton-on-Thames, 1969).
11. The details are from Barnard's engraving (fig. 31).
12. Smeaton, 'Designs', vol. IV, fols 75–7, at Royal Society.
13. *Royal magazine*, III (1760), 320. Other technical details are from this account and two engravings, one of which is fig. 33. For a general history see R. G. M. Baker, 'History

of the Thames bridges between Hampton Court and East Molesey', *Surrey archaeological collections*, LVIII (1961), 79–86.

14. Telford, 'Bridge' (1812), p. 488.

15. De Maré, *Bridges of Britain* (1954), p. 100 and pl. XLIV. See also Colvin, *Biographical dictionary* (1954), pp. 197–200.

16. Swan, *Designs in architecture* (1757), vol. II, pls XLIII–XLVII, and *Designs in carpentry* (1759), pls XLII–L. The former contains the Blair Atholl designs, pl. XLIII, and the Dunkeld design, pl. XLVI.

17. See the 16-page description published with Taylor's prints of Schaffhausen Bridge (1800) and based on J. G. Andreae, *Briefe aus der Schweiz in dem jahre 1763* (1776) and J. G. Ebel, *Schilderung der gebergvölfer der Schweiz* (2 vols, Leipzig, 1791), vol. I, pp. 388–93. Figs 36 and 37 are from the latter but fig. 37 was drawn by John Soane!

18. W. S. Childe-Pemberton, *The earl bishop* (2 vols, London, 1924). Details of the Derry bridge story are from P. J. Rankin, *Irish building ventures of the earl-bishop of Derry* (Belfast, 1972).

19. Mylne diary, 4 October 1768, in Miss Mylne's archive.

20. Shanahan's drawings were engraved at Vicenza and there is a bound set in the RIBA Library. Manuscript copies of twenty-eight of the engravings are in portfolio 3 at Sir John Soane's Museum.

21. *GM*, XLII (1772), 399–400. Details are all from this description and the engraving (fig. 38).

4. *Architecture hydraulique* and Essex Bridge, Dublin

1. Gautier, *Traité des ponts* (1714).

2. Labelye, *Short account* (1739).

3. *DNB*, vol. XIII, p. 1174.

4. F. Duncan, *History of the Royal Regiment of Artillery* (2 vols, London, 1872), vol. I, p. 113.

5. For brief biography and notice of Bélidor's writings see Hamilton, 'French civil engineers of the eighteenth century' (1941–2), and Heyman, *Coulomb's memoir* (1972), pp. 84–8, 192.

6. Bélidor, *Architecture hydraulique*, vol. II (1739), pp. 53–233.

7. *Ibid*, vol. I (1737), pp. 360–88.

8. *Ibid*, vol. III (1750), pp. 106–28.

9. Gayfere's notebook, at InstCE.

10. Telford, 'Bridge', pp. 425–6 and pl. XCVIII. This article was written in 1812.

11. Bélidor, *Architecture hydraulique*, vol. IV (1752), pp. 441–56. The title page of this volume bears the date 1753 but the censor's approval is dated 25 November 1752 and George Semple received his copy in Dublin during the second half of 1752 (see below).

12. Bélidor, *Science des ingénieurs* (1729), book II, pp. 45–64.

13. Bélidor, *Architecture hydraulique*, vol. IV, pp. 457–64.

14. Perronet, *Description des projets* (1783).

15. Semple, *Building in water* (1776) is the chief source for the remainder of this chapter and individual references to it are not separately noted. For the context of Semple's work, see M. J. Craig, *Dublin 1660–1860* (Dublin, 1969), especially pp. 170–5.

16. Semple, *Building in water*, 2nd edn (1780), preface, p. viii.

17. Probably the view of the work on 28 July 1752 in Perronet, pl. XXXI.

18. Bélidor, *Architecture hydraulique*, vol. III, pl. XIII, fig. 3, and Gautier, *Traité des ponts*, pl. XXI, fig. 1.

19. Bélidor, *Architecture hydraulique*, vol. IV, pp. 454–6 and pl. LVIII.

20. *Ibid*, p. 446.

21. Hawksmoor, *Account of London-Bridge* (1736), pp. 34–6.

22. Craig (see note 15), pp. 172–4.

23. See chapter 1.

24. *GM*, XXI (1761), 609–10.

25. Semple, 2nd edn (1780), preface, p. vi.

26. *ICJ*, VII (1761), 800.

27. J. Purser Griffith, 'The lowering and widening of Essex Bridge', *Transactions of institution of civil engineers of Ireland*, XIII (1880), 32–50.

5. Theory of arches and Pontypridd

1. R. Hooke, *A description of helioscopes, and some other instruments* (London, 1676), p. 31.

2. D. Gregory, 'Catenaria' (in Latin), *Philosophical transactions of the Royal Society*, no. 231 (1697), p. 637.

3. S. Ware, *A treatise of the properties of arches, and their abutment piers* (London, 1809), p. 6.

4. Langley, *Design for the bridge at Westminster* (1736). See above, fig. 4 (chapter 1).

5. James, *Short review* (1736), pp. 37–9.

6. P. de la Hire, *Traité de mécanique* (Paris, 1695). Heyman, *Coulomb's memoir* (1972), pp. 75–88 and 162–89, traces the sequence of developments in arch theory after la Hire.

7. Gautier, *Dissertation* (1717), p. 6.

8. W. Bowe, *Some account of the life and writings of Mr William Emerson*, prefixed to W. Emerson, *Tracts* (London, 1793).

9. Emerson, *Mechanics* (1758), p. 262.

10. The sources for the following paragraphs are Williams, 'Pont-y-ty-pridd' (1945), Ruddock, 'William Edwards's bridge' (1974), and Ruddock, 'Hollow spandrels' (1974), with the references quoted in the three papers.

11. Telford, 'Bridge' (1812), p. 486.

12. Design drawing signed 'J.B.', RHP 971/1, at SRO.

13. G. H. Jack, 'John Gethin, bridge builder, of Kingsland, Herefordshire. 1757–1831', *Transactions of Woolhope naturalists' club* (1931), pp. 86–97.

14. An inscription on Pont ar Darrell, in Brecon, dates its construction at 1829.

15. T. M. Smith, 'Account of Pont-y-tu-Prydd', *Minutes of proceedings of institution of civil engineers*, V (1846), 474–7.

16. Brecons. QS records, at NLW.

17. *Ibid*.

18. *Ibid*.

19. Pont ar Ithon in Radnorshire was rebuilt with tunnels through the abutments in 1934.

6. Four ancient bridges

1. Except where otherwise noted, the facts given here are from M. J. Becker, *Rochester Bridge: 1387–1856: A history of its early years* (London, 1930), and quoted by Becker mostly from records at the Bridge Chamber in Rochester.

2. E. Hasted, *History of Kent* (4 vols, Canterbury, 1788–99), vol. II, p. 18.

3. Hawksmoor, *Account of London-Bridge* (1736), p. 41.

4. Thomson, *Chronicles of London Bridge* (1827), is the best history of the old bridge, quoting many of the sources noted below and others for earlier events.

5. The reports, apparently quoted in full and with a plate, are in Maitland, *History of London* (1760), vol. II, pp.

827–33. Dance's report is also in Hutton, *Tracts* (1812), vol. I, pp. 115–20, and in several nineteenth-century parliamentary reports.

6. Maitland, plate at p. 828, and Dance and others, 'Report to select committee of Bridge House lands, November 1814'. This report, at p. 290, takes Hawksmoor's figure of 194 ft to be an error, but he may have been making the same allowance for the dripshot piles as did Robertson in 'The fall under bridges' (1758) (see appendix 2).

7. Journal of Common Council, 2 February 1754, at LRO.

8. Thomson, p. 497.

9. Journal of Common Council, 26 September 1754, at LRO. A printed copy of the report is BM 816.l.5/20.

10. *Ibid*. November 1754.

11. *Ibid*. 18 December 1755.

12. *HCJ*, 13 January 1756.

13. *HCJ*, 12 March 1756.

14. Dance and others, pp. 85, 98.

15. Thomson, p. 548.

16. *Universal chronicle*, 30 July 1759.

17. *LM*, XXVIII (1759), 672.

18. See chapter 8.

19. Printed *Report of committee of Common Council, 10 October 1763*, A-8-5/31 at Guildhall Library.

20. 29 Geo II c.40.

21. For history and description of the waterworks, see Thomson, pp. 555–66. Risk of reducing the water supply is the main recorded objection to Labelye's scheme of improvement in 1746.

22. R. Mylne, 'Report on London Bridge, 1800', with pl. IA (fig. 56).

23. Knight, 'Observations on old London Bridge' (1830), and Thomson, pp. 535–9.

24. Evidence of James and Ralph Walker in 'Report of committee on the state of London Bridge (1821)', pp. 56–9, 63–5.

25. J. Bridges, *Four designs for Bristol Bridge* (Bristol, 1760), pp. 56–8.

26. The collections of Bristol Central Library and Bristol University Library together include almost all, if not all, of the pamphlets. The description below is a digest of facts from many of the pamphlets, from W. Barrett, *History and antiquities of Bristol* (Bristol, 1789), pp. 73–82 and 95–7, and from S. Seyer, *Memoirs of Bristol* (2 vols, Bristol, 1821–3), vol. II, pp. 28–43. Illustrations of the old and new bridges are in both books.

27. Barrett, p. 97.

28. 'Minutes of the trustees for Bristol Bridge 1763–94', 5 December 1763, B.5014 at Bristol Central Library.

29. Most of the facts about Tyne Bridge are collected from printed reports by John Smeaton, 18 October 1769 and 6 May 1771, Smeaton and J. Wooler, 4 January 1772 (all three in Smeaton, *Reports* (1812), vol. III, pp. 252–66), R. Mylne, 12 March 1772, and J. Wooler, 17 March 1772 (both Newcastle, 1772); and from J. C. Bruce, 'The three bridges over the Tyne at Newcastle', *Archaeologia Aeliana*, new series, X (1885), 1–11.

30. Dance and others, p. 95.

31. 'Minutes of bridge committee 1771–7', 23 July 1772, document no. 45 at Newcastle City Archive.

7. Blackfriars Bridge

1. See letters by 'A.B.', a member of Common Council, in *Royal magazine*, I, 65–6 (August 1759) and *London chronicle*, 23 June 1759.

2. 'Case' of R. Mylne for payment of fees, 30 January 1776, box 12, no. 9, at LRO. A printed copy is A-8-5/42, at Guildhall Library.

3. Bridge committee report in Journal of Common Council, 18 January 1771, at LRO. A printed copy is A-8-5/35 at Guildhall Library.

4. *Ibid*.

5. *LM*, XXIX (March 1760), 115–20, letter of Thomas Simpson.

6. J. Boswell, *Life of Johnson*, ed. G. B. Hill (6 vols, London, 1934), vol. I, pp. 351–2.

7. First letter, printed 1 December 1759. All three letters are in *The architect*, 7 January 1887, pp. 13–15, and in R. W. Chapman (ed.), *Letters of Samuel Johnson* (3 vols, Oxford, 1952), vol. I, pp. 446–52.

8. *London chronicle*, 4 December 1759.

9. *GM*, XXIX (1759), 611–12, and *Royal magazine*, II, 57–8 (February 1760).

10. Letter, R. Mylne to W. Mylne, August 1759, in Miss Mylne's archive.

11. Colvin, *Biographical dictionary* (1954), p. 399.

12. This and subsequent paragraphs are based on the collection of Robert's letters in Miss Mylne's archive. Some are quoted at length by C. Gotch, 'The missing years of Robert Mylne', *Architectural review*, CX (1951), 179–82.

13. *London chronicle*, 4 December 1759.

14. For examples see J. Harris, *Sir William Chambers* (London, 1970), pls XVI, XIX and XXI.

15. G. Piranesi, *Opera varie de architettura* (Rome, 1750), pl. VIII, and *Le antichita Romane* (Rome, 1756), vol. IV.

16. Dance's design is 'Surveyor's bridges plan no. 195', at LRO. The middle of Chambers's design is in R. S. Mylne, *Master masons* (1892), p. 264. A drawing in the print room of the Guildhall Art Gallery is identified tentatively as part of one version of Gwynn's design (fig. 61).

17. Robison, 'Arch' (1801), pp. 23–4 and pl. III. The letter from Mylne to Robison, commenting on this article and dated 12 May 1799, was printed in a magazine much later, but neither the date nor the magazine has been identified. A copy is in Miss Mylne's archive.

18. Letter, R. Mylne to W. Mylne, August 1759, in Miss Mylne's archive.

19. House of Lords committee book, 20 May 1756, at House of Lords Record Office.

20. Journal of Common Council, 7 April 1756, at LRO. A copy of the design is in Smeaton, 'Designs', vol. IV, fol. 175v, at the Royal Society.

21. J. Smeaton, *Narrative of the building of the Eddystone lighthouse* (London, 1791), p. 38.

22. See biographical memoir in Smeaton, *Reports* (1812), vol. I, pp. xiii–xxiv.

23. Smeaton, *Mr. Smeaton's answer* (1760). Dated 9 February.

24. Smeaton, 'Designs', vol. IV, fols 176, 177v, 178, 178v.

25. Here and in the following table, and in fig. 65, Smeaton's arches are considered to spring from the level at which the 120° arc meets the face of the pier. The extra curve of the springer-stones (fig. 64) is ignored.

26. *LM*, XXIX (March 1760), 115–20.

27. J. Muller, *Practical fortification* (London, 1755).

28. *LM*, XXIX (April 1760), 190.

29. *London chronicle*, 21 February 1760.

30. C. Hutton, 'Memoir of the life and writings of the author', prefixed to T. Simpson, *Select exercises in mathematics*, new edn (London, 1792).

31. *LM*, xxix (1760), 117–20.

32. Hutton (see note 30).

33. Bridge committee report (see note 3).

34. The folio volume, entitled 'Surveyor's private copy . . . Entries of the several orders drawn out by Mr Robert Mylne . . . for the excavation, masonry, carpentry and smith's work . . .', and the diary are in Miss Mylne's archive. There are microfilm copies of the diary at SRO and RIBA, and Richardson, *Robert Mylne* (1955) contains an incomplete and inaccurate transcript. The engravings are Baldwin, *Plans of the machines and centring* (1787). Facts from these sources are not separately referenced in the subsequent narrative.

35. Letters, R. Mylne to his father, 15 May and 15 November 1760 in Miss Mylne's archive.

36. Colvin, p. 401.

37. Letter, R. Mylne to his father, 15 November 1760.

38. 'Case' for payment of fees at LRO (see note 2).

39. Minutes of Westminster Bridge Commissioners, 4 December 1759, WORK 6/35, at PRO.

40. Letter, R. Mylne to J. Robison, 12 May 1799 (see note 17).

41. *Ibid.*

42. £1,600 was required for repair of masonry, mostly in the piers above low-water, in 1784. See app. E.1.VI to 'Third report of select committee on the improvement of the port of London' (1800), p. 589.

8. John Smeaton and Robert Mylne

1. Smeaton, *Reports* (1812). Much of this chapter is based on these reports and obvious references are not separately noted. Details about the editing and publication are given by Skempton, 'Publication of Smeaton's reports' (1971).

2. Smeaton, 'Designs' (MS) in the library of the Royal Society of London, bound in six volumes (by John Farey, jnr). Dickinson and Gomme edited a catalogue of these designs in 1950 (but with many errors in the dating of bridge designs) and the designs have also been published on microfilm by the E. P. Group, Wakefield. The bridge designs and engravings, etc., are in vol. iv, fols 60–188. Obvious references are not noted.

3. In the case of Hexham Bridge, rebuilt by others to the same design. See below.

4. Skempton, pp. 140, 153n.

5. Smeaton, *Mr. Smeaton's answer* (1760), p. 3.

6. [Mylne], *Observations on bridge-building* (1760), p. 39.

7. Smeaton, *Mr. Smeaton's answer*, p. 2.

8. *Ibid.*

9. Smeaton, 'Designs', vol. iv, fol. 172.

10. M. Clare, *On the motion of fluids* (London, 1737).

11. Smeaton, 'Designs', vol. iv, fol. 94.

12. Smeaton, *Mr. Smeaton's answer*, p. 2.

13. *Ibid.*

14. Trevelyan Papers, WCT311, no. 69, at the University Library, Newcastle-upon-Tyne.

15. 'Minutes of the trustees for Coldstream Bridge, 1762–72', CS. Misc. 3 at Northumberland CRO, p. 2.

16. T. Pennant, *A second tour in Scotland* (2 vols, London, 1772), vol. ii, p. 114.

17. Contract drawings for widening of Banff Bridge, 1881, in custody of County Surveyor of Banff.

18. Drawings for widening of Coldstream Bridge and Hexham Bridge, in custody of County Surveyor of Northumberland.

19. In Maitland, *History of London* (1760), vol. ii, plate at p. 828. See chapter 6.

20. Shown on a 'plan for widening Stockton Bridge', 1857, at North Yorkshire CRO. Robson's engagement to design the bridge is recorded in 'Stockton Bridge order book 1762–1810' at the same archive.

21. 'Bridge Book', *c.* 1805, at North Yorkshire CRO, fol. 22.

22. Three designs are in Smeaton, 'Designs', vol. iv, fols 111–13. The fourth is in the Wentworth House muniments at Sheffield Central Library.

23. Illustrated in H. Repton, *Theory and practice of landscape gardening* (London, 1805), p. 40. Arthur Young, *Six months tour through the North of England* (4 vols, London, 1769), vol. i, p. 298, noted that a 'magnificent bridge' was about to be built.

24. Smeaton, *Reports*, vol. i, pp. 99–102; but Stratford's observations are only given in full in a printed broadsheet, *A true state of the several matters . . . with Mr. Stratford's observations for Mr. Smeaton's opinion thereon* dated 24 November 1762. See Bristol Bridge pamphlets.

25. *HCJ*, xxix (1762), 132. Reid's drawing is in Smeaton, 'Designs', vol. iv, fol. 150.

26. Unless otherwise noted, details concerning Coldstream Bridge are from Smeaton, *Reports*, vol. iii, pp. 235–51, 'Designs', vol. iv, fols 150–7 and 160v., or 'Minutes of the trustees' (see note 15).

27. Smeaton, 'Designs', vol. iv, fol. 80, and 'Minutes of the commissioners for Northesk Bridge' (MS, 2 vols), at the Town Clerk's office, Montrose.

28. Smeaton, 'Designs', vol. iv, fol. 155v.

29. 'Minutes of the commissioners for Perth Bridge, 1765–1868', at the Town Clerk's office, Perth.

30. Smeaton, 'Designs', vol. iv, fol. 176v.

31. Contract drawings for Dubh Loch Bridge, signed 10 August 1785, in Duke of Argyll's muniments, Inveraray Castle. Copies at NMRS.

32. Mylne diary, 16 December 1765, in Miss Mylne's archive.

33. Young (see note 23), vol. i, p. 361. There are entries in Mylne's diaries between August 1764 and September 1767. Thirty-two letters and other documents are in the Portland Papers, PwF 7087–7129, at the University of Nottingham.

34. Design drawing in the custody of the Surveyor, Tonbridge UDC, dated 1774 but unsigned.

35. Design and specification for the southern part of Tyne Bridge, Newcastle, 1773, Church Commission Durham Bishopric Estates Deposit, box 134, nos. 220997 and 221008, at the Department of Palaeography, Durham University.

36. Letter, Duke of Argyll to Mylne, 25 January 1775, in archive of Miss Mylne, and Mylne diary, 5 October 1775.

37. Mylne diary, 16 October 1786.

38. Mylne diary, 4 March 1782–5 October 1784, and QS records, 1782–4, at Hampshire CRO.

39. Jervoise, *Ancient bridges of the south of England* (1930), pl. xxx, gives a photograph shortly before it was reconstructed and Johnson and Scott-Giles (eds), *British bridges* (1933), p. 117, gives another shortly after the reconstruction.

40. The details are from Smeaton, *Reports*, vol. i, p. xix, vol. ii, pp. 1–26, and Mylne's report, 1800. Smeaton's draw-

ings are in Smeaton, 'Designs', vol. IV, fols 181v–186, some of them printed in *Commons Reports 1715–1801* (1803), vol. XIV, pls XIX–XXI.

41. Smeaton, *Reports*, vol. III, pp. 243, 347–8.

42. Smeaton, 'Designs', vol. IV, fols 94v–96, 101 and 104.

43. For sources see note 26.

44. Smeaton, 'Designs', vol. IV, fols 172 and 177v, *Mr. Smeaton's answer*, p. 3, and [Mylne], *Observations on bridge-building*, p. 41.

45. Perth Bridge papers, no. 18, at Town Clerk's office, Perth.

46. W. Cramond, *Annals of Banff* (2 vols, Aberdeen, 1891–3), vol. I, pp. 321–2.

47. See note 34.

48. Printed reports listed in note 29 to chapter 6. Other facts given here are from Sykes, *Local records* (1833), pp. 283–94, and from minutes of Common Council of Newcastle, 1766–85, Bridge Committee minutes, 1772–9, and other documents at Newcastle City Archive.

49. See note 35.

50. J. C. Bruce, 'The three bridges over the Tyne at Newcastle', *Archaeologia Aeliana*, new series, X (1885), pl. II.

51. Plan 6/926 and documents 48, 49 and 55 at Newcastle City Archive.

52. Contract drawing in Duke of Argyll's muniments, Inveraray Castle. Copy at NMRS and reduced copy in Lindsay and Cosh, *Inveraray* (1973), p. 235.

53. Sources additional to Smeaton, *Reports* and 'Designs', are used and listed in Ruddock, 'Foundations of Hexham Bridge'. See especially the Errington papers at Northumberland CRO, ZAN M17/53 (quoted by courtesy of the Society of Antiquaries of Newcastle-upon-Tyne), and Anon., *Narrative of proceedings re Hexham Bridge* (1788).

54. Smeaton, *Reports*, vol. III, p. 240.

55. Drawing no. B/A6079/1/3 at County Surveyor's office.

56. Additional sources and longer treatment of this section are in Ruddock, 'Hollow spandrels' (1974) and 'The building of North Bridge' (1974–6). See, in particular, Clerk of Penicuik Muniments at SRO, letters of William Mylne and Mylne (R.) diaries in Miss Mylne's archive, and minutes of bridge committee and Town Council at Edinburgh City Archive.

57. Mylne diary, 22 July 1768, and Smeaton, *Reports*, vol. I, p. 339.

58. W. Halfpenny, *Perspective view of the sunk pier* (London, 1748). Engraving.

59. Smeaton, *Reports*, vol. III, p. 294.

60. Contract drawing at Strathclyde Regional Archives, S.R.A., T-CN 14/89. Signed 16 September and 5 October 1768. A similar drawing in Robert Mylne's hand at RIBA Drawings Collection bears a note suggesting that he made the first design.

61. See notes 52 and 60.

62. Commissioners' minutes (see note 29), 15 August 1770.

63. Report re Coldstream Bridge, 10 August 1828, in Sir John Rennie's reports, vol. V, pp. 11–20, at Inst CE.

64. Records exist for Newcastle, Romsey and Tonbridge Bridges; see notes 35, 38, and bills c. 1775 in Q/AB 58, at Kent CRO.

65. Records of payments, etc., exist for many of Smeaton's works, including bridges, in the surviving papers of the clients. See also Smith, 'The professional correspondence of John Smeaton', in *Trans. Newcomen soc.*, XLVII (1974–6).

66. See T. Turner, 'John Gwyn and the building of North Bridge, Hull', *Transport history*, III (1970), 154–63.

67. 'Minutes of the trustees' (see note 15), pp. 135–42.

9. Bridges by architects

1. *ICJ*, vol. VII (1761–4), app. ccxxxviii, and W. Tighe, *Statistical observations on the county of Kilkenny* (Dublin, 1802), pp. 131–2.

2. Tighe, p. 563, and V. T. H. and D. R. Delany, *The canals of the south of Ireland* (Newton Abbot, 1966), p. 140.

3. *ICJ*, vol. VII, app. xl.

4. P. Smithwick printed a photograph of St John's Bridge in 'Georgian Kilkenny', *Quarterly bulletin of the Irish Georgian Society*, IV (1963), 75–96.

5. G. Taylor and A. Skinner, *Maps of the roads of Ireland* (London, 1778), shows a bridge too substantial to be of timber.

6. P. O'Leary, *Half hours with the old boatmen* (Wexford, 1895), p. 18.

7. Semple, *Building in water* (1776).

8. Price, *Some considerations* (1735).

9. Details supplied by Mr P. Doran, Assistant Surveyor of Kilkenny County Council.

10. G. Taylor and A. Skinner, *Maps of the roads of Ireland*, 2nd edn (London, 1783), shows no bridge, but C. Coote, *Statistical survey of Queens County* (Dublin, 1801), p. 127, mentions it.

11. *DNB*, vol. XX, pp. 82–3.

12. Drawings which appear to be those of the original design are in the records of the Dublin Port and Docks Board.

13. Paine, *Noblemen and gentlemen's houses*, vol. II (1783), pl. LXXXVI. This book is the chief source for Paine's bridges and details from it are not separately referenced.

14. Called 'Welden Bridge' in Paine, vol. I (1767), pls X and XI, and p. 3. The one-arch bridge is called 'Edensor Bridge' and 'Beeley Bridge' in modern books. Photographs of both bridges are in Johnson and Scott-Giles, *British bridges* (1933), pp. 57 and 158, and Jervoise, *Ancient bridges of mid and eastern England* (1932), figs 5 and 15.

15. Colvin, *Biographical dictionary* (1954), pp. 429–34, gives other biographical details.

16. Minutes of Chertsey Bridge committee, 5 July 1779 and 25 November 1785, in minute books MA/C1 and C2 at Middlesex Branch, GLCRO. Other facts concerning this bridge are from contracts for first construction, 1780, and major repairs, 1821 and 1842, MA/D/Br 4, 8 and 23, respectively; a contract for the south approach, 1782, Acc. 938/2 at Surrey CRO, and a longitudinal section of half the bridge in portfolio 3, no. 47, at Sir John Soane's Museum.

17. For Richmond Bridge, see Paine, vol. II, pls LXXXII and LXXXIII, and p. 29. The contracts for Kew Bridge, 24 June and 25 August 1784, with drawings, are BM Add. Charter 16155. Walton Bridge is shown in two paintings dated 1784 and 1859 at Walton and Weybridge Public Library and Museum. See also J. A. Stonebanks, *The Thames bridges at Walton*, Walton and Weybridge Local History Society paper no. 4 (Walton-on-Thames, 1969).

18. The best biographical notice of Gwynn is by W. Papworth in *The builder*, XXI (1863), 454–7, and XXII (1864), 27–30, but see also Colvin, pp. 254–6, and *DNB*, vol. VIII, pp. 848–50.

19. W. Owen, in J. Chambers, *Biographical illustrations of Worcestershire* (Worcester, 1820), pp. 504–6.

20. Ward, *Bridges of Shrewsbury* (1935), pp. 30–40. Further details of the English Bridge are from this book and Ward, 'Reconstruction of English Bridge' (1928), unless otherwise noted.

21. These and subsequent facts concerning Atcham Bridge are from the MS 'Articles of agreement' between Gwynn and the Justices, 15 October 1772, DP6 at Salop CRO.

22. *Mr. Gwynn's report to the trustees for a new bridge at Worcester* (Worcester, 1781). Dated 5 March. Copy at Town Clerk's office, Worcester.

23. T. Nash, *Collections for a history of Worcestershire* (2 vols, London and Worcester, 1781–2), vol. II, app. p. cxv. An engraving of the finished bridge is attached. Ten of the original design drawings for the bridge and approaches are in the Prattinton Collection, nos. 70/1–9, at the Society of Antiquaries, London.

24. *The builder*, XXI (1863), 456.

25. Shown by a drawing in a collection of 'Engineer's reports to the English Bridge Committee (1925–30)', MS 385 at Shrewsbury Public Library.

26. BM, Kings Maps, vol. XLIII, fol. 67-aa, design signed 'J. Gwynn Archt. 24 July 1770'.

27. Nash (see note 23), vol. II, app. p. xcvi.

28. *The builder*, XXI (1863), 456.

29. Johnson and Scott-Giles, pp. 240 and 328–9.

30. Contract for Maidenhead Bridge, 1 July 1772, and seven drawings, M/AB5 at Berkshire CRO.

31. Letter from Staines Bridge Commissioners to Lord Chancellor in acc. 809/BR/17 at Middlesex Branch, GLCRO, and Colvin, p. 279. For a photograph of the bridge see Johnson and Scott-Giles, p. 245.

32. Berkshire CRO, D/EE E22, fols 9–12, and printed report by J. M. Davenport in *Oxfordshire bridges* (1869), copy at Oxfordshire CRO.

33. Details extracted by Professor A. W. Skempton from Quarter Session records at West Yorkshire CRO.

34. York Georgian Society, *Works of John Carr* (1973), p. 47. Information on Carr's bridges is from this book and Quarter Session records at West and North Yorkshire CROs.

35. Bound as the 'Bridge book', c. 1805, at North Yorkshire CRO. There are plan and elevation drawings of 113 bridges.

36. See de Maré, *Bridges of Britain* (1954), pp. 100–13, for a well-illustrated survey of estate bridges.

37. Among the Adam drawings at Sir John Soane's Museum. Most of the finished drawings are in vol. LI, fols 1–44. Bolton, *Architecture of Robert and James Adam* (1922), vol. II, gives an almost complete topographical index of the collection, but an important series of sketches in vol. IX, fols 72–96, are not listed. All the bridge designs mentioned are in the collection unless otherwise noted.

38. Fleming, *Robert Adam and his circle* (1962), pp. 109–244.

39. Adam drawings, vol. IX, fols 72–96.

40. De Maré, pls 51 and 61, are the Bowood design and Kedleston Hall bridge.

41. J. Lees-Milne, *Earls of creation* (London, 1962), p. 97.

42. Lindsay and Cosh, *Inveraray* (1973), p. 125. All the references to Inveraray bridges are from this work, unless otherwise noted.

43. Dimensions and other factual data have been provided by Messrs R. H. Cuthbertson and Partners, who were responsible for a major repair of the bridge in 1970. The estimate and accounts are in the Dumfries House estate papers, box 9, bundle 30, seen by courtesy of the Marquess of Bute.

44. First published 1786, see J. A. Morris, *The brig of Ayr*, 7th edn (Ayr, 1912), pp. 19–28.

45. Committee minutes of trustees for Ayr Bridge, 1787–1831, B6/29/7 at SRO. The payment was made in 1788. The contract drawing, signed by Stevens and the magistrates but not by Adam, hangs on the wall at Ayr Public Library (copy at NMRS).

46. The elevation is shown in a sketch by John Rennie in notebook 23, box 25, Rennie papers, at NLS, and on a design for rebuilding in 1795, drawings D-TC 13/293–4, at Strathclyde Regional Archives.

47. In RIBA Drawings Collection.

48. Telford, 'Bridge' (1812), pp. 524–5 and pl. XCVI.

49. See A. J. Youngson, *The making of classical Edinburgh* (Edinburgh, 1966), pp. 113–17, and A. Rowan, 'After the Adelphi: forgotten years in the Adam brothers' practice', *Royal society of arts journal*, CXXII (1974), 659–710, especially pp. 670–4.

50. Morris, *The brig of Ayr*, and trustees minutes (see note 45).

51. Prestonhall estate papers, GD 244, box 18, at SRO. Box 19 contains copies of the contracts and lists of payments for Drygrange Bridge and Bridge of Dun.

52. *OSA*, vol. xv, p. 26, and T. Reid, 'Fords, ferries, floats and bridges near Lanark', *Proceedings of society of antiquaries of Scotland*, 4th series, XI (1912–13), 209–56.

53. RIBA Drawings, K 12/10–11, n.d. Rennie papers, acc. 5111, box 26, notebook 36, at NLS, notes that a bridge by Stevens had failed after fourteen years.

54. Details from contract, see note 51, and recent repairs and inspections of which records are at the Scottish Development Department, Bridges Section.

55. Letters to Wm Ker, 28 September and 24 November 1784, in Duchess of Roxburgh's muniments, section 3, box 3/18, bundle 46. All the information on Stevens's work at Ancrum and Teviot Bridges is from this bundle.

56. Telford, 'Bridge', p. 486, and report by John Rennie to Sir J. B. Riddle, 14 May 1804, in Rennie reports, vol. III, p. 179–80, at InstCE.

57. See note 51. Other information is from Bridge of Dun papers and town council minutes, at Town Clerk's office, Montrose.

58. Minutes of Northesk Bridge Commissioners, 1765–1880, at Town Clerk's office, Montrose, and printed *Memorial and petition to General Assembly of Church of Scotland* (1773). Copy at NLS.

59. D. Mitchell, *History of Montrose* (1886), p. 69.

60. *GM*, LXIII (1793), 311.

61. *OSA*, vol. XIV, pp. 396–400.

62. Copy of letter to 'Mr Gregson', n.d., in Rennie reports, Lancaster Canal vol., at InstCE.

63. Copy of specification for Lune Aqueduct, signed 26 July 1793, in W. Radford's commonplace book, QAR/5/39 at Lancashire CRO.

64. Letter, Rennie to Gregson, 16 January 1795, in Lancaster Canal papers, HL2 at PRO (Transport).

65. Letter, Rennie to Gregson, 6 February 1796, in Lancaster Canal papers, HL2 at PRO (Transport).

66. Colvin, pp. 268–70. See also J. M. Crook, 'The architecture of Thomas Harrison', *CL*, CXLIX (1971), 876–9, 944–7, 1088–91.

67. Recited in contract for Skerton Bridge, 21 February

1783, QAR/2/5 at Lancashire CRO.

68. The notes are in QAR/5/30 at Lancashire CRO, together with two designs (not those referred to in the notes) probably by Harrison.

69. Minutes of trustees for St Mary's Bridge, at Derby Central Library.

70. Crook, *CL* (see note 66), CXLIX, 1088, and Jervoise, *Ancient bridges of the north of England* (1931), pp. 125–6 and fig. 67, which shows the façade.

71. Crook, *CL*, CXLIX, 1088, and Jervoise, *Ancient bridges of mid and eastern England*, p. 7.

72. Colvin, pp. 404–10, and Summerson, *John Nash* (1935), pp. 17–54.

73. List of payments for designs, etc., Q/CB 001/2, at Gwent CRO.

74. Extracts from letter, 'Mr Millne' (i.e. Robert Mylne) to 'R.B.', 1791, in Fox papers, no. 167, at Plymouth and West Devon Record Office.

75. J. Baker, *Picturesque guide through Wales* (1792).

10. Navigable aqueducts

1. Early printed sources for the story of the Bridgewater Canal are the anonymous *History of inland navigations, particularly those of the Duke of Bridgewater* (London, 1766), often attributed to James Brindley; F. H. Egerton, *First part of a letter to the Parisians* (Paris, 1818) and *Second part* (Paris, 1820); and *Biographia Britannica*, 2nd edn (5 vols, London, 1778–93), vol. II, pp. 591–604. Details from estate and canal company papers are added by Malet, *The canal duke* (1961), and Boucher, *James Brindley* (1968). Smiles, *Lives* (1861), vol. I, pp. 307–476, adds further details without revealing his sources.

2. Malet, pp. 22–7.

3. Described by Boucher, *James Brindley*, and Smiles.

4. Malet, p. 54.

5. Brindley diary, 1 July and 14 October 1759, at InstCE.

6. *Biographia Britannica* (see note 1), vol. II, p. 595.

7. S. Hughes, 'Memoir of James Brindley', in *Quarterly papers on engineering* (London: John Weale), vol. I (1843), p. 47.

8. *HCJ*, 7 March 1760.

9. These and subsequent dimensions and details are from notes and a sketch made by John Rennie on 15 May 1784 in notebook 1, box 25, Rennie papers, at NLS.

10. See Malet, p. 87.

11. *Manchester mercury*, 21 July 1761.

12. Egerton, *Second part* (see note 1), p. 62.

13. 2 Geo III c.11.

14. Brindley diary, 11 October 1763, at InstCE.

15. Sir Bosdin Leech, *History of the Manchester Ship Canal* (2 vols, London, 1907), vol. II, pp. 19–20, and Malet, p. 159.

16. Smiles, vol. I, p. 384.

17. J. Farey, *Treatise on the steam engine* (London, 1827), pp. 257ff., and Smeaton, *Reports*, vol. I, p. 223.

18. Illustrated in Boucher, *James Brindley*, p. 98.

19. *Ibid.* p. 74.

20. *Ibid.* p. 73.

21. Copies of MS notebooks of John F. Green, entitled 'Orders by Mr Brindley', covering the periods 17 March 1767–15 June 1768, and 8 January 1769–9 August 1771, at Staffordshire CRO. The remainder of the paragraph is taken from these books.

22. See V. I. Tomlinson, 'The Manchester, Bolton and Bury Canal', *Trans. Lancashire and Cheshire archaeological society*, LXXV–LXXVI (1965–6), 231–99.

23. Minutes of Scottish Committee of the Forth and Clyde Canal Company, BR/FCN/1/37, pp. 213–14, at SRO.

24. Notebook 4, box 25, Rennie papers, at NLS.

25. Except where noted otherwise, details are from William Radford's copies of specification, estimates and observations, QAR/5/39, at Lancashire CRO, and a construction drawing printed in Burton, *The canal builders* (1972), p. 167.

26. Burton, p. 195, quoting Lancaster Canal Company records.

27. Reports of John Rennie, Lancaster Canal vol., at InstCE.

28. Burton, p. 167.

29. Letter, Rennie to Lancaster Canal Company committee, 20 February 1795, in collection HL2, Lancaster Canal papers, at PRO (Transport). The rest of the paragraph is derived from later letters in the same collection.

30. Lancaster Canal reports (see note 27).

31. Notes by W. Johnson, copied in Radford's commonplace book (see note 25). See chapter 13 and fig. 164 for a similar arrangement.

32. Paybills for foundation work, CAN 30, HL2/3, at PRO (Transport).

33. K. R. Clew, *The Kennet and Avon Canal*, 2nd edn (Newton Abbot, 1968), pp. 64–8.

11. The first iron bridges

1. Copies of the drawings are at NMRS. See also Lindsay and Cosh, *Inveraray* (1973), pp. 232–3.

2. Swan, *Designs in architecture* (1757), vol. II, pls 43–4, and *Designs in carpentry* (1759), pls 45–7, 50.

3. White, 'Mr Pritchard's progress' (1832).

4. Except for the subsequent note, this paragraph is taken from the 'Minute book of the proprietors of the Iron-bridge, 1775–98', Shrewsbury MSS 337A at Shrewsbury Public Library. Longer narratives from this source are in Raistrick, *Dynasty of ironfounders* (1953), pp. 193–202, and Maguire and Matthews, 'Ironbridge at Coalbrookdale' (1958).

5. Lady Labouchere collection, no. 1987, at Salop CRO, quoted in G. R. Morton and A. F. Moseley, 'An examination of fractures in the first iron bridge at Coalbrookdale', *West Midlands Studies Special Publication no. 2* (Wolverhampton, 1970), pp. 2–3.

6. See Maguire and Matthews for details of all intersections.

7. Rickman, *Life of Telford* (1838), p. 29. Maguire and Matthews discuss the authorship of the design more fully.

8. Raistrick, pp. 201–2, and Telford, 'Bridge' (1812), pp. 488, 539.

9. A. T. Bolton, *Portrait of Sir John Soane* (London, 1924), pp. 17, 22, 26–9, 305–7, etc.

10. Colvin, *Biographical dictionary* (1954), p. 557.

11. Soane's notes and sketches are in portfolio 3 at Sir John Soane's Museum.

12. P. Foner, *The complete writings of Thomas Paine* (2 vols, New York, 1945), vol. II, pp. 1026–8. The narrative following is derived from this volume, pp. 1026–57, 1255–1322, and 1411–12, except where otherwise noted.

13. Portfolio 3 (see note 11 above), no. 64, inscribed: 'Copy, Great Scotland Yard Nov 9th 1791', and on the back, 'Jan 1792'. No. 63 is a better-drawn copy, but undated.

14. Transcripts of five letters, Paine to Thomas Walker,

April–25 September 1790, acc. 30215, Sheffield Central Library.

15. Letter, Thomas Walker to Earl Fitzwilliam, 28 September 1790, F 127/29 in Wentworth Woodhouse Muniments, at Sheffield Central Library. This letter establishes the span as 110 ft, against ambiguity in Paine's writings.

16. F. Oldys, *Life of Thomas Pain*, 5th edn (London, 1792), p. 63.

17. Foner (see note 12), vol. II, pp. 1054, 1411.

18. Letter of Smeaton, dated 18 August 1791 without direction, in Smeaton letter-books, vol. III, fol. 107, at InstCE. Nash's own account of the affair, published after his death in *Mechanics magazine*, XXIV (1836), 26–9, is very detailed but so obviously prejudiced that it has to be discounted.

19. Notes of contents of a letter from Mylne to Burdon in November 1791, no. 167, Fox Papers, at Plymouth and West Devon Record Office. Mylne's diary records letters sent to Burdon on 8 November and 6 December 1791.

20. Bolton (see note 9), p. 202.

21. Copy of letter, Hutton to Burdon, 18 June 1792, no. 164, Fox Papers (see note 19).

22. Letter, Burdon to Soane, 13 August 1836, in Bolton, p. 532.

23. Description of Sunderland Bridge by R. Burdon, with specification of Patent no. 2066 (18 September 1795), in *Repertory of arts*, 1st series, V (1796), 361–9.

24. Rees, *Cyclopaedia* (1819), vol. V, article 'Bridge'. This volume was actually issued in 1805.

25. Stephenson, 'Iron bridges' (1856), p. 579.

26. 'Minutes of the commissioners for Staines Bridge', 1 April 1806, at GLCRO Middlesex Branch.

27. *Commons reports, 1846* (407.) XLV. 493.

28. Hutton, *Tracts* (1812), vol. I, p. 147.

29. Stephenson, p. 579.

30. Soane's portfolio 3 contains copies of five designs of centring by different people, with notes and estimates.

31. Hutton, *Tracts*, vol. I, p. 156.

32. *Ibid.*

33. Engraving, dated 1797, in Rennie drawings collection, vol. I, no. 7, at InstCE.

34. Date and foundry inscribed on bridge.

35. Date of erection, 1800, inscribed on both bridges. Foundry not known.

36. *Shrewsbury chronicle*, 30 October 1795, quoted by Raistrick, p. 204.

37. Prattinton collection, box 4, folder 2, no. 5, at Society of Antiquaries, a MS landscape view with note: 'Original plan for Stanford Bridge by Mr Nash. This fell down.' For other references see Summerson, *John Nash* (1935), pp. 43–4.

38. Prattinton collection, vol. XIII, p. 117. Apparently a transcript of a newspaper report.

39. Patent no. 2165, dated 7 February 1797, published in *Repertory of arts*, 1st series, VI, pp. 361–8.

40. Summerson, p. 53. A drawing of the bridge is in the Prattinton collection, box 4, folder 2, no. 6.

41. Minutes of Committee of Peak Forest Canal Co., 22 April 1795, RAIL 856/1 at PRO (Transport).

42. M. D. Conway, *The life of Thomas Paine* (2 vols, London, 1892), vol. II, pp. 443–5.

43. R. Fulton, *A treatise on the improvement of canal navigation* (London, 1796), and Telford, 'Canals' (1797), pp. 308–10.

44. Rolt, *Thomas Telford* (1958), pp. 17–18.

45. Copy of letter, Telford to Messrs Walker, 21 April 1801, T/LO. 29, Telford Manuscripts, at InstCE.

46. Telford 'Bridge', pp. 488, 539, and Rickman, p. 29.

12. Rennie and Telford – early years

1. Biographical details are from Smiles, *Lives* (1861), vol. II, and Boucher, *John Rennie* (1963). A manuscript life of Rennie by his son John is at InstCE.

2. Notebook 1, dated 1784, in box 25, Rennie papers, at NLS.

3. Report of Rennie and John Wilson to Lord Provost of Glasgow, in Rennie reports, vol. I, pp. 335–9, at InstCE. R. Renwick (ed.), *Burgh records of Glasgow*, vol. IX (Glasgow, 1914), p. 40, shows that the report was submitted in September 1796.

4. Rennie reports, vol. II, pp. 105–11, at InstCE. Dated 8 May 1799. Facts about the construction are confirmed by reports of the resident engineer John Duncan in 1801–3, in box 1, Rennie papers, at NLS. A plan and elevation is in the Rennie drawings collection, vol. I, no. 19, at InstCE.

5. Notebook 31, box 25, Rennie papers, dated 19 November 1798, at NLS.

6. Boucher, *John Rennie*, p. 22.

7. In Sir David Brewster's review of Rickman (ed.), *Life of Thomas Telford* (1838), in *Edinburgh review*, LXX (1839–40), 46. The following pages are derived from the biographies by Rickman, Smiles, Gibb and Rolt, with the additional references noted for individual events.

8. Papers re widening of Meole Brace Bridge in 1788, in DP108 at Salop CRO.

9. St Mary Magdalene, Bridgnorth (1792) and St Michael, Madeley (1793–6).

10. St Leonard, Dawley (1805). See Telford, 'Civil architecture' (1813), p. 644.

11. Orders and contracts re Montford Bridge, 1790–4, in QS Order books and DP114 at Salop CRO.

12. QS records at Salop CRO.

13. Records re Bewdley Bridge are in acc. 4600, nos. 303 and 765, at Worcestershire CRO.

14. The sequence of dates for the aqueducts is discussed by C. Hadfield, 'Telford, Jessop and Pont Cyssylte', *Journal of the railway and canal history society*, XV, no. 4 (October 1969), and also by Rolt, *Thomas Telford* (1958), pp. 44–50.

15. Telford, 'Canals' (1797 and 1800), p. 300.

16. Additional references are in QS records and DP113 at Salop CRO. See also M. C. Hill, 'Iron and steel bridges in Shropshire 1788–1901', *Transactions of Shropshire archaeological society*, LVI, part II (1959), 104–24.

17. See note 13. The patent is no. 2109.

18. Rolt, pp. 44–50.

19. Telford, 'Canals' (1797 and 1800), p. 300. Paxton, 'Influence of Thomas Telford' (1976), pp. 71–4 and 339–40, analyses the structure in detail.

20. Having searched all the documents in vain, I quote from the words of an octogenarian met on the tow-path of Pont Cyssylte, who claimed a family connection in that his great-grandfather worked on the construction!

21. Rickman, p. 682.

22. See Telford, 'Bridge' (1812), pp. 488–9 and 539–40, and Hill (see note 16), pp. 106–12. The description given is from these and Rickman, pp. 29–30.

23. Telford, 'Canals' (1797 and 1800), p. 316.
24. See note 13.
25. Telford, 'Bridge', p. 524 and pl. xcvi.

13. Iron and timber

1. R. Dodd, *Letters to a merchant, on the improvement of the port of London* (London, 1798). Dated 9 and 28 November. The design was also published as pl. ii of appendix to the third report of the select committee on the improvement of the port of London (1800). The committee's reports with appendices are in *Commons reports 1715–1801* (1803), vol. xiv, pp. 461–635.
2. *HCJ*, 2 May 1799.
3. For biography and career see J. G. James, 'Ralph Dodd, the very ingenious schemer', in *Trans. Newcomen soc.*, 1974–6.
4. Third report of select committee (see note 1). The following paragraphs are derived from this report and all the designs mentioned are printed or described in it.
5. In fourth report of select committee (1801), pp. 604–5. Except where otherwise noted, the following discussion is all based on this report.
6. There are four successive drafts in Telford's hand in Telford MSS, T/LO.4–8, at InstCE.
7. Letters to and from Wm Reynolds, John Wilkinson, Neville Maskelyne, Abraham Robertson, Charles Hutton, Charles Bage and Joshua Walker, are in Telford MSS, T/LO.11–38.
8. *Edinburgh review*, LXX (1839–40), 3.
9. Telford MSS, T/LO.19, 24, 26, 30.
10. *HCJ*, 3 June 1801.
11. Gibb, *Telford* (1935), p. 51.
12. *Ibid*, pp. 43–4, 52.
13. Report of Rennie to Charles Abbot, 16 February 1802, appendix D to second report of committee on Holyhead roads and harbour, *Commons papers 1810* (352.) IX. 41–51. Hutton's and Jessop's reports are appended. There are drafts and correspondence in box 1, F1 and F2, Rennie papers, at NLS.
14. See H. R. Davies, *Conway and Menai ferries* (Cardiff, 1942), pp. 252–60.
15. Specification of patent to Thomas Wilson, 23 July 1802, in *Repertory of arts*, 2nd series, III (1803), 87–8. Description of the Staines arch is from this and Rennie's sketches and notes, 25 April 1802, in notebook 51, box 26, Rennie papers, at NLS.
16. J. W. Wardell, *History of Yarm* (Sunderland, 1957), pp. 121–2, gives full details from the QS records of Durham and the North Riding. His pl. 15 is the landscape view.
17. The main sources are a report of the commissioners to the Lord High Chancellor of 17 October 1803, the 'Minutes of commissioners for Staines Bridge 1804–28', an estimate for the work done in 1806–7, acc. 809/BR/17, 15 and 18 respectively at GLCRO Middlesex branch, and reports by Rennie dated 10 October 1804 and 17 June 1805, in Rennie reports, vol. III, pp. 219–20, 318–22, at InstCE.
18. Colvin, *Biographical dictionary* (1954), pp. 520–3.
19. Hutton, *Tracts* (1812), vol. I, p. 150.
20. *Ibid*, pp. 153–5.
21. Telford, 'Bridge' (1812), pp. 541–2 and pl. xciv. R. A. Buchanan and N. Cossons, *Industrial archaeology of the Bristol region* (Newton Abbot, 1969), pp. 51–2, gives dates and other details.

22. Letter and estimates, 19 November 1800, in Rennie reports, vol. III, pp. 1–4, at InstCE. There is one MS design, for a stone bridge, in Rennie drawings collection, vol. I, at InstCE.
23. The fullest description is in Telford, 'Bridge', p. 541, but see also Hutton, *Tracts*, vol. I, pp. 152–3, and Stephenson, 'Iron bridges' (1856), p. 583. P. A. Townley, 'Boston's iron bridge', *Lincolnshire life*, VI, no. 8 (October 1966), 52–55, quotes bridge committee records from 1799 to 1816.
24. Paxton, 'Influence of Thomas Telford' (1976), p. 55 and fig. 11, shows twelve built in 1806–10, compared with thirty-one in 1796–1805.
25. R. Southey, *Journal of a tour in Scotland in 1819* (London, 1929), pp. 128–9.
26. Telford, 'Bridge', pp. 539–45. Extra details concerning Bonar Bridge are from 'Sixth report of commissioners for Highland roads and bridges' (1813), *Commons reports 1812–13* (110.) V. 1.
27. W. W. Lowson, 'The reconstruction of the Craigellachie Bridge', *Structural engineer*, XLV (1967), 23–28.
28. Gibb, pp. 155–6.
29. 'Seventh report of commissioners for Highland roads and bridges' (1815), *Commons papers 1814–15* (205.) III. 19.
30. Johnson and Scott-Giles, *British bridges* (1933), p. 330.
31. Rolt, *Thomas Telford* (1958), p. 73n. Paxton, pp. 90–1, 284, gives details of this bridge.
32. Johnson and Scott-Giles, p. 379.
33. Rolt, pp. 72, 115. A list of Hazledine's works is in his obituary in *Civil engineer and architect's journal*, IV (1841), 48–9.
34. Paxton, pp. 90–100, traces the developments in the design after Bonar Bridge.
35. Telford, 'Bridge', pp. 542–5.
36. James (see note 3) gives a summary of events concerning these bridges. Details of Vauxhall Bridge as built are given by Stephenson, p. 583.
37. The story is condensed from Rennie reports, vols VIII–XI, at InstCE, Rennie papers, boxes 2 and 3, at NLS, Smiles, *Lives* (1861), vol. II, pp. 327–34, Stephenson, pp. 583–4, Cresy, *Encyclopaedia* (1847), vol. I, pp. 498–503, Sir J. Rennie, 'Presidential address' (1846), pp. 29–31, and *Autobiography* (1875), pp. 7–26, and T. Young, 'Bridge', in *Supplement to Encyclopaedia Britannica*, 4th edn (1817).
38. Rennie drawings collection, vol. II, fols 26–30, at InstCE.
39. Sir J. Rennie, 'Presidential address', p. 30.
40. Sir J. Rennie, *Autobiography*, p. 8.
41. G. Rennie, 'On the expansion of arches', *Trans. InstCE*, III (1840–2), 201–18.
42. F. W. Simms, *Public works of Great Britain* (London, 1838), division II, pp. 23–7, contains the specification for High Bridge, and pls IC–CI the plan and details. Brief notes on both High Bridge and Chetwynd Bridge are given by R. J. Sherlock, 'Industrial archaeology in administrative Staffordshire', *North Staffordshire journal of field studies*, II (1962), 96–107.
43. Notes by H. P. Forge in discussion of Buckton and Fereday, 'The demolition of Waterloo Bridge'.
44. J. Guest, *Historic notices of Rotherham* (Worksop, 1879), p. 495.
45. A. H. John, *The Walker family, 1741–1893* (Council for preservation of business archives, 1951), pp. iii, 25–8.
46. Except where otherwise noted, the source is Telford, 'Bridge', pp. 537–8 and pl. LXXXIII.

47. *Aberdeen journal,* 11 September 1850, 15 January and 17 September 1851.

48. J. Milne, *Aberdeen* (Aberdeen, 1911), p. 259.

49. RHP 4299, at SRO. Signed and dated 1797.

50. RHP 11714, at SRO.

51. 'Fifteenth report of commissioners for roads and bridges in Scotland', *Commons papers 1829* (114.) v. 135, and Rickman, *Life of Telford* (1838), pl. 50.

52. Twentieth report of commissioners, *Commons papers 1834* (164.) XL. 165.

53. Contract for Warrington Bridge dated 23 November 1812, QAR 18, at Cheshire CRO.

54. G. A. Carter, 'History of Warrington to 1847', in *Warrington hundred* (1947).

55. Wiebeking, *Traité de construire les ponts* (1810). For an assessment of Wiebeking's work see Booth, 'Development of laminated timber arch structures' (1971).

56. Green's bridges are surveyed in detail by Booth, 'Development of laminated timber arch structures', L. G. Booth, 'Laminated timber arch railway bridges in England and Scotland', *Trans. Newcomen soc.,* XLIV (1971–2), 1–21, and H. Hagger, 'The bridges of John Green', *Northern architect,* new series, no. 8 (April 1976), 25–31.

57. Stevenson, *Life of Robert Stevenson* (1878), p. 162.

58. Mitchell, *Reminiscences* (1883), vol. I, p. 161.

59. These and subsequent details are from the Spey Bridge papers 1804–35 in Gordon Castle papers, GD44/53, box 1, at SRO. The specification is dated 16 December 1831 and the contract 28 February and 4 April 1832. The drawing (fig. 171) is attached to the printed Act of 1832.

60. *Elgin courant and courier,* 12 August 1912.

14. Climax in masonry

1. Except where otherwise noted, the source is the article 'Telford did not design Union Bridge', *People's journal* (Aberdeen), 19 November 1904.

2. Rennie reports, vol. I, pp. 414–18, at InstCE.

3. Telford, 'Bridge' (1812), p. 486.

4. N. Murray, 'Union Bridge widening, Aberdeen', *Minutes of proceedings of Aberdeen association of civil engineers,* VIII (1907–8), 69–75.

5. For instance, W. Hosking in Weale (ed.), *Architecture of bridges* (1843), vol. II, p. 41.

6. C. Smith, *The county and city of Cork* (2 vols, 1893), vol. I, p. 417, and watercolour of 'St Patrick's Bridge as it appeared 3 November 1853', at Cork City Library.

7. *Freeman's journal* (Dublin), 4 and 6 December 1802.

8. *Commons reports 1715–1801* (1803), vol. XIV, p. 603 and pl. 51.

9. Letters dated 1800 in Rennie papers, box 34, at NLS, show that the bridge was then under construction.

10. Rennie reports, vol. III, pp. 1–4, and Rennie drawings collection, vol. I, fol. 5, at InstCE.

11. The narrative is based on the MS journal of the construction by George Rennie in Rennie papers, box 3, F2, at NLS, supplemented by letters and reports in Rennie reports, vols V–IX, at InstCE, and Buckton and Fereday, 'The demolition of Waterloo Bridge'.

12. Engraved 1808. Copy at the Guildhall Library.

13. Letters of 14 and 20 February 1809, in Rennie reports, vol. V, pp. 356–62, at InstCE.

14. Estimate of 4 December 1811, in Rennie reports, vol. VI, p. 407.

15. Full specification and drawings printed in Cresy, *Encyclopaedia* (1847), vol. I, pp. 445–7.

16. W. Muirhead, discussion on Buckton and Fereday (see note 11), *Journal of InstCE,* III (1935–6), 515–16.

17. Cresy, vol. I, pp. 447–8.

18. H. W. Dickinson, 'Jolliffe and Banks, contractors', *Trans. Newcomen soc.,* XII (1931–2), 1–8.

19. See C. D. Brown, 'London Bridge: planning, design and supervision', *Proceedings of InstCE,* LIV (1973), part 1, 25–46, and discussion by J. K. T. L. Nash, 726–32.

20. Memoir of George Rennie, *Minutes of proceedings of InstCE,* XXVIII (1868–9), 610–15.

21. 'Old London Bridge', in Cooke, *Views of the old and new London Bridges* (1833), p. 9.

22. J. W. Clarke, 'The building of Grosvenor Bridge', *Chester archaeological society journal,* XLV (1958), 43–55.

23. An engraving published in 1839 is inscribed 'G. Rennie const.', and reports to the commissioners dated 1834–8, acc. 809/BR/185–93 at GLCRO Middlesex Branch, were all written by him. Cresy (1847) also ascribed it to him.

24. Sir J. Rennie reports, vol. II, pp. 292–3, at InstCE, and Sir J. Rennie, 'Presidential address' (1846), p. 27.

25. Drawing in Telford drawings at InstCE, assumed to be Harrison's design.

26. 'Account of the new bridge at Chester', *Trans. InstCE,* I (1836), 207–14, pl. XXII. The following paragraphs are derived from this paper, Clarke (see note 22), and manuscripts quoted by Clarke at Chester City Record Office.

27. Telford, 'Practice' in article 'Bridge' (1812), pp. 519–45.

28. Telford, 'Civil architecture' (1813).

29. A. Alison, *Essays on the principles of taste* (Edinburgh, 1790). Alison was a close friend of Telford, see Gibb, *Telford* (1935), pp. 8, 157–8.

30. Contract and other papers are DP36 and DP113 at Salop CRO.

31. See Rickman (ed.), *Life of Telford* (1838), pp. 38–41 and pl. 13, and Gibb, pp. 30–5.

32. Telford, 'Bridge', p. 487.

33. *Ibid.*

34. Rickman, pl. 47.

35. *Ibid.* pls 49 and 51.

36. *Ibid.* pp. 423–6, has the specification for bridges on the Speyside road.

37. Summarised by Haldane, *New ways through the glens* (1962). See especially pp. 122–30.

38. *Ibid.* p. 125.

39. Rickman.

40. Details from DP180 at Salop CRO; vol. 32, pp. 1–6, of Prattinton MSS, and Prattinton collection, box IV(5), at the Society of Antiquaries, London.

41. Rickman, pp. 190–1 and pl. 60.

42. *Ibid.* pp. 258–67 and pls 63 and 82.

43. *Ibid.* pp. 202–3 and pl. 65.

44. Sources for Dean Bridge are Rickman, pp. 195–201 and pls 62 and 63, and original drawings by Telford at the Architect's Department, Edinburgh District Council; for Pathhead Bridge, Rickman, pp. 201–2 and pl. 64. The drawing for Stonebyres Bridge is in the Telford drawings at InstCE.

45. Drawing in Telford drawings at InstCE, and Gibb, pp. 254–5.

46. Rickman, pp. 191–5 and pl. 61.

47. I am grateful to the Lothian Regional Council for permission to publish details of the inspection and to Mr

R. Paxton for inviting me to be present.

48. Paxton, 'Influence of Thomas Telford' (1976), pp. 31–5, describes Telford's part in the development of concrete road bases.

49. Description by Charles Atherton in *Encyclopaedia Britannica*, 7th edn (London, 1842), vol. v, pp. 283–5, also printed in Rickman, pp. 197–201.

50. See *Commons papers 1836* (418.) xx. 65.

51. See *Commons papers 1844* (477.) vi. 663, and *Commons papers 1846* (574.) xv. 277.

52. Stevenson, *Life of Robert Stevenson* (1878) is the source, except for Hutcheson Bridge, for which specification, description and drawings were published in Weale, vol. i, pp. 97–140, and vol. iii, pls 27–33.

53. Colvin, *Biographical dictionary* (1954), pp. 528–30, and 'Minutes of the corporation for the port of Dublin', 1805–16, at the Ballast Office, Dublin.

54. *Freeman's journal* (Dublin), 16 March 1816.

55. 'Minutes of . . . port of Dublin', 1816 (see note 53).

56. QS records at Bedford CRO contain minutes, contracts, drawings and correspondence (QBM1, QBP8/1–7, PB5/5–6). See also *The lock gate*, 1 (1961), 190–4.

57. Weale, vol. ii, p. cxxxii.

58. Sources for Nimmo's biography include *DNB*, vol. xiv, p. 512; minutes of the presbytery of Chanonry, 1802–5, at SRO; Rickman, pp. xiii–xiv; Mitchell, *Reminiscences* (1883), vol. i, p. 44; various parliamentary reports; and manuscript collections in Dublin archives.

59. Illustrated in Mr and Mrs S. C. Hall, *Ireland, its scenery and character*, new edn (3 vols, London, c. 1840), vol. iii, p. 480, and V. T. H. and D. R. Delany, *Canals of the south of Ireland* (Newton Abbot, 1966), p. 166.

60. S. Lewis, *Topographical dictionary of Ireland* (2 vols, 1837), vol. i, pp. 152–3.

61. The sources are Weale, vol. ii, pp. cxxx–cxxxii, and vol. iv, pls 54 and 55, and M. Lenihan, *History of Limerick* (Dublin, 1866), pp. 98, 469–71.

BIBLIOGRAPHY OF CHIEF PRINTED REFERENCES

Adam, R. See Bolton, A. T., and Fleming, J.

Anon. *A short narrative of the proceedings of the gentlemen, concerned in obtaining the Act for building a bridge at Westminster.* London, 1737.

A narrative of the proceedings relative to Hexham Bridge. London, 1788.

(attrib. W. Adam). *Vitruvius Scoticus.* 3 vols. n.d.

Baldwin, R. *Plans, etc., of the machines and centring used in erecting Blackfriars Bridge.* London, 1787.

Bélidor, B. F. de. *La science des ingénieurs.* 6 vols. Paris, 1729.

Architecture hydraulique. 4 vols. Paris, 1737–52.

Bolton, A. T. *The works in architecture of Robert and James Adam.* 2 vols. London, 1922.

Booth, L. G. 'The development of laminated timber arch structures in Bavaria, France and England in the early nineteenth century', *Journal of the institute of wood science*, no. 29 (July 1971).

Boucher, C. T. G. *John Rennie, 1761–1821.* Manchester, 1963.

James Brindley, engineer, 1716–1772. Norwich, 1968.

Brindley, J. For biographies, see Smiles, S., Boucher, C. T. G., and Malet, H.

Bristol Bridge pamphlets. Collections of pamphlets, all dated 1759–63, at Bristol Central Library and Bristol University Library.

Buckton, E. J., and H. J. Fereday. 'The demolition of Waterloo Bridge', *Journal of the institution of civil engineers*, III (1935–6), 472–522.

Burton, A. *The canal builders.* London, 1972.

Carr, J. See York Georgian Society.

Carson, P. M. 'Provision and administration of bridges over the lower Thames 1701–1801, with special reference to Westminster and Blackfriars'. Unpublished MA thesis, University of London, 1954.

'The building of the first bridge at Westminster, 1736–1750', *Journal of transport history*, III (1957–8), 111–22.

Colvin, H. M. *Biographical dictionary of English architects 1660–1840.* London, 1954.

Cooke, E. W. *Views of the old and new London Bridges.* London, 1833.

Cresy, E. *Encyclopaedia of civil engineering.* 2 vols. London, 1847.

Dance, G. (the elder). See London Bridge reports, 1746.

Dance, G. (the younger) and others. See London Bridge reports, 1814.

Dickinson, H. W., and A. A. Gomme, eds. *Catalogue of the civil and mechanical engineering designs 1741–1792 of John Smeaton, F.R.S.* London, 1950.

Dictionary of national biography. 22 vols. London, 1908–9.

Emerson, W. *The principles of mechanics.* 2nd edn. London, 1758.

Fleming, J. *Robert Adam and his circle.* London, 1962.

Gautier, H. *Traité des ponts.* Paris, 1714. 3rd edn, 1728.

Dissertation sur les culées, voussoirs, piles et poussées des ponts. Paris, 1717.

Gibb, A. *The story of Telford.* London, 1935.

Haldane, A. R. B. *New ways through the glens.* Edinburgh and London, 1962.

Hamilton, S. B. 'The French civil engineers of the eighteenth century', *Trans. Newcomen soc.*, XXII (1941–2), 149–59.

Hawksmoor, N. *A short historical account of London-Bridge, with a proposition for a new stone-bridge at Westminster.* London, 1736.

Heyman, J. *Coulomb's memoir on statics.* Cambridge, 1972.

Hutton, C. *The principles of bridges.* Newcastle-upon-Tyne, 1772.

Tracts on mathematical and philosophical subjects. 3 vols. London, 1812.

James, J. *Short review of the pamphlets in relation to the building of a bridge at Westminster.* London, 1736.

Jervoise, E. *The ancient bridges of the south of England.* London, 1930.

The ancient bridges of the north of England. London, 1931.

The ancient bridges of mid and eastern England. London, 1932.

The ancient bridges of Wales and western England. London, 1936.

Johnson, S. M., and C. W. Scott-Giles, eds. *British bridges: an illustrated technical and historical record.* London, 1933.

Journal of the institution of civil engineers. London, annual, 1935/6–51.

Knight, W. 'Observations on the construction of old London Bridge', *Archaeologia*, XXIII (1830), 117–19.

Labelye, C. *A short account of the methods made use of in laying the foundations of the piers of Westminster Bridge.* London, 1739.

A description of Westminster Bridge. London, 1751.

See London Bridge reports, 1746.

Langley, B. *A design for the bridge at New Palace-Yard, Westminster.* London, 1736.

Lindsay, I., and M. Cosh. *Inveraray and the dukes of Argyll.* Edinburgh, 1973.

London Bridge reports (all printed):

1746   Reports by C. Labelye, G. Dance (the elder) and B. Sparruck, printed in Maitland, *History of London*, 3rd edn (1760), vol. II, pp. 827–33. Dance's report is also in Hutton, *Tracts* (1812), vol. I, pp. 115–20.

1754   Report of committee of Common Council, 26 September. A copy is at BM, 816.l.5/20.

1763   Report of committee of Common Council, 10 October. Guildhall Library, A-8-5/31.

1763   Report by J. Smeaton in Smeaton, *Reports* (1812), vol. II, pp. 1–17.

1766–7 Reports by J. Smeaton in Smeaton, *Reports* (1812), vol. II, pp. 18–26.

1790   Report of the committee for letting the Bridge House lands, 2 June. Guildhall Library, A-8-5.

1799   Second report of the select committee on the improvement of the port of London, 11 July (with appendices B.1–11 and G.5–7), in *Reports of committees of the House of Commons 1715–1801* (1803), vol. XIV, pp. 461–84.

1800   Third report of select committee, 28 July (with appendices), in *Commons reports*, vol. XIV, pp. 543–630b. Robert Mylne's report is appendix 1A.

1801   Fourth report of select committee, 3 June (with appendices), in *Commons reports*, vol. XIV, pp. 604–35.

1814   Report of G. Dance (the younger), W. Chapman, D. Alexander and J. Montague to select committee of Bridge House lands, November (with appendices). Printed as appendix no. 2 to 1821 report.

1821   Report of committee on the state of London Bridge, 25 May (with appendices). *Commons reports 1821* (569.) v. 281 and (609.) v. 415.

1831   Reports presented to the City of London by T. Telford, W. T. Clarke, J. Walker, Sir J. Rennie and Sir E. Banks. *Commons papers 1831–2* (61.) XLV. 99.

Maguire, R., and P. Matthews. 'The ironbridge at Coalbrookdale: a reassessment', *Architectural association journal*, LXXIV (1958), 30–45.

Maitland, W. *History of London*. 3rd edn. 2 vols. London, 1760.

Malet, H. *The canal duke*. London, 1961.

Maré, E. de *The bridges of Britain*. London, 1954.

Mitchell, J. *Reminiscences of my life in the Highlands*. 2 vols. 1883–4. Reprint, Newton Abbot, 1971.

Minutes of proceedings of the institution of civil engineers. London, annual, 1842–1935.

Mylne, R. For biography, see Richardson, A.
    See London Bridge reports, 1800.

[Mylne, R.] *Observations on bridge-building and the several plans offered for a new bridge*. London, 1760.

Mylne, R. S. *Master masons to the crown of Scotland*. Edinburgh, 1892.

Nash, J. For biography, see Summerson, J.

Nimmo, A. 'Theory', in article 'Bridge' (1812), in *Edinburgh encyclopaedia*, vol. IV, pp. 479–545. Edinburgh, 1830. See also Telford.

Paine, J. *Plans, elevations and sections of noblemen and gentlemen's houses*. 2 vols. London, 1767 and 1783.

Palladio, A. *I quattro libri dell' architettura*. Venice, 1570. English editions entitled *Four books of architecture* published by G. Leoni (1715) and I. Ware (1738).

Facsimile of Ware's edition by Dover Publications, New York, 1965.

Paxton, R. A. 'The influence of Thomas Telford on the use of improved construction materials'. Unpublished MSc thesis, Heriot-Watt University, 1976.

Perronet, J.-R. *Description des projets*. 2 vols. Paris, 1783.

Price, J. *Some considerations for building a stone-bridge from Westminster to Lambeth*. London, 1735.

Raistrick, A. *Dynasty of ironfounders*. London, 1953.

Rees, A., ed. Article 'Bridge' (1805), in *The cyclopaedia: or universal dictionary of arts, sciences and literature*, vol. V. 39 vols. London, 1819.

Rennie, J. For biographies see Boucher, C. T. G., and Smiles, S.

Rennie, Sir J. 'Presidential address, 1846', *Minutes of proceedings of InstCE*, V (1846), 19–122.
    Autobiography of Sir John Rennie, F.R.S. London, 1875.

Repertory of arts and manufactures. 1st and 2nd series. London, monthly.

Richardson, A. *Robert Mylne, architect and engineer*. London, 1955.

Rickman, J., ed. *Life of Thomas Telford, civil engineer, written by himself*. 2 vols. London, 1838.

Robertson, J. 'Concerning the fall of water under bridges', *Philosophical transactions of the Royal Society*, L (1758), 492–9.

Robison, J. Article 'Arch' in *Supplement to Encyclopaedia Britannica 3rd edn*, vol. I, pp. 14–31. Edinburgh, 1801.

Rolt, L. T. C. *Thomas Telford*. London, 1958.

Ruddock, E. C. 'Hollow spandrels in arch bridges: a historical study', *Structural engineer*, LII (1974), 281–93.
    'William Edwards's bridge at Pontypridd', *Industrial archaeology*, XI (1974), 194–208.
    'The building of North Bridge, Edinburgh, 1763–75', *Trans. Newcomen soc.*, XLVII (1974–6).
    'The foundations of Hexham Bridge', *Géotechnique*, XXVII (1977), 385–404.

Salmond, J. B. *Wade in Scotland*. 2nd edn. London and Edinburgh, 1938.

Semple, G. *A treatise on building in water*. Dublin, 1776. 2nd edn, London, 1780.

Shanahan, M. Engravings of bridges, c. 1770–80. Published at Vicenza. Collection in RIBA Library and copies at Sir John Soane's Museum.

Skempton, A. W. 'The publication of Smeaton's reports', *Notes and records of the Royal Society of London*, XXVI (1971), 135–55.

Smeaton, J. *Mr. Smeaton's answer to the misrepresentations of his plan for Black-Friars Bridge*. London, 1760.
    Reports of the late John Smeaton, F.R.S. 3 vols. London, 1812. See also Skempton.
    Designs by the late John Smeaton, F.R.S. Microfilm, Wakefield, 1972. See also Dickinson and Gomme.

Smiles, S. *Lives of the engineers*. 3 vols. London, 1861–2.

Stephenson, R. Article 'Iron bridges', in *Encyclopaedia Britannica*, 8th edn, vol. XII. London, 1856.

Stevenson, D. *The life of Robert Stevenson*. Edinburgh, 1878.

Summerson, J. *John Nash, architect to King George IV*. London, 1935.

Swan, A. *Collection of designs in architecture*. 2 vols. London, 1757.
    Designs in carpentry. London, 1759.

Sykes, J. *Local records*. Newcastle-upon-Tyne, 1833.

Taylor, J., ed. *Schaffhausen Bridge*. 2 engravings with descriptive pamphlet. London, 1800.

Telford, T. 'Canals' (1797 and 1800), in J. Plymley, *General view of the agriculture of Shropshire*. London, 1803.

'History' and 'Practice', in article 'Bridge' (1812), in *Edinburgh encyclopaedia*, vol. IV, pp. 479–545. Edinburgh, 1830. See also Nimmo.

'Civil architecture' (1813). Article in *Edinburgh encyclopaedia*, vol. VI, pp. 521–663. Edinburgh, 1830.

For biographies see Gibb, A., Rickman, J. (ed.), Rolt, L. T. C., and Smiles, S.

Thomson, R. (pseud. 'An antiquary'). *Chronicles of London Bridge*. London, 1827.

Transactions of the institution of civil engineers. 3 vols. London, 1836–42.

Transactions of the Newcomen Society. London, annual.

Ward, A. W. 'The reconstruction of the English Bridge, Shrewsbury', *Selected engineering papers*, no. 66, InstCE, 1928.

The bridges of Shrewsbury. Shrewsbury, 1935.

Weale, J., ed. *The architecture of bridges*. 5 vols. London, 1843.

White, J. 'Notice of Mr Pritchard's gradual progress in the application of iron to the erection of bridges', *Philosophical magazine and annals of science*, 2nd series, XI (1832), 81–2.

Wiebeking, C. F. von. *Traité contenant une partie de la science de construire les ponts*. Munich, 1810.

Williams, E. I. 'Pont-y-ty-pridd: a critical examination of its history', *Trans. Newcomen soc.*, XXIV (1945), 121–30.

York Georgian Society. *The works in architecture of John Carr*. York, 1973.

INDEX OF BRIDGES

Only bridges built are included; for unexecuted designs consult the general index. Gaps occur where no information has been found. The strength of the evidence for individual facts varies greatly, but the information given is considered to be the most reliable available. Where documentary evidence is lacking, a few author's suppositions are included and are followed by a question mark. See general index for bridges not mentioned here.

Column 1 Place name precedes bridge name.

Column 3 The reference is to the British Ordnance Survey National Grid and, for bridges in Ireland, the Irish Grid.

Column 5 Technically, a bridge built under Act of Parliament was owned and built by commissioners or trustees named by the Act. These bodies were usually derived more or less exclusively from the members of a more permanent institution, such as a bench of justices or a borough council, and ownership and responsibility for maintenance passed to the permanent body after a few years. In such instances the permanent body is named as the owner.

Column 6 The source of capital is specified by the following abbreviations:

CC Funds of a canal company
L Lotteries
O Owner's general wealth
JT Funds of joint-stock bridge company, with perpetual right of profit from toll on finished bridge
OT Owner's general wealth with right of repayment by toll on the finished bridge
P Grant(s) from Parliament
R Rates levied on property in borough, city or county
S Subscriptions (i.e. donations) from individuals and/or statutory bodies
ST Subscriptions to be repaid (i.e. loans) from toll on the finished bridge, and usually bearing interest until repaid

Unless otherwise noted the cost quoted is the actual cost of construction.

Where this is not recorded, the agreed contract price (cont) or the designer's estimate (est) is sometimes given; both these figures dating from before construction.

Column 7 Small land arches and arches through approach embankments are not counted.
Abbreviations:
circ semicircles
CV *cornes de vâche*
elpse semi-ellipses
segt segments of circles

Column 8 Abbreviations:
Arch Architect undertaking both design and supervision of construction
Eng Engineer undertaking both design and supervision of construction
Surv Surveyor undertaking both design and supervision of construction
Des Designer without responsibility for supervision of construction
Cont Main, general or sole contractor
Carp Contractor for carpentry only
Iron Contractor for supply (and sometimes erection) of ironwork
Mason Contractor for masonry only
Smith Contractor for blacksmith work only
Ovsr Overseer, supervisor or resident engineer, representing the architect, engineer, surveyor and/or owner on the site

Column 9 Abbreviations:
Coll collapsed
Dem demolished

Column 10 Numbers in italics refer to figures. References to notes are in the form 225(12.7), where 225 is the page number, 12 the chapter number, and 7 the number of the note to chapter 12.

Place and name	River	Grid reference	Date of construction	Owner or promoter	Source of finance and cost	Number, shape and material of arches	Persons in design and construction	Present condition	Page refs.
		3	4	5	6	7	8	9	10
Aberdeen, Don	Don	NJ945095	1827–30	Burgh (and Co.?) of Aberdeen	O, £17,000	5, segt, stone	Eng T. Telford Cont J. Gibb & Son	Widened 1960s	193
Aberdeen, Grandholm	Don	NJ897113	c. 1810	Leys, Masson and Co. (mill owners)	O, £1,200	2, segt, timber	Des (& cont?) J. Burn	Dem	171
Aberdeen, Union	Dean Burn	NJ939061	1802–5	Burgh of Aberdeen	O, £13,000	1, segt, stone	Surv T. Fletcher Cont W. Ross	Widened 1906–7	175; 172
Aberfeldy	Tay	NN852494	1733–5	Board of Ordnance	P, £4,095	5, segt, stone	Des W. Adam	As built	22–5, 96, 118; 20, 21; see general index, Aberfeldy profile
Amesbury, ornamental bridge	Avon	SU150418	1776	Duke of Queensberry	O	3, elpse, stone	Des J. Smeaton	As built, disused	82, 86, 90; 79, 84
Ancrum	Teviot	NT638237	1784	Co. of Roxburgh?	R?	3, segt, stone	Des & cont A. Stevens	As built, disused	121; 120
Atcham	Severn	SJ541093	1768–76	Co. of Shropshire	R, £7,239 (cont)	7, circ, stone	Arch & cont J. Gwynn	As built, disused	111–13; 105
Avoncliffe aqueduct	Avon	ST805600	1795–8	Kennet and Avon Canal Co.	CC	1 elpse, 2 circ, stone	Eng J. Rennie Cont J. McIlquham	Distorted, disused	131
Ayr, New	Ayr	NS337223	1786–9	Burgh of Ayr	£4,063	5, segt, stone	Des & cont A. Stevens	Dem 1878	118–20, 122; 117
Aysgarth (widening and improvement)	Ure	SE010886	1788	Co. of York, N. Riding	R, £420	1, segt, stone	Surv J. Carr	As built	115; 110
Banff	Deveron	NJ695638	1772–9	Board of Ordnance	P, £8–10,000	7, segt, stone	Des J. Smeaton Ovsr W. Kyle	Widened 1881	81–4, 87–8, 90, 96, 103; 87
Barton aqueduct	Irwell	SJ767976	1760–1	Duke of Bridgewater	CC	3, segt, brick	Eng J. Brindley (with J. Gilbert?)	Dem 1893	125–7, 147; 125
Bath, Pulteney	Avon	ST752649	1769–74	Sir J. W. Pulteney	O	3, segt, stone	Arch R. Adam	As built	120; 118
Bettws-y-coed, Waterloo	Conway	SH799557	1815	Commissioners for Holyhead road	P	1, segt, iron	Eng T. Telford Iron W. Hazledine	Strengthened and widened 1923	165; 162
Bewdley	Severn	SO787754	1798–9	Borough of Bewdley	OT, £8,512	3, segt, stone	Surv T. Telford Cont J. Simpson Ovsr? J. Cargill	As built	154,190,194; 151
Birmingham, Galton	Birmingham Canal	SP015893	1829	Birmingham Canal Co.	CC	1, segt, iron	Eng T. Telford Iron Horseley Bridge Co.	As built, bypassed	164–5

1	2	3	4	5	6	7	8	9	10
Coalbrookdale	Severn	SJ672033	1777–80	Proprietors of Ironbridge	JT, c.£6,000	1, circ, iron	Des A. Darby III? Cont Coalbrookdale Co.	As built	134–5, 138–9, 150–1; 132
Coldstream	Tweed	NT848401	1762–7	Turnpike trust	P and ST, c.£6,000	5, segt, stone	Eng J. Smeaton Ovsr R. Reid	Widened 1960	81–4, 86–90, 94–6, 100–4, 147; 76, 77, 78
Cork, St Patrick's	Lee	W6772	1789	City of Cork		3, elpse, stone	Arch M. Shanahan	Coll 1853	177
Craigellachie	Spey	NJ285452	1814–15	Highland Roads Commissioners	P and R, £8,000 (est)	1, segt, iron	Eng T. Telford Mason J. Simpson Iron W. Hazledine	Restored 1963, disused	162–4, 190; 158, 159, 160
Crubenbeg	Truim	NH680923	1730	Board of Ordnance	P	1, circ, stone		As built?	22; 17
Cumnock, Dumfries House	Lugar Water	NS537207	1760–2	Earl of Dumfries	O, £543	3, elpse, stone	Des & cont Jn Adam	As built	118; 115
Dalkeith House	Esk	NT334682	c.1793	Duke of Buccleuch	O	1, circ, stone	Arch R. & Jas Adam	As built	117; frontispiece, 113
Dalwhinnie, 'Wade Bridge'	Truim	NN639828	1730	Board of Ordnance	P	2, segt, stone		As built?	22; 18
Darlaston	Trent	SJ886355	1805	Co. of Stafford	R?	1, segt, stone	Eng J. Rennie	As built	147, 178; 141
Derby, St Mary's	Derwent	SK354367	1789–94	Borough of Derby	£2,900 (cont)	3, elpse, stone	Arch T. Harrison Ovsr C. Moore Cont S. Lister and J. Hanley	As built	124
Drygrange	Tweed	NT575346	1779–80	Turnpike trust	ST, £2,100 (cont)	3, segt, stone	Des & cont A. Stevens	As built, disused	121; 119
Dublin, Essex	Liffey	O1534	1753–5	City of Dublin	P, £20,661	5, circ, stone	Des & cont G. Semple	Dem 1873–4	40–5, 54, 56, 98, 105–7; 42, 43, 44, 45, 46
Dublin, Queen's (now Queen Maeve)	Liffey	O1434	1764–8	Port of Dublin?		3, elpse, stone	Eng C. Vallancey	As built	106–7, 196; 98
Dublin, Richmond (now O'Donovan Rossa)	Liffey	O1534	1813–16	Port of Dublin	R, £25,950	3, elpse, stone	Des J. Savage Cont G. Knowles	As built	196; 198
Dublin, Ringsend	Dodder	O1834	1803	City of Dublin?		1, elpse(CV), stone		As built	177; 175
Dublin, Sarah's	Liffey	O1334	1791–3	Dublin Pavings Commissioners		1, elpse, stone	Des (& cont?) A. Stevens	As built	122
Dublin, Whitworth (now Father Mathew)	Liffey	O1534	1816–17	Port of Dublin	R	3, elpse, stone	Des & cont G. Knowles Ovsr G. Halpin	As built	196
Dunkeld	Tay	NO026425	1805–9	Highland Roads Commissioners	P and S, c.£14,000	7, segt and circ, stone	Eng T. Telford	As built	190

1	2	3	4	5	6	7	8	9	10
Dyce	Don	NJ889141	1803	Turnpike trust		1, segt, timber	Des (& cont?) J. Burn	Dem 1851	169-72
Edinburgh, Dean	Water of Leith	NT243740	1829-31	J. Learmonth and J. Paton and district road trustees	O, £18,556	4, segt, stone	Eng T. Telford Ovsr C. Atherton Cont J. Gibb	As built	193-5, 200; 192, 193, 194
Edinburgh, North	—	NT258738	1763-75	City of Edinburgh	O, £15,897	5, circ, stone	Des & cont W. Mylne	Dem 1895	81-3, 96, 100-2; 93
Edinburgh, South	—	NT260736	1786-8	City of Edinburgh	O	19, circ, stone	Des & cont A. Laing	As built	120
Fochabers	Spey	NJ349595	1802-6	Duke of Gordon	P and ST, £12,000 (est)	4, segt, stone	Des & cont G. and J. Burn	Half coll 1829	173; 171
Fochabers (reconstruction)	Spey	NJ349595	1832	Duke of Gordon's trustees	O, £4,986	1, segt, timber	Arch A. Simpson Cont W. Minto and W. Leslie	Dem 1853	173-4; 171
Garstang, Wyre aqueduct	Wyre	SD490448	c. 1798	Lancaster Canal Co.	CC	1, elpse, stone	Eng J. Rennie	As built	129, 131
Glasgow, Broomielaw I	Clyde	NS587647	1768-72	Burgh of Glasgow	O, £8,000	7, segt, stone	Des W. Mylne Cont J. Adam	Dem 1833	83, 102
Glasgow, Broomielaw II	Clyde	NS587647	1833-5	Burgh of Glasgow	O	7, segt, stone	Eng T. Telford Ovsr C. Atherton Cont J. Gibb & Son	Dem 1899	193-4
Glasgow, Hutcheson	Clyde	NS594645	1829-32	Burgh of Glasgow	£23,000	5, segt, stone	Eng R. Stevenson	Dem 1868	195-6; 195, 196
Glasgow, Hutchesonstown	Clyde	NS594645	1793-5	Burgh of Glasgow	ST, £5,260	5, segt, stone	Cont J. Roberton	Coll 1795	118-19, 147
Glasgow, Kelvin aqueduct	Kelvin	NS563690	1787-9	Forth and Clyde Canal Co.	CC	4, circ, stone	Eng R. Whitworth	As built	128-9; 127
Glasgow, Old (widening)	Clyde	NS592645	1820	Burgh of Glasgow	O	6, segt, iron	Eng T. Telford	Dem 1847	192-3; 190
Gloucester, Over	Severn	SO816195	1826-30	Co. of Gloucester	R, £43,500 (inc. approaches)	1, elpse(CV), stone	Eng T. Telford Ovsr T. Fletcher and J. Hall Cont J. Cargill	As built, disused	193, 195; 191
Godalming	Wey	SU974441	1782	Co. of Surrey	R	3, segt, brick	Arch G. Gwilt	As built	109; 102
Graiguenamanagh	Barrow	S7224	1770s?			7, circ, stone		As built	105-6; 97
Great Haywood aqueduct	Trent	SJ994230	1771	Staffs. and Worcs. Canal Co.	CC	4, segt, brick and stone	Eng J. Brindley Ovsr? T. Dadford	As built	127-8
Greta	Greta	NZ086131	1773	Co. of York, N. Riding	R, £850	1, segt, stone	Surv J. Carr Cont W. & J. Peacock	As built	85?, 115

GENERAL INDEX

Bridges are indexed by place name, followed by the name (if any) of the bridge. Page numbers are in ordinary type, figure numbers in italics.

References to notes are in the form 225(12.7), where 225 is the page number, 12 the chapter number and 7 the number of the note to chapter 12.